Advance Praise for

Contemporary Voices from Anima Mundi

"This is and will be a necessary book, due to the fundamental awakening of our troubled world's beating heart. Perhaps the prefix 'post-' betrays our way of life, with the continuous talk surrounding post-positivism, post-modernism and everything else that emerged as a result of post-materialistic science. It is true that profitable mechanical matter has flooded Western society and signaled a dogmatic 'language game' in the mouth of the false myth, whose concept of the 'objective observer' has relegated the spirit to pathology in too many cases. However, that spirit is being recovered thanks to those who have kept listening to Nature's subtlety, in the jungles, the rivers, mountains and monasteries, all without falling into infantile justifications. Science today also unites the awakening of the 'spiritual intelligence' that penetrates everything that exists and, what is more, it has set off on the path towards the 'only search' that, in the end, the divine breath can offer us. An ontology of BEING is co-created, as well as an epistemology of our 'inner world' that allows us to lose our fear and be able to map the numinous cosmos. You will find this and much more among these pages. May this book become an itinerant encyclopedia in space and time as successive contributions are made."

—Manuel Almendro, PhD, clinical psychologist, member of European Federation of Psychologists' Associations, Specialist Certificate in Psychology and Psychotherapy-EuroPsy, Director of Oxigeme, Center for a Psychology of Consciousness, Barcelona, Spain

"Tribes and Ancient Societies have always negotiated with animal spirits their right to hunt for food, asked mother earth and Anima Mundi for rain if there was drought, and resolved their problems of social and ethnic justice through rituals of mediation with the Spirits of Nature and the Universe with the support of master plants and other techniques. What is the logic of ancient societies maintaining rituals and balance with supernatural forces and laws for millennia? That scientific-materialistic rationalism does not have the tools to explain this cannot delegitimize or reduce these spiritual practices, an authentic technology of the quantic or spiritual World, to a simple irrational folkloric belief. We have a big blind spot in our western society, and we don't want to see it."

—Lluís M. Tomas, MD, PhB, Associate Professor at Health Sciences Foundation University, Bogota, Colombia, co-author and coordinator of published peer-reviewed articles in medicine and alternative medicine, founder of Nuna Hampi Retreat Center, Bogota, Colombia

Contemporary Voices from Anima Mundi

This book is part of the Peter Lang Regional Studies list.
Every volume is peer reviewed and meets
the highest quality standards for content and production.

PETER LANG
New York • Bern • Berlin
Brussels • Vienna • Oxford • Warsaw

Contemporary Voices from Anima Mundi

A Reappraisal

Edited by Frédérique Apffel-Marglin
and Stefano Varese

PETER LANG
New York • Bern • Berlin
Brussels • Vienna • Oxford • Warsaw

Library of Congress Cataloging-in-Publication Data

Names: Apffel-Marglin, Frédérique, editor. | Varese, Stefano, editor.
Title: Contemporary voices from anima mundi: a reappraisal / edited by
Frédérique Apffel-Marglin and Stefano Varese.
Description: New York: Peter Lang, 2020.
Includes bibliographical references.
Identifiers: LCCN 2019036655 | ISBN 978-1-4331-6337-1 (hardback: alk. paper)
ISBN 978-1-4331-6975-5 (ebook pdf) | ISBN 978-1-4331-6976-2 (epub)
ISBN 978-1-4331-6977-9 (mobi)
Subjects: LCSH: Spirituality. | Anima (Psychoanalysis) | Animism. |
Shamanism. | Cosmology.
Classification: LCC B105.S64 C66 2019 | DDC 204—dc23
LC record available at https://lccn.loc.gov/2019036655
DOI 10.3726/b15663

Bibliographic information published by **Die Deutsche Nationalbibliothek**.
Die Deutsche Nationalbibliothek lists this publication in the "Deutsche
Nationalbibliografie"; detailed bibliographic data are available
on the Internet at http://dnb.d-nb.de/.

Cover drawing by Randy Chung Gonzales, Amazonian painter and curandero/shaman:
<www.urkumamanwasi.com>

The paper in this book meets the guidelines for permanence and durability
of the Committee on Production Guidelines for Book Longevity
of the Council of Library Resources.

Printed in the United States of America

Table of Contents

Figures

Acknowledgments

This book is the result of many conversations among friends and colleagues over several decades: Jacques Mabit, Guillermo Delgado, Robert Tindall, Inés Hernández-Ávila, D. Ahmed, Neela Saxena, and Fern Feldman. As editors, the two of us are deeply grateful for this illuminating and creative process that has resulted in this collection of essays. All of us want to thank the many indigenous communities that have hosted us and initiated us into different ways of knowing and experiencing the spiritual in everyday life. We are particularly in debt to Randy Chung Gonzales and all the staff at Sachamama Center for Biocultural Regeneration, and the late Eduardo Grillo, founding member of the Peruvian intellectual activist organization Proyecto Andino para las Tecnologías Campesinas (PRATEC, Andean Project for Peasant Technologies). We also wish to express our gratitude to the many friends and colleagues who accompanied us throughout the process of conceptualizing and producing this volume. We owe an especially profound gratitude to Catherine V. Howard for exceptionally careful, accurate, and wise editing of the whole manuscript. This book could not have come about without her.

Frédérique Apffel-Marglin and Stefano Varese
Cambridge, Massachusetts, May 24, 2019

Introduction

Cosmic Dialogues

FRÉDÉRIQUE APFFEL-MARGLIN AND STEFANO VARESE

For the Anishnaabe people of northeastern United States and Canada, as for many other Indigenous peoples of the Americas, silence and alertness are the necessary conditions that allow the receptive quietness into which the world spirit can be heard. This state of quietness of the individual mind is consciousness, the rejoining of the personal mind with the cosmic mind. The Amazonian shaman of the Asháninka people—the *shiripiari*—can listen to the stories that plants, trees and animals can tell him and learn how to treat people with them. When European anthropologist Jeremy Narby asked *shiripiari* Don Carlos how he learned the properties of the different plants of the Amazon rainforest, the answer was as simple as enigmatic: "The plants talk to me" (Narby 1998, 38). Similar humble and still indecipherable answers given by Indigenous intellectuals and spiritual elders to Euro-American researchers reveal the fundamental separation of the modern materialist reductionist theory of knowledge and the Indigenous concept and practice of knowledge, that is, the "coming-to-knowing" as a constant process of the individual coming "into relationship with the energy and animating spirit of the universe" (Peat 1994, 55). Along thousands of miles of Andean and Amazonian mountains, the Yatiris and Yachak of each community are charged with maintaining the cosmic dialogue and keeping the Apus and Wak'as alive and satisfied of being a part of the larger sacred kinship. No human would be able to converse with rocks, mountains, and caverns unless the language of the spirits is learned and practiced in reiterated opportunities of communion between the place and the mind, the space and the timeless coming-to-knowing.

With the extraordinary new discoveries in the twentieth century about our universe initiated by Edwin Hubble's discovery in 1929 that the universe is expanding, our view of the universe has dramatically changed. These observations have revealed a living, expanding, evolving, sentient universe. This means, as cosmologist Brian Swimme (1996, 99) expresses it, that "we now realize we are all embedded in a living, developing universe and thus cousins to everything in the universe. The universe is not a collection of dead objects but a seamless whole community made up of cosmos-creating subjects" (1996, 103). The phrase "cosmos-creating subjects" underlies the living, evolving and sentient character of the cosmos where nonhuman as well as human subjects play an active role in its ongoing evolution.

As Swimme and Thomas Berry (a Catholic priest *cum* cultural historian) formulate it, "This story, as we know it for the first time through empirical observation and critical analysis, brings us back to the fifteenth-century Renaissance world of intimate presence of all things to each other" (Swimme and Berry 1992, 228). This fifteenth-century world was called Anima Mundi. It did not survive beyond the seventeenth century for reasons detailed by Apffel-Marglin in her essay in this volume, "Western Modernity and the Fate of Anima Mundi: Its Murder and Transformation into a Postmaterial Ecospirituality" (Chapter 1). As she points out, however, the primal peoples of all the continents share an

> insistence on establishing a close relationship with the psychic depths of the universe . . . . The drum was part of the sacred techniques for orchestrating the unity of the human/universe dance. The drum beat . . . the songs, chants, and dances . . . expressed the visions . . . awakened in them by the spirit world, by those dimensions of nature beyond the phenomenal world, but integral with materiality. (Swimme and Berry 1992, 44)

The essays gathered in this volume speak *from within* these worlds rather than *about* them in the classical anthropological manner. They come from many different parts of the world, some from Indigenous North and South America, and others from religious traditions in Bengal and Pakistan, as well as one from a mystical Jewish tradition. They speak of the spirit world as being both beyond the phenomenal world as well as integral with materiality since it brings about real changes on the material, phenomenological plane. This means that the spirit world is not spoken of as either a facet of the collective human unconscious *à la* Jung or as projections of the inner world of humans as many varieties of psychology maintain. Nor is the spirit world seen as somehow expressing some facets of the sociocultural worlds of humans in the manner of much anthropology. Rather, the spirit world is spoken of as an experienced tangible reality resonating beyond the world of the one experiencing the spirits. The spirit world has been denied in modernity, but it is a real world beyond the empirical world which ever since the Age of Reason has been called simply "reality." Jacques Mabit, a physician and shaman who apprenticed in the Upper Peruvian Amazon for many years, and who has thirty years of clinical experience at his center in the Peruvian Upper Amazon healing drug addicts using Indigenous healing practices, knows that the external invisible world is inhabited by spirits and other invisible entities. In his essay in this collection, "The Sorcerer, the Madman and Grace: Are Archetypes Desacralized Spirits? Thoughts on Shamanism in the Amazon" (Chapter 4), he states:

> these are objective realities, which will later, of course, be interpreted in varied ways according to the cognitive capacity, cultural background, and personal outlook of each individual subject. From this point of view, "animism," in the sense of the attribution of a living soul to natural phenomena, is not merely a belief but, rather, an acknowledged, verifiable fact, established by those who have taken the trouble to explore this world—the "otherworld." (Mabit, Chapter 4, this volume)

The casting of this "otherworld" as an unreal one, the fruit of projections from humans' inner world onto the outer world or simply as hallucinations or superstitions is dominant in modernity today. It is the result of the murder of the Soul of the World in sixteenth- and seventeenth-century Europe and the propagation of such views outward through the slave trade, conquests, colonialism and globalization.

The new cosmological and other scientific revelations of the late twentieth and twenty-first centuries impel us to leave behind our European colonialist mindset born at the turn of the fifteenth and sixteenth centuries, a mindset that has cast the spirit worlds encountered in those continents as a mark of backwardness and inferiority. Such a mindset, born of the murder of Anima Mundi in Europe, exported its view of the spirit world as an irrational, superstitious and backward mode of knowing. As the philosopher of science Isabelle Stengers put it:

> "We" on our side presume to be the ones who have accepted the hard truth that we are alone in a mute, blind, yet knowable world—one that is our task to appropriate ....
>
> Science, when taken in the singular and with a big S, may indeed be described as a general conquest bent on translating everything that exists into objective, rational knowledge .... What is called Science, or the idea of a hegemonic scientific rationality, can be understood as itself the product of a colonization process. (Stengers 2012)

What Stengers—following Bruno Latour (1995)—calls Science with a capital S has been called by others materialist reductionist science or simply classical and/or Newtonian science. The science emerging especially since the twenty-first century has been called by some "post-materialist science" (Beauregard, Dossey, and Miller 2014) and by others "New Materialism."[1] As the manifesto for a post-materialist science referred to in the endnote makes clear, this label is meant to convey a transcendence of the dominant materialist reductionist classical science paradigm and not a rejection of the relevance of matter. As Swimme and Berry (1992, 44) put it so felicitously, the spirit world refers to those dimensions of nature beyond the phenomenal world but integral with materiality. In other words, what they are saying is that although the spirit world is invisible, it is real since it is integral with materiality.

In the Amazonian region, every aspect of existence has a "mother/spirit" (*madre*), the life-giving soul that has form, energy, and intelligence and that communicates with humans (Mabit, Chapter 4, this volume). With his long clinical experience in his center for the treatment of addiction in the Peruvian Upper Amazon, where he combines ancestral Amazonian shamanism with a Western form of psychotherapy—Mabit bases his assertions on this long clinical experience. However, too often such views are understood either as superstitious beliefs or as superseded prescientific notions. If we contemplate them in the light of the kind of post-materialist science exemplified by Swimme, they take on a completely different significance.

Swimme and Berry (1992) offer a radically different view with what they call the "cosmogenetic principle." The first feature of this principle is that form-producing powers are latent throughout the universe, and the second is the relationship between such powers through time. For example, a star formation cannot be activated in the primeval fireball. Only a coordinated sequence of transitions makes possible the emergence of new realities. This contrasts with an indifferent universe in either a chaotic or an equilibrium state, in which the chances that a galactic structure will evolve in a billion years are negligible—even in one hundred billion years (1992, 70). It is well to recall that our universe began some fourteen billion years ago, which is far short of the one-hundred-billion-year threshold of chance.

> The cosmogenetic principle has three aspects: (1) differentiation (diversity, variation, disparity, heterogeneity); (2) autopoiesis (subjectivity, self-manifestation, sentience, self-organization, voice, interiority, identity); (3) communion (interrelatedness, interdependence, mutuality, kinship, reciprocity, complementarity).
>
> Were there no differentiation, the universe would collapse into a homogenous smudge; were there no subjectivity, the universe would collapse into inert, dead extension; were there no communion, the universe would collapse into isolated singularities of being. (Swimme and Berry 1992, 73)

In other words, the cosmogenetic principle speaks of a diverse, sentient, and interrelated universe. It is not a far stretch to recognize the Indigenous and other non-Western cosmovisions of a universe filled with a diversity of sentient beings, spirits, demons, or deities with different identities interdependent among themselves as well as with humans as portrayed in the essays gathered in this volume, as particular expressions of this cosmogenetic principle. It is also our act of poeisis to stretch the discursive reach of Anima Mundi beyond its original home in

Renaissance Europe and have it poetically evoke this cosmogenetic principle of Swimme and Berry with its universe-wide reach.

What the new science of cosmology has also shown through empirical observation is that these three aspects of the cosmogenetic principle are embedded in a universe that is unbelievably finely tuned for the emergence of structures, life and sentience. In the Orchestrated Objective Reduction theory (ORCH OR) of the physicist and mathematician Sir Roger Penrose and his collaborator, the anesthesiologist Stuart Hameroff (Penrose and Hameroff 2017), they name this sentience "consciousness" and the principle that makes it possible the "anthropic principle." This principle refers to

> The 20 or so fundamental constants which govern the universe (e.g., the mass of the proton, the gravitational constant, etc.) are all precisely, exactly what are needed for stars, life and consciousness—a coincidence of astronomically unlikely probability ... consciousness is intrinsic to reality, as suggested in ORCH OR, its quality woven into the 20 or so fundamental constants which regulate the universe. (Hameroff 2016, 370–71)

We must immediately point out that this view is rather controversial and that among professional physicists it has become standard practice to ignore the seeming encounter between quantum mechanics and consciousness.[2] However Penrose and Hameroff are not alone in making this link.[3]

Although Brian Swimme and Mary Evelyn Tucker (2011) do not speak of cosmic consciousness, their concept of autopoiesis imply it, covering as it does subjectivity, sentience, and interiority.[4] In fact, all three aspects of the cosmogenetic principle together amount to the recognition of consciousness in the universe. As the essays in this book make abundantly clear, the invisible beings with which humans interact are at once different, sentient, and in a relation of reciprocity or mutuality with humans. What makes Swimme and Tucker's new story of the universe quite different from the ORCH OR theory is their recognition of, and emphasis on, the numinous quality of this universe revealed by the latest scientific discoveries.[5] This is not surprising coming from scholars of religion like Berry and Tucker, but it is more unusual for those working in the sciences. Swimme, an evolutionary cosmologist and mathematician, does not shy away from spirituality; in fact, he insists that it is not only impossible not to recognize it in the new vista opened by the latest cosmological scientific discoveries but, rather, such a lack of recognition "is to live a life that is vulnerable to fundamental distortions" (Swimme 1996, 48). He and Tucker assert that "we have identified a nonmaterial realm suffusing not only the great macrocosm of the universe but suffusing as well the microcosm of the human and of every being of the Earth and universe" (2011, 104). In this their work converges with that of the post-materialist science manifesto and their many signatories.

In the context of this volume, primal and other non-Western traditions recognize the reality of spirits, deities, demons, and many other invisible beings and interact with them as possessing the same type of consciousness as we humans do. In his contribution to this volume, "Andean Entifications: Pachamamaq Ajayun, The Spirit of Mother Earth" (Chapter 7), Guillermo Delgado-P. suggests that what he witnesses in his native high Andes could be named after Pachamama, the Earth Mother of Andean peoples:

> Their ritual and sustained spiritual activity could also be named the Pachacene, just as some Native peoples in the Maya area propose the Mayacene, both pointing to the renewed cycles of spiritual regeneration. This regeneration implies the restoration of the multiple forms of intricate life: nature, humans, and other-than-humans acknowledging each other as equal, possible "alter-Natives" to the Anthropocene. (Delgado-P., Chapter 7, this volume)

Similarly, in her essay in this collection, "Spirit Crossings: A Nimipu/Tejana Cultural Perspective on Mortality and Death" (Chapter 3), Ines Hernandez-Avila, quotes Gloria Anzaldúa (2002, 541) on the Nahuatl concept of *Nepantla*, through which "you glimpse *el espíritu*—see the body as inspirited. *Nepantla* is the point of contact where the 'mundane' and the 'numinous' converge." Hernandez-Avila's Indigenous voice grounds her words in the immediate reality of life with relatives, both the living and the dead, both human and nonhuman. In her narrative, she conjures the reality and immediacy of an integral web of life, what was known in premodern Europe as Anima Mundi. The integrality of her cosmovision is clear, just as it is in the words of another Indigenous contributor to this volume, Guillermo Delgado-P. In their voices, there is no dualism between nature and culture, between matter and spirit.

So, too, Neela Bhattacharya Saxena, in her contribution to this volume, "Mapping the Chiasmus: Liberating Patterns in a Planetary Mandala" (Chapter 8), writes of the sacred geometries linking the human body and the universe, microcosm and macrocosm, and the way the various Indic Tantric traditions of meditation on the interior geometries bring one in resonance with cosmic geometries. In this she points out a parallelism with the ORCH OR theory of Penrose and Hameroff. It is also strikingly similar to what Swimme asserts in the above citation.

In his autobiographical essay (Chapter 5), Stefano Varese explains that the Ashéninka people conceive of the loss of the sacred cosmic interrelatedness as the result of an original human error, which must be corrected to re-establish the lost familial dialogue with all cosmic relatives.

The same theme is echoed in Rabbi Fern Feldman's essay in this volume, "To Dwell in the Thick Darkness: The Sacred Dark in Jewish Thought" (Chapter 6), where human interiority and embodiment are integral with transcendence: "Darkness can be source, essence, innermost being, transcendence, embodiment, nothingness, emptiness, mystery." She herself points out the similarity between this cosmic darkness and the "fecund nothingness" of cosmic black holes that are at once vacuum and generators of being.

D. Ahmed's narrative in this collection, "Lost and Found: Gifts, Dreams, and Sanity" (Chapter 2), will detain us here, since it highlights an aspect of colonization seldom acknowledged—namely that of the upper classes in formerly colonized countries—as well as giving us a vivid example of how the spirits intervene in this life. Ahmed narrates the story of a woman who came to her psychotherapeutic office with strange symptoms. Her patient was hearing voices, having strange visions and felt she was going crazy. The woman came from the same social class as Ahmed, educated and running her own successful business.

Ahmed, as a scientist—with an Ed.D. in psychology from Columbia University, as well as three M.A.s also from Columbia University—was faced with what seemed to her (and her patient) to be "magical" actions they could not explain. They both relied on modern secular assumptions about such beliefs, which they ascribed to the ignorant lower classes. The narrative takes us on their journey towards very different views, ones that reveal the deep erosion of the mystical traditions of Pakistan and, more generally, of Indo-Persian Islam. Their journey shines a bright light on how it is possible to overcome the colonization of the minds of English-educated upper classes in subaltern countries.

What we wish to emphasize in this example is that sorcery and healing are both holotropic practices. "Holotropism" is a term coined by transpersonal psychologists such as Stanislav Grof and Richard Tarnas. What distinguishes black magic from healing practices is the intention and the character of the practitioner. The town of Lamas in the Peruvian Upper Amazon, where Frédérique Apffel-Marglin founded the Sachamama Center for Biocultural Regeneration, is renowned for being a town where both sorcery and shamanism are practiced. Having been associated with this place for more than two decades, she has become aware of what distinguishes healing from sorcery, namely the intention and character of the practitioner and not the healer/shaman's recourse to invisible beings. What unites them is the knowledge that the cosmos is inhabited by invisible/nonmaterial entities, invisible to the normal human senses as well as the ability to have recourse to such entities for either healing or harmful purposes. She has also learned how strong the impacts can be from such practices, both sorcery and shamanism. In this respect, her experience is very similar to that of Ahmed. Furthermore, as in Ahmed's case, the educated upper classes prefer not to consult a healer (shaman/*curandero/yachak*)—or at least they do not want others to know they have done so.

Jacques Mabit's essay in this volume (Chapter 4) is based on a long clinical experience effectively treating drug addicts using the shamanic practices and ceremonies taught to him over many years by local shamans in the same Upper Peruvian Amazon region as Apffel-Marglin's center. He writes about the reality of these nonhuman entities, including spirits of the dead and their interactions with humans. According to him, good and evil spirits coexist in this invisible cosmos. He points the finger at the "masters of suspicion" that for contemporary Westerners have erased the existence of a coherent cosmic vision:

> The universal nature of these figures from the invisible world has led some rationalist Westerners to interpret them as facets of the collective unconscious, projections of the inner world common to all human beings, denying them all objective external reality. This kind of logic is no more consistent than one that would deny the existence of trees, the sun, or the sea simply because these elements are often employed symbolically in dreams, in psychotherapeutic processes, or in artistic creations. (Mabit, Chapter 4, this volume)

The recognition of the actual existence of incorporeal cosmic beings by all the contributors to this volume resonates with Berry, Swimme, and Tucker's cosmogenetic principle: "Differentiation, autopoiesis, and communion which together spell the cosmogenetic principle, are what prevents the universe from being a smudge or blob, from collapsing into an inert, dead mass, or from splintering into isolated singularities." Since, as they repeatedly assert, we humans have emerged from this universe and are inextricably part of it, in all our dimensions, both bodily as well as psychically/mentally, it would seem irrational to consider humans to be estranged from the many other entities inhabiting the universe. This is made clear by the communion aspect of the cosmogenetic principle. That violence, destruction and suffering are part of the universe is fully recognized. It is what Swimme refers to as the "sacrificial" character of the universe where the death of one is the life of another whether in the birth and death of stars or galaxies or the predator/prey relationship here on earth. That immense beauty, generosity, love, and sacrality also suffuse the universe is abundantly made visible and affirmed.

However, as we are so vividly aware, all of this is highly controversial and too often dismissed as fringe or even worse. This is the direct result of the murder of Anima Mundi, the sacred, living and communally bonded universe of the Renaissance as well as of those who lived within such a universe, the occult philosophers and the wise women and men that were labeled heretics and declared by the Pope to be heretics towards the end of the fifteenth-century.[6] Many of them were shamanesses and shamans using different European psychotropic plants such as belladonna, some psychotropic mushrooms or ergot.[7] There is no doubt that some and perhaps many among them practiced black magic, harming others. There is also no doubt that many of them were genuine healers as in the case described in Ahmed's essay or in Apffel-Marglin's personal experience in Lamas, Peru or in Mabit's essay. In any case our collective memory about this category of person has completely erased their positive healing powers and selectively retained only a memory of them as perpetrating evil.

As Apffel-Marglin details in the first part of her essay (Chapter 2), the worldview of these people, whether belonging to the literate occult philosophers or the oral peasant shamanesses and shamans, needed to be destroyed. The murder of Anima Mundi was enacted by the inquisitions of both Protestant and Catholic churches with the encouragement of the natural philosophers who later were called scientists. These practices were forever branded as evil, satanic, and superstitious to be destroyed. Belief in their worldview referred to by historians as hylozoism, invoking a living, integrated and sacred cosmos, is viewed as a regression to a less enlightened obscurantist past to which we must never regress. We are heir to this legacy whether in a variety of religious traditions or in the materialist reductionist scientific tradition.

What has been forgotten and erased is that hylozoism needed to be replaced by a dead mechanical nature, a neutral domain required as a new and firm basis upon which to reconfigure the certainty that the wars of religion thoroughly fractured.[8] Such a neutral terrain became the "object of study" of science thus insulating this new science from the bloody religious conflicts decimating Europe at the time.[9] To build this new certainty, the new paradigm needed to be absolutely separate from what both the church and the Protestants considered heretical, namely hylozoism. Catholics and Protestants violently disagreed on just about everything except about the need to eradicate the witches and other "heretics." On their part, natural philosophers needed a neutral terrain where religion and metaphysics could not intrude and risk creating political murderous mayhem, a terrain upon which to recreate certainty on novel bases which could only be claimed outside of religion and politics. This division—*and hidden partnership*—between a supernatural domain and a dead mechanical material one is still dominant today and has become global. What continues to enthrall us all to its regime is the certitude taught from kindergarten to PhD that religion is about a nonmaterial "supernatural" realm located above an insentient, dead, mechanical material world.

The period of these transformations in Western Europe was also the beginning of European expansion begun with the slave trade and the invasion of the Americas at the turn of the fifteenth to sixteenth centuries. This was followed by colonization and later by globalization. The expansion of this modern Western worldview continues apace through the hegemony of its type of education, its form of production, its global financial system, its consumerism and its technological wizardry.

Varieties of holotropic, shamanic and other related practices are viewed as ignorant and superstitious heir to this history. This is particularly relevant to the editors who work among Indigenous peoples of the Amazon region since there the dominant mestizo and *criollo* (descendants of European colonizers) society overwhelmingly sees Indigenous shamanism as primitive and backward, steeped in superstition. In her work with several high schools through the local provincial school board, Apffel-Marglin runs into such prejudice regularly. This history is also at the root of the resistance to a change in the materialist reductionist paradigm which Berry, Swimme, and Tucker all recognize to underlie all the institutions of modernity including education. Isabelle Stengers trenchantly captures this attitude and its political implications in the following passage:

> I received this word "reclaiming" as a gift from neo-pagan contemporary witches and other U.S. activists. I also received the shocking cry of neo-pagan Starhawk: "The smoke of the burned witches still hangs in our nostrils." Certainly, the witch hunters are no longer among us, and we no longer take seriously the accusation of devil worshipping that was once leveled at witches. Rather, our milieu is defined by the modern pride in being able to interpret both witchery and witch hunting in terms of social, linguistic, cultural, or political constructs and beliefs. What this pride ignores, however, is that we are the heirs of an operation of cultural and social eradication—the forerunner of what was committed elsewhere in the name of civilization and reason . . . . In this sense, our pride in our critical power to "know better" than both the witches and the witch hunters makes us the heirs of witch hunting. (Stengers 2012)

The murder of Anima Mundi and the witches' burning stakes that went with it began before conquest, slave trade, colonization and their heirs. This legacy is very much with us. We have all been colonized by this dominant version of history born in Western Europe. Coloniality has existed in Europe since those times and has spread everywhere there are schools around the world.[10] The oral ones, or those who somehow evaded or resisted what Bayo Akomolafe (2018) calls "the One Tongue" in their education, are the ones least in need of decolonizing their minds.

In our academic experience, whether on campus or in study abroad situations, we have encountered a deep-seated suspicion about any form of Native American shamanism. In our study abroad courses, we have had to scrupulously keep such practices away from our students' experience. The most that could be done is to bring them to the Takiwasi Center, founded by Jacques Mabit, for lectures *about* the therapeutic protocol, based on shamanic practices that his center follows. Academic administrations across the board seem to view the experience of such practices as not only illegal in the United States (although they are not illegal in Peru and hence not so for U.S. citizens while visiting Peru) but as forbidden to their students anywhere. In those offices, the legacy of the murder of Anima Mundi is alive and well and the "smoke of the burned witches" can still be smelled. This is particularly ironic given the relentlessly growing epidemic of drug addiction in a country like the United States. Apffel-Marglin found a New York Times article (2017) that details with precise figures the terrible price modern consumer society is paying for the devastating opioid epidemic sweeping the fifty states. Recent reports show that the epidemics of opioid addiction and mental illness are getting worse, even becoming a leading cause of shortened life expectancies in the United States.

More recent news informs us that now the number of opioid deaths has surpassed that of car crashes and homicides. Well, the problem will continue to be inescapable as long as our collective ego defenses repress the depth source of our murder some four hundred years ago in Western Europe of the experience and celebration of this numinous world from which we sprang and of which we are made, a murder necessary to the emergence of this consumer society where just about everything is a commodity for sale. The fact that such an epidemic crosscuts social classes makes it clear that addiction can with difficulty be correlated with a lack of material goods.

The irony is that mainstream addiction treatment institutions mostly using methadone are known to have a low rate of success. Furthermore, recent research into the effectiveness of certain psychotropic substances for the treatment of a variety of mental illness, especially post-traumatic syndrome disorder (PTSD), have contributed to a groundswell of interest in different types of psychotropic substances, including the plants used in Amazonian shamanism.[11] Ayahuasca has become very popular and given rise to what critics have labeled "shamanic tourism." The steep rise in popularity of ayahuasca in North America in the last few decades is sending an increasing number of people to South America in search of this brew. Inevitably, this has engendered the rise of charlatans wanting to cash in on that trend. Several scholars have condemned what they label "shamanic tourism" and warned of its dangers. Indeed, the rise of charlatan *curanderos* has occasioned not only financial despoliation and sexual exploitation but even some deaths. However, this phenomenon cannot be used to indict Amazonian shamanism per se. Such ever-growing stream of ayahuasca seekers from the north is also part of the search for alternatives to allopathic medicine as well as liberation from the monopolistic reign of a narrow bandwidth of "normal" consciousness. This search has led to the creation of the National Center for Complementary and Integrative Health (NCCIH) part of the National Institute for Health. Some observers claim that complementary and alternative medicine nowadays draw as much as 70% of the U.S. population. The popularity of Amazonian shamanism is directly related to the fact that for the great majority of allopathic medical practitioners, the body is an insentient material object part of Descartes's *res extensa* along with the material world and like the latter, devoid of mind.

There is now a growing body of research on the therapeutic efficacy of non-material treatments such as the placebo/nocebo effect, the role that prayer plays in recovery and other such phenomena and includes the whole field of psychoneuroimmunology. University of Arizona's neuroscientist Mario Beauregard's book *Brain Wars* (2012) is devoted to an in-depth study of such phenomena. The founders and directors of the Princeton Engineering Anomalies Research Laboratory (PEAR), Robert Jahn and Brenda Dunne, who have devoted some thirty years to the rigorous study of anomalous phenomena in the laboratory at the Princeton University's School of Engineering and Applied Sciences, write the following in their book-length report on their research concerning medical mysteries:

> Disillusionment with the perceived inaccessibility, impersonality, cost and in many cases, ineffectiveness of the contemporary allopathic approaches has driven growing communities of consumers to explore a burgeoning array of alternative health care options, many of which entails reinvestment in ancient or remote cultural traditions . . . and includes such practices as homeopathy, acupuncture, yoga, meditation, herbal remedies, prayer, and contemporary shamanism. (Jahn and Dunne 2011, 15)

However, the most detailed and thorough reporting on the therapeutic effectiveness of psychedelics/psychotropics comes from Michael Pollan (2018). He quotes the former director of the National Institute of Mental Health (NIMH) as saying that the U.S. mental health field is broken. Pollan reports on the last twenty-five years of scientific research on the effectiveness of psychedelics for treating addiction and most forms of mental illness. He also states the following, referring to quantum physics, "matter might not exist as such in the absence of a perceiving subject" (2018, 413). This sentence comes from his new book on psychedelics. On that same page, he also states the following:

> One of the gifts of psychedelics is the way they reanimate the world, as if they were distributing the blessings of consciousness more widely and evenly over the landscape, in the process breaking the human monopoly on subjectivity that we moderns take as a given . . . . Psychedelic consciousness overturns that view, by granting us a wider, more generous lens through which we can glimpse the subjecthood—the spirit!—of everything, animal, vegetal, even mineral, all of it somehow returning our gaze. Spirits it seems are everywhere. New rays of relation appear between us and all the world's Others. (Pollan 2018, 413)

Given this growing body of research—experimental, empirical and rigorous—it is no longer possible to dismiss shamanism or the many examples of so-called paranormal experiences that Robert Tindall gives us of his visit to the ancient temple of Chavin de Huantar in Peru. In his essay in this collection, "Shamanic Archaeology at Chavín de Huántar" (Chapter 9), Robert Tindall gives us telling evidence of archaeologists using states of altered consciousness during shamanic ceremonies to identify specific features of pre-Columbian sites but do so under a pseudonym or anonymously given the mainstream rejection of the validity of such methods. The lived experience of most of the contributors to this volume, similarly, is overall considered out of bounds by the academic mainstream, and a form of superstition. Such name-calling is the historical legacy of how a past age judged such phenomena in Western Europe for very specifically Western-European cultural, political, intellectual, and religious reasons. Those reasons no longer exist and never existed outside of Western European societies and clinging to the materialist reductionist scientific paradigm begins to look more like what Stengers (2012) suggests, namely that of "a hegemonic scientific rationality . . . understood as itself the product of a colonization process."

In addition to these political ramifications, one of the most crippling implications of the rejection of the reality of this living, sentient and numinous cosmos and our integrality with it is for finding our way out of the present global ecological and climate crisis that threatens the very survival of countless species, including our own, and of the planet as we know it. Today, what Graham Harvey (2005, 31) calls "the monopolistic dominance of the alert-problem-solving state of consciousness" is protected by laws in most nation states while sacred, nonaddictive consciousness expanding plants are demonized and illegal. This has led to the gruesome contemporary landscape of totally legal consciousness-altering drugs such as Prozac, Seroxat, Ritalin, Oxycontin, and, of course, alcohol, all of which have generated untold harm. The recent exposé by Patrick Keefe (2017) of the case of Purdue Pharma, owned by the famous philanthropic Sackler family, and its development of Oxycontin in full knowledge of its addictive nature starkly reveals the enormity of the destruction deliberately unleashed by the opioid epidemic. It is totally baffling that such drugs are legal while those that bring us the voices of the Soul of the World are illegal and viewed as unreal, destructive and superstitious. More than ever we need the voices of the "Greater Anima Mundi" to awaken us to other intelligences in the cosmos that have healed, taught, and inspired us to find our way out of this catastrophe we find the earth and ourselves in.

This does not mean that we need to give up or eradicate the alert, problem-solving state of consciousness, only that we need to recognize its proper place as well as open ourselves to the humbling realization that our everyday problem-solving mind is not the only or even the best mode to respond to our present condition. We are not advocating an either/or paradigm, only an enlargement of what is considered mind and consciousness and who or what possesses them.

Notes

1. See especially Dolphijn and van der Tuin (2012) and Keller and Rubinstein (2017).
2. In what is known as the Copenhagen Interpretation, led by Niels Bohr, the observer/measurer plays a fundamental role in the collapse of the wave function in quantum mechanics. However, what the ORCH OR theory posits is much more radical, namely, a cosmic consciousness in resonance with human consciousness, a domain that Bohr, like most other physicists, banished as metaphysical and outside the purview of science.
3. This interpretation is followed by several leading physicists, among them Paul Davies, Bernard d'Espagnat, Freeman Dyson, Henry Stapp, and Andrei Linde (see Nelson 2015, 52).
4. See also the extraordinary Emmy-award-winning film based on this work, *Journey of the Universe* (Northcutt and Kennard 2011).
5. This is something they share with Beauregard, Dossey, and Miller (2014).
6. The pope declared witches to be heretics to be burned at the stake in 1484.

7. The famous Harvard ethnobotanist Richard Evans Schultes has cowritten a beautiful book on European psychotropic plants (see Schultes, Hoffman, and Rätsche 2001).
8. Recall that for many centuries, the Church had a monopoly on knowledge and education and thus on certainty until the advent of the Reformation. On this, see especially Toulmin (1990).
9. This is only a partial view; the emergence of a different economy—mercantilism—replaced the manorial system, and its attendant enclosure movement was equally implicated. See Apffel-Marglin (2011).
10. We are well aware that a group of Latin American intellectuals argue that coloniality—that is, the colonizing of the mind—began with the Spanish invasion of the Americas, a point Stengers (2012) implicitly rejects when she asserts that Europeans are heirs to an operation of eradication, "the forerunner of what was committed elsewhere in the name of civilization and reason." Hylozoism was being eradicated in Europe before Europeans sailed to other worlds.
11. In his latest book on psychedelics, Michael Pollan (2018, 335) cites Tom Insel, former director of the National Institute of Mental Health (NIMH), as saying that the U.S. mental health system is broken. Pollan's book details recent scientific research about the effectiveness of psychedelics in successfully treating both addiction and mental illnesses.

References

Akomolafe, Bayo. 2018. *Those Wilds beyond Our Fences*. Berkeley, CA: North Atlantic Books.

Anzaldúa, Gloria. 2002. "Now Let Us Shift . . . . The Path of Conocimiento . . . . Inner Work, Public Acts." In *This Bridge We Call Home: Radical Visions for Transformation*, edited by Gloria E. Anzaldúa and Analouise Keating, 541–77. London: Routledge.

Apffel-Marglin, Frédérique. 2011. *Subversive Spiritualities: How Rituals Enact the World*. Oxford: Oxford University Press.

Beauregard, Mario, Larry Dossey, and Lisa Miller. 2014. "Manifesto for a Post-materialist Science." http://opensciences.org/about/manifesto-for-a-post-materialist-science (Accessed October 3, 2017).

Dolphijn, Rick, and Iris van der Tuin. 2012. *New Materialism: Interviews and Cartographies*. Ann Arbor, MI: Open Humanities Press.

Hameroff, Stuart. 2016. "Is Consciousness Guiding the Universe?" In *Consciousness: Integrating Eastern and Western Perspectives*, edited by Prem Saran Satsangi and Stuart Hameroff, 351–73. New Delhi, IND: New Age Books.

Harvey, Graham. 2005. *Animism: Respecting the Living World*. London: Hurst.

Jahn, Robert, and Brenda Dunne. 2011. *Consciousness and the Source of Reality: The PEAR Odyssey*. Princeton, NJ: ICRL Press.

Keefe, Patrick Radden. 2017. "Empire of Pain: The Family That Profited from Oxycontin." *The New Yorker* (October 30). https://www.newyorker.com/magazine/2017/10/30/the-family-that-built-an-empire-of-pain.

Keller, Catherine, and Mary-Jane Rubenstein, eds. 2017. *Entangled Worlds: Religion, Science, and New Materialisms*. New York: Fordham University Press.

Latour, Bruno. 1995. *We Have Never Been Modern*. Translated by Catherine Porter. Cambridge, MA: Harvard University Press.

Narby, Jeremy. 1998. *The Cosmic Serpent: DNA and the Origin of Knowledge*. New York: Jeremy P. Tarcher/Putnam.

Nelson, Adrian David. 2015. *Origins of Consciousness*. Nottingham, UK: Metarising Books.
New York Times. 2017. "Inside a Killer Drug Epidemic: A Look at America's Opioid Crisis." *New York Times* (January 6). https://www.nytimes.com/2017/01/06/us/opioid-crisis-epidemic.html.
Northcutt, Patsy, and David Kennard, dirs. 2011. *Journey of the Universe*. Film, 56 mins. San Francisco, CA: Northcutt Productions.
Peat, F. David. 1994. *Blackfoot Physics*. Boston, MA: Red Wheel/Weiser.
Penrose, Sir Roger, Stuart Hameroff, and Subhash Kak, eds. 2017. *Consciousness and the Universe: Quantum Physics, Evolution, Brain and Mind*. Cambridge, MA: Cosmology Science.
Pollan, Michael. 2018. *How to Change Your Mind: What the New Science of Psychedelics Teaches Us about Consciousness, Dying, Addiction, Depression, and Transcendence*. London: Penguin Random House.
Schultes, Richard Evans, Albert Hoffman, and Christian Rätsche. 2001. *Plants of the Gods: Their Sacred, Healing and Hallucinogenic Properties*. Randolph, VT: Healing Arts Press.
Stengers, Isabelle. 2012. "Reclaiming Animism." *e-flux* 36 (July). https://www.e-flux.com/journal/36/61245/reclaiming-animism/.
Swimme, Brian Thomas. 1996. *The Hidden Heart of the Cosmos: Humanity and the New Story*. New York: Orbis Books.
Swimme, Brian Thomas, and Thomas Berry. 1992. *The Universe Story: From the Primordial Flaring Forth to the Ecozoic Era*. San Francisco, CA: Harper.
Swimme, Brian Thomas, and Mary Evelyn Tucker. 2011. *Journey of the Universe*. New Haven, CT: Yale University Press.
Toulmin, Stephen E. 1990. *Cosmopolis: The Hidden Agenda of Modernity*. New York: Free Press.

CHAPTER ONE

Western Modernity and the Fate of Anima Mundi

Its Murder and Transformation into a Postmaterial Ecospirituality

FRÉDÉRIQUE APFFEL-MARGLIN

> There exist no occult forces in stones or plants. There are no amazing and marvelous sympathies and antipathies, in fact there exists nothing in the whole of nature which cannot be explained in terms of purely corporeal causes totally devoid of mind and thought. (Descartes [1641] 1998, *Pr. Phil.*, Pt. 4, § 187)

> The veneration wherewith men are imbued for what they call nature has been a discouraging impediment to the empire of man over the inferior creatures of God. (Boyle [1685] 2012, 15)

The Historical Demise of Anima Mundi

During the Enlightenment, the fathers of Western modernity and science, such as Descartes, Boyle, and Newton, actively argued against a prior Renaissance view that the world had a soul, that it was alive, and that all things in it—both human and nonhuman—were connected in an enormous web that was named *Anima Mundi*, the "Soul of the World." This ensouled world was alive, and divinity pervaded it. The literate "occult philosophers," as they are called, wrote on these topics. One of them, for instance, the fifteenth century philosopher Picco de la Mirandola (1463–1494), stated the following:

> All this great body of the world is a soul, full of the intellect of God, who fills it within and without and vivifies the All .... The world is alive, all matter is full of life .... Matter and bodies or substances ... are energies of God. In the All there is nothing which is not God. (Picco de la Mirandola, quoted in Potter 2001, 89)

Paracelsus (1494–1541), a famous sixteenth century physician and occult philosopher, held that God as prime matter is the invisible substance that originates, sustains, and exists in all things. He held that God is not outside the world and that the human soul is divine. For Paracelsus, each planet crowns a hierarchy of people, animals, plants, minerals, and elements, all of them being bound together so that the action of one affects all the others. All things—be they natural, human, or made by humans—are connected through the Anima Mundi and are sacred, since divinity pervades everything.

The fathers of Western modernity—the natural philosophers—marshaled their arguments and their experiments against the occult philosophers. Simultaneously, the Catholic church and the new Protestant ones marshaled their respective inquisitions against the occult philosophers and their allies, the peasant "witches" and sorcerers who were the teachers of Paracelsus and, very likely, of other occult philosophers as well. All of them were declared heretics by both versions of Christianity, which were successful in eradicating most of them. By the end of the seventeenth century, the notion of Anima Mundi had become thoroughly tainted by the allegations of heresy and superstition. Today it is relegated to the status of quaint notions of an earlier unenlightened European age and gets dismissed along with the allegedly exotic and superseded spiritual beliefs of societies deemed to be primitive, backward, and underdeveloped.

The campaign against Anima Mundi was not a mere sideshow, a minor issue for the fathers of Western modernity; it was at the heart of the emergent worldview they were championing. The need to destroy Anima Mundi was perhaps the only issue on which Protestants and Catholics agreed. Almost immediately after Martin Luther's expulsion from the Catholic church in 1521 and the creation of the rival Protestant church, conflicts erupted between these two branches of Christianity. The sixteenth century saw eight civil wars in France. These conflicts led to the thirty-year war in the early seventeenth century that engulfed Western Europe, when 35% of the people perished. There was only one issue on which these two sworn enemies agreed and that was the need to eradicate the occult philosophers and their oral peasant allies, the so-called witches.[1] Such eradication was carried out successfully during the sixteenth and seventeenth centuries, a period known as "the burning times." Yet another name for these centuries is the "Age of Reason"—an era that could only be established on the ashes of the stakes where witches and occult philosophers were burned alive.

The worldview of Anima Mundi clashed powerfully with two major trends of the times. First was the belief shared by both Protestants and Catholics that God transcended this world, that he was outside of His Creation. The immanence—as well as the femininity—of Anima Mundi had an aura of paganism and nature worship surrounding it, too much so for both varieties of Christianity. The Jewish Kabbalah and its Christian version, Cabala, had appeared in thirteenth century

Castille (on the heels of the Zohar, the most sacred text for the Kabbalists) and became influential among occult philosophers, such as Pico della Mirandola, Marsilio Ficino, and others (Yates 1979).[2] As the essay by Rabbi Fern Feldman in this volume shows so profoundly, Kabbalah is a nondualist, mystical version of Judaism in which darkness, the feminine, and the depths are as impregnated with divinity as are light and the heights. Thus, occult philosophy garnered the opprobrium of both the Protestant and Catholic churches for its links to paganism and Judaism, both of which the churches considered to be superseded by Christianity.[3]

Secondly, the worldview of Anima Mundi clashed as well with the emerging trend of transforming land into a commodity through what is known as the enclosure movement. The commoditization of land—and, along with it, of labor—was a slow but crucial process in the formation of modern Western Europe, accompanying the development of capital. These trends led to the emergence of capitalism, also known as the market economy.[4] Such changes were long in the making: it was not until the eighteenth century that Britain passed laws that gave legitimacy and full legal recognition to the enclosure movement, already underway for centuries (see Apffel-Marlin 2011, chap. 3).

Anima Mundi succumbed to the overwhelming combination of stronger forces triggered by the hostility of the Protestant and Catholic churches and the emerging capitalist economy. These forces were irrepressible in eradicating Anima Mundi. Ever since the second half of the seventeenth century, it has been tainted by the brush of heresy. It is now part of the quaint attic—or perhaps, better said, the "basement" with its resonance with the unconscious—in the imagination of Europeans and their far-flung descendants, having lost its heretical sting but none of its nonrational, even irrational, aura. Renaissance Anima Mundi was a numinous phenomenon, steeped in sacrality. It spoke of an enchanted world, harboring no nature/culture dualism in its breast. It included the earth and everything in, on, or around it: animals, plants, rocks and minerals, human-made things, planets and constellations. If we were to revive such enchantment, it would put a serious brake on those who are exploiting the world—both natural and cultural—for profit.

In fact, Descartes's radical separation between *res cogitans* and *res extensa* transformed all that the latter encompasses—including the human body along with the bodies of every earthly being, rock, plant, animal—into an inert mechanism without sentience, without agency. This was a prerequisite for the Industrial Revolution, which treated every earthly being as a natural resource to be exploited or, in Brian Swimme's ironic phrase, as "premanufactured consumer stuff" (Swimme 1996, 18).

The nonrationality or irrationality attributed to Anima Mundi stems from the view that rationality is the distinguishing, defining characteristic of humans, a view with a long pedigree in the West and one made into a quasi-dogma by Descartes. Human reason—this *res cogitans*—has metamorphosed into the view in fully fledged capitalism that such rationality represents the ability to calculate one's self-interest. The dogma that human beings "by nature" pursue their self-interest through using their calculative rationality is enshrined in introductory economics textbooks.[5] Hence the deep-seated resistance to drawing what to many seems to be the rational conclusion when faced with ecological degradation and the contemporary exploitation of both human and natural resources. These processes have gone to extremes, involving the mass extinction of species, the climate crisis, and the income inequality in which 1% of the U.S. population owns over 60% of the nation's total wealth. These are the signs of the corrosive as well as ethically perverse nature of late industrial capitalism. As Patrick Curry puts it,

> The idea of an ideally disembodied, nonemotional, analytical, calculative reason in the service of self-interest terminates in an ideology, invented and spread to justify exploiting some other men, nearly all women, virtually all nonhuman animals, and the Earth itself: most recently and "successfully" through industrial capitalism. It doesn't actually describe human beings as such (apart from those few who have succeeded, asymptotically, in turning themselves into quasi-machines), so it also can't be used to distinguish us from other animals. (Curry 2011, 154)

The murder of Anima Mundi was a prerequisite for the emergence of both the scientific revolution and the capitalist economic system. For reasons I have detailed more fully elsewhere, the conflicts engulfing Western Europe in the sixteenth and seventeenth century created a crisis of law and order, which emerged from a crisis of certainty.[6] The Catholic church had exercised a monopoly over education for centuries; the Reformation shattered the Catholic monopoly on certainty. It was essential to re-establish certainty on a new foundation that had to be separated and insulated from the raging politico-economic-religious conflicts of the times (Toulmin 1990). Those new foundations can be summarized as consisting in the invention of a nature thoroughly cleansed of any religious traces of numinosity, divinity, and sacrality, on the one hand, and on the other, of anything labeled "metaphysical" that could lead down dangerous paths to religio-political conflicts. Hence there was a necessity to invent a purely "material" nature, which in turn required the murder of Anima Mundi.

This purely material nature was to be the "object of study" of a group of learned men—women were explicitly excluded from the new domain—who were to gather in a new space called "the laboratory." This was a public space not in the sense that it was open to anyone but, rather, in the sense that it was the opposite of the occult philosophers' cabinet of experiments, which were private and even secret. In this new space, the new material nature was to be interrogated through the use of experiments, often involving the use of measuring devices and other types of apparatus. The rules of behavior in that space were strictly enunciated in order to protect it from any potential conflict arising from political or religious disputes. Robert Boyle's invention of the scientific experimental method and of the laboratory in mid-seventeenth century England still dominates scientific practice today; this approach has enshrined the complete separation of the sacred from a nature that is viewed as purely material. It has also enshrined the division between a supposedly inert, mechanical nature and anything reminiscent of the mind or emotion—in other words, of the soul. Boyle's experimental scientific method and the public laboratory where it is practiced enact the radical separation that Descartes posited between the mind and nature, the latter including the human body.[7]

Ironically, various kinds of scientific research conducted over the past century are revealing that the paradigm created by the fathers of the scientific revolution is sprouting more and more leaks. However, since all the institutions of modernity rest on the demise of Anima Mundi and on the new view of nature, intellectual breakthroughs, even those made in institutions dedicated to scientific pursuits such as universities, do not contain the necessary vital momentum to transform the various institutions of the nation state to ensure the survival of the world. Educational institutions cannot lead this transformation because their organization and functioning are too enmeshed with most of the other institutions of the nation state.

The nation state itself was an invention to seal the end of the wars of religion in mid-seventeenth century Western Europe.[8] James Scott (1998) richly details how the new materialist worldview was completely entangled with the emergence of the nation state in the mid-seventeenth century, a result of the treaty of Westphalia, which put an end to the thirty years' war. All the central institutions of the nation state (not all of them emerging at the same historical period), namely, the political, economic, educational, and financial institutions, rest upon the foundation of this new view of nature as pure materiality that is completely separate from the human mind.

Such a view was able to accommodate itself to a Western Christian understanding of the divinity as transcending creation. This statement oversimplifies the much more complex and often conflictual relationship between the two main forms of Christianity (Protestant and Catholic) and between both forms and the new worldview ushered in by the scientific revolution, but nonetheless, the statement contains a core of truth. The new mechanistic view of nature could not have triumphed if, in some fundamental way, it had contradicted Christian belief that creator and creation are distinct from each other; the very transcendence of the Godhead implied and necessitated such a separation. Indeed, it was the lack of such a separation in Anima Mundi that rendered it heretical to both forms of Christianity as well as unacceptable to the creators of the scientific revolution.

A Contemporary Claim That Anima Mundi Has Returned

In an essay first published in 1982, the neo-Jungian archetypical psychologist James Hillman called attention to the anthropocentric (human-centered) nature of the field of psychology. He pointed out that dictionaries of psychology and schools of psychology of all orientations agree that reality is of two kinds:

> First, the world [reality] means the totality of existing material objects or the sum of conditions of the external world. Reality is public, objective, social, and usually physical. Second, there is a psychic reality, not extended in space, the realm of private experience that is interior, wishful, imaginational. (Hillman 1992, 95)

In this view, the external world has no psyche, no soul; the soul has migrated and shrunk to the interiority of human beings. Since psychology is a modern discipline, its view could not have been otherwise. It was born thanks to the murder of Anima Mundi. Hillman continues to be impressed by the soul's sophistication that he saw in his patients, which he attributed to a hundred years of psychoanalysis. He notes that during that same period, psychology has become increasingly individuated and intrasubjective. When the concept of mental pathology appeared, the focus of psychology shifted to the task of readjusting inner psychodynamics. As Hillman puts it: "Complexes, functions, structures, memories, emotions—the interior person needed realigning, releasing, developing." (1992, 93)

Hillman briefly reviews the more recent field of family and group therapy, in which problems are seen as intersubjective, located in the patient's close social relationships. Therapy then consists of improving interpersonal psychodynamics. However, the world remains "external reality," an objective backdrop to human action and subjectivity but still one without its own subjectivity, its own psyche.

Hillman argues that although this intersubjective perspective acknowledges that the dynamics of the psyche are influenced by the small social group surrounding the patient, it still denies that such dynamics are part of the outside world. It does not view the outside world, whether it be human-made or natural, as having a psyche itself. This external world does not suffer nor does it communicate with the individual human psyche. Some varieties of social psychiatry admit that the built environment, such as cities, buildings, agriculture, mining, and the like, may be possible objective causes of psychopathology, but they are nonetheless considered radically different from subjectivity.

> This was especially the American dream, an immigrant's dream: change the world and you change the subject. However, these societal determinants remain external conditions, economic, cultural, or social; *they are not themselves psychic or subjective*. The external may cause suffering but it does not itself suffer. For all its concern with the outer world, social

> psychiatry too works within the idea of the external world passed to us by Aquinas, Descartes, Locke, and Kant. (Hillman 1992, 94; emphasis added)

It would appear that social psychiatry recognizes the built environment as a possible cause of psychopathology but, like the "natural" environment, it is not considered to possesses subjectivity. In other words, the *res extensa*—whether built or natural, whether bodies or buildings—has no subjectivity, no psyche, no soul. This soulless *res extensa* therefore cannot be diagnosed with any sort of psychopathology of its own.

Hillman goes on to suggest that the external world exhibits serious, deep psychopathologies and that people, including his patients in particular, are suffering from psychopathologies originating in the external world. Hillman sees this causal relationship as being of a different order than is the one posited by social psychiatry. He reviews a long list of the psychopathologies of the external world, all of which are excluded from psychological etiology and therapy. All of his examples of psychopathologies in the external world involve humanly created disasters, such as wars, bank scandals, or ecological pollution.

Reading Hillman's 1992 essay in 2018, I was struck by how these psychopathologies in the world have become much more pervasive, grave, and threatening in the intervening thirty-five years. Hillman mentions Vietnam, Watergate, Three Mile Island, and Chernobyl, as well as bank scandals, government collusion, pollution, street crime, and the increase of junk consumer goods and deceit, among many other pathologies. For me, his labeling of aspects of the human-made environment with psychopathological terms triggers a plethora of revelatory perspectives on modernity. Let me quote some of his especially vivid observations:

> To call a business "paranoid" means to examine the way it presents itself in defensive postures . . . [with] its delusional relations between its product and the speaking about its product . . . . To call a building "catatonic" or "anorexic" means to examine the way it presents itself, its behavioral display in its skinny, tall, rigid, bare-boned structure, trimmed of fat, its glassy front and desexualized coldness and suppressed explosive rage . . . . To call consumption "manic" refers to instantaneity of satisfaction, rapid disposal, intolerance for interruption . . . the euphoria of buying without paying (credit cards), and the flight of ideas made visible and concrete in magazine and television advertising. To call agriculture "addictive" refers to its obsession with ever-higher yields, necessitating ever more chemical energizers (fertilizers) and mass killers (pesticides, herbicides) at the expense of other life forms and to the exhaustion of agriculture's earthen body. (Hillman 1992, 104–5)

Of course, when reading these lines, we tend to understand the pathology labels as literary devices, as poiesis. Hillman calls such a response an "ego-defense" that is unwilling to acknowledge an unconscious move to protect the Western modern dualist materialist paradigm. It is as if we moderns simply *know* that buildings, businesses, consumer behavior, and agriculture have no agency and no psyche;

only those humans designing and implementing those aspects of reality have these capacities. If their creations are viewed as pathological, it is because their designers and implementers are pathological; the causality runs from the humans to the external world, not the reverse. I can only agree that such a metaphorical reading of these labels is an unconscious defense of the modern Western dualistic materialist worldview in which the external world—*res extensa*—is strictly separate from mind, the latter being an exclusively human capacity.[9]

Today, some thirty-five years after Hillman wrote these lines, we constantly hear news in the media about "sick" buildings, toxic toys, carpets, and a plethora of other products, as well as descriptions of "sick" cities choked off by toxic air or reports about depressed agricultural workers.[10] These pathological labels, however, do not begin to disturb the exclusive dualism between mind (or psyche) and world, since everyone is convinced that sickness comes only from human actions. These pathologies are human-made; in fact, all of Hillman's examples are human-made ones. However, he does not seem to grasp the crucial difference between human-generated perversion, conflicts, and climate problems, on the one hand, and on the other, the world of Anima Mundi, as the following passage makes clear:

> Death lurks in things: asbestos and food additives; acid rain and tampons, insecticides and pharmaceuticals, car exhaust and sweeteners, televisions and ions . . . . The material world is inhabited again; the repressed returns from the matter declared dead by Aquinas and Descartes, now as Death itself, and because of this resurrecting ghost in matter we are aware at last again of the *anima mundi.* (Hillman 1992, 111)

Hillman makes two problematic identifications in this passage. First, he identifies his list of examples, all being polluted human-made things, with the material world. Second, he identifies *anima mundi* with death-dealing pollution as a resurrected ghost in matter, an entity that lives not in matter in general but, more specifically, in human-caused polluted or perverted things. These misidentifications then lead him to the strange conclusion that such a death-dealing resurrected ghost is now, at long last, making us aware again of Anima Mundi. I would argue that this Hillmanian *anima mundi* bears hardly any resemblance to the original Anima Mundi.

As is clear in the statement by Picco della Mirandola and the description of Anima Mundi by Paracelsus that were quoted at the beginning of this essay, Anima Mundi includes the solar system, constellations, mountains, rocks, plants, and much more, and it is thoroughly pervaded by the divinity. The beauty of Renaissance Florence—which Hillman admires, along with many other cities of the period—arose out of the spirituality that was seen as saturating both natural and human creation. The figure of Anima Mundi, I contend, is a powerful symbol of an integral reality in which the human and the natural are integrated in an overarching web. I very much doubt that these occult philosophers, including Marsilio Ficino, whom Hillman especially praises, would recognize Anima Mundi in these death-dealing human-made ecological catastrophes and other human-made perversions. Anima Mundi was an integral spiritual phenomenon devoid of the exclusive dualism between creator and creation as well as between the natural and human worlds—that is, between nature and culture—a dualism that was exacerbated by the advent of the scientific revolution and, more generally, by Western modernity.

Having said this, Hillman's argument that the pathologies of this degraded modern, human-made external world are in fact *psychopathologies* is a crucial step toward recognizing that the external world, not just individual human beings or their social contexts, possess a psyche. This step allows us to recognize the agency of this external human-made world and how it can cause pathologies in human individuals. However, it stops short of acknowledging agency in the nonhuman world—an acknowledgment that was definitely made for Anima Mundi. I would contend that our times now call for a far more robust response to the psychopathologies of the human-made external world.

Reintegrating Mind and Body, Psyche and World

As the two opening quotations to this chapter reveal, Anima Mundi succumbed to the murderous attacks in the seventeenth century of both forms of Christianity as well as to the outright denial of its validity by the fathers of the scientific revolution. Taking a quick pulse of the situation in the world today, I would suggest that the situation may be different in the specifics but not in substance. The dualistic Cartesian/Boylian/Newtonian paradigm—also described in studies of the history of science as the classical scientific paradigm—which radically separated mind/psyche from world is still dominant today in academia and all other institutions of the nation state, as well as in modern media and culture in general.

In terms of religious conflicts—centrally responsible in the birth of this paradigm—the contemporary resurgence around the world of fundamentalist varieties of religion, some of a violent nature, as well as the almost instantaneous spread of ideologies through the internet, means that religious conflicts are by no means a thing of the past. In such a troubled religious and ideological global landscape, the search for inspiration from any particular religious or spiritual tradition easily stokes the fires of most forms of fundamentalism and, perhaps, some more mainstream religious traditions as well. I wish I could follow the lead of philosopher Raimon Panikkar (2018) in celebrating a diversity of such religious traditions within himself (as a Catholic, Hindu, and Buddhist), thus enabling others to contemplate or revive such practices,[11] although I realize this would be an extremely long shot for most denizens of modernity. I would also love to find convincing arguments to legitimize the suggestion made by Jacques Vigne (2008), a French psychiatrist who became a Hindu *sadhu* (an ascetic who has renounced worldly life), to alleviate psychological suffering through the practice of "theodiversity." I delight in this term coined by him since it carries with it the positive connotations of the term "biodiversity," as well as being a cousin of Panikkar's proud ownership of a multireligious identity. Nevertheless, as much as I would love to argue that theodiversity fosters the same resilience in human communities as biodiversity does in nonhuman communities, I must restrain my urge to equate these desires with reality.

Another major impediment to such a path is that in major theistic traditions, at least, the nonhuman world tends to be seen as something apart from the Source and as a mere backdrop to the human drama between that Source and Humanity.[12] To my own astonishment, I have come to suspect that if we are to find our way out of contemporary ecological and other catastrophes, our best bet is a totally secular science, being a less conflictual source than existing religio-spiritual traditions. I must hastily add, however, that the classical scientific paradigm is seriously frayed and something else is trying to emerge to replace it. That something else *is* revising the dualism between mind and matter; it is also revising the total disenchantment of the world that science promoted. Let me offer some examples of this emergent revision.

The findings of certain recent scientific research projects are opening up new vistas through strictly empirical means. This research is being carried out entirely outside the bounds of anything remotely religious or spiritual. The most cogent of arguments advocating for the return of Anima Mundi or, alternatively, for a worldwide acceptance of Indigenous varieties of spirituality with their profound affinity to the worldview of Anima Mundi, are colliding with problems similar to those that Western Europe ran into when the Reformation unfolded. I think advocating such a return is a tempting choice but, unfortunately, also a dead end. Later in this chapter, however, I will come back to the necessity of making visible the numinosity of this postmaterialist approach. I will also argue for the necessity of embodying and enacting those new vistas, since a purely cerebral, analytical knowledge is not enough to generate actions that will bring about changes in the world. To foster these changes, it will be essential to have recourse to ancient ancestral rituals and other forms of expressions that have been tested through countless generations.

I have chosen the research of a biochemist named Candace Pert, a mainstream scientist who is utterly dedicated to her assays in the lab at Georgetown University, working feverishly and, in her own words, "accountable only to members of our highly exclusive club." What is remarkable about this researcher is that the results of her painstaking work have completely transformed her in ways she never imagined possible. Even more striking is how the transformation she underwent as a result of her findings in the lab propelled her to share what she had learned as widely as possible, beyond the usual fora of scientific conferences, meetings, and publications. I am grateful for her efforts, since I lack the specialized scientific training that would allow me to thoroughly understand the scientific communications that she and her coresearchers have published. She has also been willing to ponder the wider cultural and philosophical implications of her work, something that is discouraged, if not derided, in younger scientists by their mentors and colleagues. Pert published her findings for a wide general audience only some fourteen years after her breakthroughs in the lab and after having established a solid scientific reputation in her field (Pert 1997).[13]

In her study of peptide molecules and their receptors, Pert has discovered that all the systems of the body—the neurological, endocrine, immunological, and gastrointestinal—are interconnected through the actions of discrete messenger molecules. These different systems of the human body communicate with each other through such molecules and their receptors. For example, the immune system can communicate with the brain or nervous system through peptides issuing from immune cells, which can affect the brain through their action on peptide receptors in the brain's blood vessels, on surrounding membranes, or even on neurons (Pert 1997, 172). She has demonstrated that the nervous and immune systems are clearly in communication with each other. Peptide molecules produced by one system can travel to another system where, through their attachment to a peptide receptor in the latter, information is conveyed between the two systems. Ironically, given the tendency of academic knowledge to be confined to silos, where, for example, the specialties of immunology and those of neuroscience, endocrinology, or gastroenterology prevent efficient communication among them, the implications of Pert's discovery took a decade to have an impact on these other fields, even though she and her coresearcher Michael Ruff had published their findings in *Science*.

Her work managed to pique interest in a field that has been emerging primarily from psychiatry and psychology, a field called psychoneuroimmunology (PNI). Pert said, "By providing PNI with a clear scientific language, that of neuropeptides and their receptors, we helped legitimize it" (1997, 176). However, the term she and Ruff had coined for the broader system they were researching was the "psychoimmunoendocrine system." The reason, she explains, was that they considered the word particles "psycho-" and "neuro-" to be redundant, since the psyche (Greek for "soul") and the neurological system (or brain) are both of the same kind, namely, the mind. But as they discussed terminology, Ruff asked Pert a crucial question: why use the word particle "neuro-" to label a peptide that is also found in the immune or endocrine systems? Why speak of "neuroreceptors" if these are equally active in other body systems? Pert conceded that these were pertinent question. When they shifted their terminology and began using the terms "peptides" and "information substances," it became more obvious that they were describing a body-wide communication system that shares information across cellular barriers and among different systems of the body. Because the section of the brain where peptides and receptors are most abundant is also the section involved in the expression of emotions, Pert and Ruff concluded that the peptides and their receptors "thus join the brain, glands and immune system in a network of communication between brain and body, probably representing the biochemical substrate of emotion" (1997, 179).

Pert modulates this apparent emphasis on emotion by pointing out that what she and Ruff have been talking about all along is *information*. She speculates that the "mind" is the flow of information as it moves among the cells, organs, and systems of the body. Although the mind as humans experience it is immaterial, it has a physical substrate in the body and the brain where the immaterial flow of information occurs. She writes:

> The mind then, is that which holds the network together, often acting below our consciousness, linking and coordinating the major systems and their organs and cells in an intelligently orchestrated symphony of life. Thus we might refer to the whole system as a psychosomatic information network, linking *psyche,* which comprises all that is of an ostensibly nonmaterial nature, such as mind, emotion, and soul, to *soma*, which is the material world of molecules, cells, and organs. Mind and body, psyche and soma. (Pert 1997, 185)

Thus, we can see that Pert and her coresearchers are conducting scientific work that is strictly empirical, based on an endless series of carefully prepared lab assays, and that is producing results that upend one of the most tenacious of modernity's dualisms—that between mind and body. Since the body is part of the material external world, part of Descartes's *res extensa*, her work has far wider implications for our understanding of nature, supposedly soulless, mindless, and purely material.

Skeptical, reductionist colleagues asked Pert, "Doesn't the flow of peptide *change* the physiologic responses, which then create the feelings we experience? Doesn't the chemical release of endorphins *cause* the feeling of pain relief?" The answer she has given opens up a radically new horizon, a veritable paradigm shift. Without denying the validity of the questions, she points out that they are only half of the truth. What she has discovered—and verified biochemically—is that, for instance, changes in the rate and depth of breathing bring about changes in the quantity and kind of peptides released from the brain stem and vice versa (Pert 2007, 312). Since many of the peptides are endorphins, which are the body's natural opiates, our conscious, willed changes in breathing pattern have the capacity to diminish pain through the release of endorphins via peptide messenger molecules. Similarly, the consciously willed practice of biofeedback is able to control pain, heart rate, blood circulation, tension, and relaxation. Such discoveries lead her to the following extraordinary statement:

> We can no longer think of emotions as having less validity than physical, material substance, but instead must see them as cellular signals that are involved in the process of translating information into physical reality, literally transforming mind into matter. Emotions are the nexus between matter and mind, going back and forth between the two and influencing both. (Pert 1997, 189)

Thus, the peptides and their receptors amount to a process of information exchange, of two-way conversations between various systems of the body, all going on below the threshold of consciousness. Furthermore, these peptides and receptors amount to conversations between our conscious mind or will and the processes occurring inside the body. This bidirectional flow of information takes us out of the narrow Newtonian paradigm of unidirectional chains of cause and effect, like a trail of billiard balls.

Pert came to another profound realization via a colleague, Dr. Robert Gottesman, who brought their conversations around to the concept of information itself. He quoted Gregory Bateson's (1973) definition of information as "the difference that makes a difference." It is crucial to realize that the perception of such differences can only originate in a particular observer or perceiver: that is, a difference makes a difference only *to the observer*. This understanding constitutes a new metaphor, one that allows us to comprehend that *the observer is an active part of defining reality*.[14]

In recounting her scientific journey and insights in her book, Pert (1997) comes to an even broader conclusion: that the psychosomatic network she and her colleagues have discovered and documented should be described as a system run by an intelligence not of the individual but one that is shared among all humanity and all life. Recent research supports this view, she says, by demonstrating the existence of intelligence in mycelium, in trees and their root system, possibly in all plant life, and in animals. Humans are but nodal points in a vast, comprehensive shared network made up not only of humans but of the nonhuman world as well. She states:

> It is this shared connection that gives us our most profound sense of spirituality, making us feel connected, whole .... To think otherwise is to suffer, to experience the stresses of separation from our source, from our true union. And what is it that flows between us all, linking and communicating, coordinating and integrating our many points? The emotions! .... I believe that the receptors on our cells even vibrate in response to extra-corporeal peptide reaching, a phenomenon that is analogous to the strings of a resting violin responding when another violin's strings are played .... The oneness of all life is based on this simple reality: Our molecules of emotion are all vibrating together. (Pert 1997, 312)

In short, this dedicated biochemist, through patient, detailed, and sustained research in her laboratory, far from any church, synagogue, mosque, shrine, or other religious sites, has come to a view that powerfully evokes Anima Mundi without once mentioning the name. History and theology are not fields that biochemists typically frequent. That is precisely why the spirituality that Pert discovered and articulated at the close of her book deserves to be called a form of a postsecular spirituality. It has its roots, ironically enough, in empirical research that has been carefully insulated from anything remotely spiritual. Since her findings do not spring from any particular spiritual tradition, they cannot be misunderstood as some disguised attempt to dislodge our traditional worldview in favor of a religious creed. Conversely, it is equally impossible to accuse Pert of "unbelief," since belief nowhere plays a role in her research, her findings, or her theories. Furthermore, the spirituality she invokes is phrased in terms that are nonspecific and general enough to allow interpretation through the languages of other traditions. In today's world, where religious strife is once again front and center in the news, this is welcome news indeed.

Postmaterialist Integral Ecological Ethics and Spirituality

Let me now return to the issue of Western modernity and the classical scientific paradigm, which I earlier contended is thoroughly implicated in our present widespread destruction of both the human and the nonhuman worlds. In this context, it is worth considering what has been called "integral ecology" (Grim and Tucker 2016). If it is true that the phenomenon of emotional resonance exists, then Hillman's assertion—that psychopathologies in the external world cause psychopathologies in human individuals—takes on additional weight and substance. The daily news in the United States is saturated with reports of epidemic levels of drug and substance addictions, depression, suicide, alcoholism, and other psychopathologies. Additionally, the media gives us daily reports about new products found to be toxic to humans, animals, plants, as well as the soil, air, and waters. We can say that the psyche of the modern world is saturated with pathologies in the external human-created world, the nonhuman world ravaged by humans, and the psyche of individual humans. Our world and all of us are in dire need of healing at the collective and individual levels; to enable this to come about, it is essential that we devise new approaches to ethics and spirituality.

Although it is indubitable that many religious traditions harbor invaluable practices and ethical teachings that can help us on this path, it is also true that the spread of deeply entrenched forms of fundamentalism as well as secularism renders any possible recommendation of one particular tradition not only hopeless but dangerous, possibly abetting rage and violence from people who may interpret this as a derision of their own traditions.

Before exploring this topic further, I want to make clear what I do *not* mean by spirituality. I emphatically do not mean "belief in a supernatural domain and beings." First of all, the term "belief" is problematic because it points to adherence to a reality that one does not experience directly. The word "supernatural" assumes that this realm is not located in the world shared by humans, animals, plants, mountains, and rivers, and beyond to stars and galaxies. The term "supernatural" literally means that which is above that world, above nature. As the French anthropologist Jean Pouillon (1982, 8) remarks about the Dangaleat of Chad in Africa, the world for them—and, I would add, for Indigenous Amazonian people—is not divided between a this-worldly reality and another-worldly reality. The spirits are *experienced* rather than believed in, and this experience is, above all, a local one. Such spirits do not necessarily exist in the same way everywhere. Let me emphasize that my point is not that disembodied beings cannot be experienced; rather, in modernity, nonempirical beings are ipso facto relegated to a domain inaccessible to experience—a supernatural domain that one must *believe in*. I am *not* asserting that nonempirical entities or beings are simply the fruit of human imagination and do not "really" exist since they cannot be measured or empirically verified. I am instead insisting on the distinction between a *belief* in such entities, on the one hand, and on the other, the direct *experience* of them in an embodied manner. As Patrick Curry puts it in his commentary on religion and belief,

> Most definitions of "religion" specify belief in one or more supernatural entities, which is not only parochial (the emphasis on belief, as distinct from practices and a way of life, is a particularly monotheistic one, and there is no God in Buddhism), but prejudices matters against nature from the start, since whatever this being or power is, it cannot be in or of nature, since it is *super*natural. (Curry 2011, 139)

As our discussion of Renaissance Anima Mundi has made clear, varieties of Christianity as well as of Judaism embraced this nondualist spiritual vision that integrated the human, the natural, and the divine in one great sacred whole. The mystical esoteric traditions within the three major monotheistic faiths appear to have engendered nondualistic visions as well as experience-based practices.

However, the English language—and European languages in general—do not make it easy to articulate what I am attempting to convey here. The words "spirituality" and "spirit" commonly refer to something that is *not* matter, but our contemporary condition requires a language that communicates the integration of matter, the human world, and sacrality. As Curry points out, the *sacred* is that which is beyond human control, intrinsically valuable, and awe-inspiring, something that should not and ideally cannot be transformed into a commodity for sale. The more-than-human world—that is, nature with humans as part of it—goes beyond any particular understanding of it, any use of it, or even appreciation of it. Its value, as Curry (2011, 139–140) points out, is that of an inexhaustible mystery.

An integral ecological ethic and spirituality should be robust enough to lead us out of our modern Western anthropocentrism. To make such a massive undertaking possible, it must be translated into ways of life and daily practices that repeatedly and incessantly enact our integration with the nonhuman world, with the whole cosmos, with the sacred.

There is no doubt that the innumerable religious traditions of the world possess practices and ethics that can not only greatly help us on this path but may even be indispensable. However, given the global existence of entrenched secularism and varieties of fundamentalism, as well as militantly exclusivist or triumphalist mainstream forms of religions, advocating for any one of these traditions may seem to be an impossible endeavor, sure to be rejected—sometimes violently—by many. Paradoxically, it is from within decisively secular scientific practices in several disciplines that we can see initial steps to open up to both ecological spirituality as well as ecological ethics.

In this chapter, I have discussed the work of the biochemist Candace Pert, but in previous writings, I have focused on the work of Niels Bohr and that of quantum physicist Karen Barad, who has interpreted and extended Bohr's findings. Barad emphasizes the ethical imperative that surges forth from Bohr's breakthrough insight that the experimental scientific method is not just a method for measuring and observing the world but is part of what is being measured and observed. The scientific experimental method is not a neutral, transparent mirror of nature; rather, along with scientists, engineers, technicians, and all those who apply the scientific experimental method, it has a creative role in producing the "observed" phenomenon. We humans cannot but be responsible for a reality we co-create. One inescapable entailment of this is that there is no universal nature "out there," whether it be thought of as an insentient, mechanical, purely material object or imagined as the One and Only True Nature. Since we co-create the more-than-human world, and since we humans are extraordinarily diverse, our dream of universality must also be surrendered.

Bateson's point about the difference that makes a difference makes it clear that the reality being perceived/observed is co-created by the perceiver/observer. The classical scientific dream of a purely objective observer beholding a purely objective nature has to be abandoned. We humans are part of the cosmos, of nature; we need to shed the Western dream of our specialness, our singularity, a quality residing in our purportedly unique human faculty of reasoning, as if that gives us license to exploit, extract, and ravage the nonhuman world, using as a justification that it is good or profitable for (certain) humans.

From the science of cosmology, we learn the same lesson. Our galaxy is only one among innumerable other galaxies, even clusters of galaxies, all of which are expanding in the universe. However, no observation is possible of this expanding universe from a point external to it. As cosmologist Brian Swimme puts it:

> When I picture the cosmic birth as some kind of explosion that is taking place off in the distance, away from me, away from where I am observing it, just where am I standing? What provides the platform for my feet? How is it that I can stand outside the universe and watch its birth if I myself, from the beginning, am woven into this birth? . . . . The central archetypal pattern for understanding the nature of the universe's birth and development is omnicentricity . . . a developing reality which from the beginning is centered upon itself at each place of its existence. In this universe of ours to be in existence is to be at the cosmic center of the complexifying whole. (Swimme 1996, 85–86)

It will take a while before I (and, most likely, many others) can digest this word "omnicentricity," which sounds so much like a contradiction. A "center," by definition, is not everywhere, as the Latin word *omni* suggests. As if this were not challenging enough, the discovery of the quantum vacuum has radically displaced reductionist materialism. Swimme says the quantum vacuum is *nonvisible*, meaning that many things are invisible to us but through the use of certain apparatuses can become visible. The quantum vacuum, however, by its nature, "can never be seen, because it is neither a material thing nor an energy constellation . . . it cannot even be *pictured* . . . . Even so, it is profoundly real and profoundly powerful" (1996, 97). More astonishing yet, out of this nonmaterial, nonvisible quantum vacuum, elementary particles arise and then disappear. Where there is "no-thing," a vacuum, no atoms, no elementary particles, no protons, no photons, is where elementary particles suddenly emerge. Being itself arises out of a field of "fecund emptiness," something that takes place throughout the universe. It seems that only the paradoxical language of contradictory words can capture such mysteries. And as Swimme points out, to suggest that material reductionism is not the only foundational reality in the universe throws into an abyss of doubt modernity's justification of consumerism, its anthropocentrism and exploitative posture.

It seems that we are in a situation reminiscent of sixteenth and seventeenth century Western Europe where the radical discoveries of Copernicus, Galileo, Kepler and Newton ushered in a new age, the modern age, anthropocentric, materialist, mechanist and reductionist. The recent scientific discoveries mentioned here—and only a few could be focused upon—have already intellectually superseded the classical scientific paradigm. What is harder to accomplish is to use those as the foundations for new institutions of society, new political frameworks, and new economic structures that embody the heart of these discoveries that we humans are entirely embedded in the nonhuman world, in nature, in the cosmos. We need a cosmocentric polity, economy, society, etc., enacting our nature as entirely part of, as well as embedded in the more-than-human world.

Spirituality: Lessons from the Epidemic of Drug Addiction

Brian Swimme shares a revealing personal story about cosmological discoveries and attempts to communicate them. When describing his feelings when entering into the new cosmological findings and orientation, people often asked him if he used drugs. His initial reaction was to resent and reject such a suggestion, which he thought would lead his listeners to equate the feelings that cosmological discoveries induced in him with "tripping" and would equate it with all the negative consequences that come with drug addiction. However, upon reflection, Swimme came to very different views. He concluded that alcohol and drugs are an intrinsic feature of consumerism, necessary for its sustainability. Consumerism is based on the basic assumption of the modern worldview, namely that the world is made of dead objects. These objects are mostly "unmanufactured consumer goods." The deliriously abundant glory of the natural world, of the cosmos, is reduced to an inert mechanism. Humans are of this world, created from and with it, and this Western modern paradigm cuts us off from the extraordinary expressiveness of this living, sensuous, numinous world. We are left alone among our kind, bereft of this numinous and exuberantly varied part of ourselves. The nonhuman world, the cosmos has agency, sentience and more. Candace Pert writes that like information, emotions travel between two realms: mind and body; the peptides and their receptors in the physical, body realm and the emotions in the nonmaterial mind realm. Information belongs to neither of these realms but touches both, occupying a nonmaterial realm called by information theorists the "inforealm." She concludes thus: "Information theory releases us from the trap of reductionism, positivism, determinism and objectivism. Information theory is a rich language of relatedness, cooperation, interdependence and synergy" (Pert 1997, 261).

These and other such findings among several scientists have barely percolated within academia, let alone the wider society and culture. As argued earlier, all the institutions of modern society are based on the classical scientific paradigm, one that gives certainty and power over the nonhuman world and all those humans perceived as being close to it. All those institutions thus are not going to welcome with open arms the kind of news some scientists are bringing us lately. The reaction is what Hillman calls "collective ego-defenses" of the repressed unconscious of modernity. Modernity is still overwhelmingly in the grip of this dead world, which is also a deadening, pathological world. Swimme thinks that "hoping for a consumer society without drug abuse is as pointless as hoping for a car without axle grease." He explains why in these terms:

> When humans find themselves surrounded by nothing but objects, the response is always one of loneliness .... But isolation and alienation are profoundly false states of mind. We

> were born out of the Earth Community and its infinite creativity and delight and adventure. Our natural genetic inheritance presents us with the possibility of forming deeply bonded relationships throughout all ten million species of life as well as throughout the nonliving components of the universe. Any ultimate separation from this larger and enveloping community is impossible, and any ideology that proposes that the universe is nothing but a collection of preconsumer items is going to be maintained only at a terrible price. (1996, 33–34)

Today these ten million species have been severely diminished with the largest extinction of species since the disappearance of the dinosaurs, and one caused by humans, giving our geologic era the label of the "Anthropocene." We are all in deep mourning, depressed and bereft whether we are aware of it, or of its deep underlying causes or whether we have repressed all of this to our collective unconscious.

In terrible synchronicity, after writing these words, I opened my online New York Times and in it, I found an article on the devastating drug addiction epidemic sweeping the fifty states, entitled "Inside a Killer Epidemic: A Look at America's Opioid Crisis," detailing with precise figures the terrible price that consumer society is paying:

> Opioid addiction is America's 50-state epidemic. It courses along interstate highways in the form of cheap smuggled heroin, and flows out of "pill mill" clinics where pain medicine is handed out like candy. It has ripped through New England towns, where people overdose in the aisles of dollar stores, and it has ravaged coal country, where addicts speed dial the sole doctor in town licensed to prescribe a medication.
>
> Public health officials have called the current opioid epidemic the worst drug crisis in American history, killing more than 33,000 people in 2015. Overdose deaths were nearly equal to the number of deaths from car crashes. In 2015, for the first time, deaths from heroin alone surpassed gun homicides.
>
> And there's no sign it's letting up, a team of New York Times reporters found as they examined the epidemic on the ground in states across the country. From New England to "safe injection" areas in the Pacific Northwest, communities are searching for a way out of a problem that can feel inescapable. (New York Times 2017)

The problem will continue to be inescapable as long as our collective ego defenses repress the depth source of our murder some four hundred years ago in Western Europe of the experience and celebration of this numinous world from which we sprang and of which we are made, necessary to the emergence of this consumer society where just about everything is a commodity and thus for sale.

Meanwhile, there is hopeful news from Lisa Miller, the director of the Clinical Psychology Program at Columbia University Teachers College and also director of the Spirituality Mind-Body Institute. In her book (2015), she marshals an impressive amount of scientific psychological research that shows unambiguously that there is only one thing that can protect up to 40% of adolescents and young adults from drug addiction, 60% of them from depression, and 80% from risky sex, and that is spirituality. She echoes the words of physician Jacques Mabit, founder and director of Center Takiwasi in Tarapoto, Peru, which is having remarkable success treating people with drug addictions, using a combination of Western psychotherapy and Indigenous ancestral plant medicine and shamanism.[15] Miller states that the escape and connection that teens experience with drug use "needs to be understood as a spiritual quest, inherently good and important" (2015, 43). Such a view is also shared by Swimme, who writes:

> It is simply not human finally to live a life sealed off from all conscious contact with those powers at work throughout the Earth and universe and within every one of our cells. So intolerable is this sense of being out of it, of being left out, of being without central meaning for the world, we will resort to any route to ease the pain. And the quick and mindless way ... is to ingest mind-altering chemicals that dissolve the thin veneer of consumer culture ... spiritually desiccated ... [and] out of touch with the numinous powers pervading each being in the universe .... Thus, if only for a moment, and sometimes at a horrible cost ... one can be at home again in the great flood of beauty. (Swimme 1996, 35)

Miller's book focuses on bringing to parents the results of rigorous and multiple scientific studies, which amount to showing that lived spirituality is the only preventive measure that works. She does not spend her energies on tracing the deep collective roots of the drug epidemic in modern consumer society, although she repeatedly points to the problem as the result of a misplaced emphasis on material science. She emphasizes the split that occurs early in children's development between "logic-based learning and direct experience and inner heart knowing" (Miller 2015, 169). Modern cultures, particularly the American, lack the value given especially to children and adolescents' inner voices and inner wisdom as "not real" and "not scientific" (2015, 74).

Educational institutions from kindergarten through graduate programs overwhelmingly emphasize the rational, analytical mind, leaving to the arts such concerns pertaining to the creative imagination but not to "reality." She calls this inner wisdom "heart knowing" and adds: "Due to socialization in our current society, heart knowing is often blocked, denied, or disintegrated. This leads to enormous suffering, as we can become cut off from other people, our higher selves, and even our transcendent relationship" (2015, 78).

Miller refers to a fresh kind of scientific research, a new generation of empirically based *postmaterial* scientific research. Although she makes no direct reference to Candace Pert's research or the new cosmology of Swimme and others, these would certainly qualify for the label of postmaterial scientific research. According to Miller, children are born with a natural sense of being that is related to everything:

> Children are entranced by nature . . . . All things in nature are assumed part of the family. This natural curiosity is human, and the sense of a caring relationship with all living beings is spiritual. Children have a natural spiritual attunement with the world around them: they relate to animals, trees.
>
> The young child is born with this assumed relationship with all of nature, from goslings to galaxies. (Miller 2015, 125–5)

Miller bases these assertions on several scientific studies of twins, both identical and nonidentical, that show unambiguously that we are inherently, genetically spiritual (especially Miller 2015, 54–64). She refers to a landmark 1997 twin study focusing on inner personal spirituality by Kenneth Kendler. The terminology used in this study, like others, speaks of spirituality as varieties of a *transcendent* relationship. Of course, the word "transcendent" is used in these studies is different from the way I have used it in this essay. It does not refer particularly to a dualist notion of the ultimate, or God, as being outside of creation. To my ears, this use of the term "transcendence" is almost synonymous with either the term "numinous" or "sacred." The three basic varieties of transcendence according to these psychology laboratory studies are: (1) a "transcendent dialogue" with God, referring principally to the Judeo-Christian traditions; (2) "transcendent oneness," a sense of being one with the universe, found among Eastern traditions and religions; and (3) the "transcendent other" known through nature, animals, forces of nature, the universe, and other people, as found within Indigenous traditions.

Since this essay is centrally concerned with integral ecology and the deep roots in modernity of the current integral—ecological and social—crises, my choice of terms is guided by that. Miller's central focus is guiding parents toward helping the development of lived spiritual reality in their children as the most and perhaps only effective prevention against the most devastating epidemics of addiction, mental illness and other destructive behaviors.

In her laboratory research on adolescent individuation, Miller makes a significant distinction between lived reality and borrowed reality. In their individuation process during the decade from fourteen (puberty) and twenty-five (early adulthood), teens and young adults develop their natural, genetic spirituality and learn to distinguish it from the one received from their immediate social context and culture. Such a process takes a great deal of emotional investment as well as serious intellectual work. Puberty not only unlocks the process of sexual maturity but is also a time of "a biologically primed tidal surge in natural spirituality" (2015, 64).

Miller draws the conclusion, based on her studies and those of others, that for spiritual individuation to successfully protect against the most destructive behaviors, the process of individuation of this natural spirituality must be successful. However, she and her coresearchers found that adolescents and young adults were faced with a curious silence: in fact, *nobody has talked to them about this experience.*

"Without supported and guided spiritual awakening in adolescence, our teens are left to fend for themselves. The cost is high" (2015, 71). She states that a successful process of spiritual individuation may or may not include the family received religion. The process may replace the received family religious tradition with the young person's own lived spirituality, or it may replace it with a mix of both. Nevertheless, Miller is clear that for a successful process of individuation to take place, it has to go through a personally lived experience in which spirituality is embodied and intensely felt, whatever its characteristics might end up being. Such a process should be impregnated with what she calls *heart knowing*, a kind of knowing that is rarely, if ever, cultivated in modernity.

Similarly, Brian Swimme points out that the dry, rational, abstract language pervading classical science (hopefully, I might add, less so in what Miller calls "postmaterial" science) cannot lead us to change our everyday behaviors: "Facts by themselves are not enough; what is needed is embodiment .... What we need is just the simple recognition that as we deprive ourselves and our children of direct contact with the numinous powers that fill the universe, we are choosing a diminished existence" (Swimme 1996, 45–46).

Conclusion: Healing the Split in the Peruvian Upper Amazon

For me, the imperative of our times is the need to heal ourselves and to help our children and our ravaged earth to heal—in other words, the need for integral ecological healing that will mend and overcome the split between nature and culture, between mind and body, and between mind and heart. In this endeavor, we need to avoid the Charybdis of fundamentalist rationality and materialism and the Scylla of unquestioningly accepting the received wisdom through parents, school and community. We also need to recognize that spirituality is at the very core of such an endeavor. The following anecdote reported by Miller could have happened in my own institution, Smith College, where I taught anthropology for twenty-six years:

> A few months into her first year at an all-women's college, [Marin] knew she was no longer unique. Her initial visit to the campus health clinic resulted in a brief chat, followed by a diagnosis of depression and an immediate prescription for an antidepressant; there were no offers for extended conversation beyond management of the medication. Her inner experience was not the topic at hand. Developmental depression wasn't mentioned. Her depression was viewed strictly as an illness. (Miller 2015, 31)

During my academic career as an anthropologist at Smith, many students came to me with similar stories. I remember vividly being told by deans at orientation workshops for faculty that we were never to engage students about their personal lives. We were told we were teachers, not psychologists or psychotherapists. The purpose of education, the deans said, is not to transform our students but to educate them with the accumulated knowledge housed in libraries and other repositories. I found myself frustrated and alienated with such dictums.

However, what finally drove me to retire early and run from academia was the sense that spirituality was considered taboo as a personal experience outside of attending church, temple, or mosque on the weekend. Anthropologists, of course, study the spiritual traditions of other folks, but to be personally touched, inspired, or even transformed by such traditions is a taboo that is strictly enforced within the tribe of anthropologists. We study *about* other people's customs and practices, just like classical natural scientists study *about* a mindless, inert, soulless nature. Indeed, the discipline of anthropology was modeled in the second half of the nineteenth century on that of the natural sciences. It asserted that social and cultural phenomena were facts just like those in nature and could be studied with analogous methods. However, the new postmaterialist science—and the new materialism in science[16]—is showing us that it is impossible to remain outside of what we observe and learn *about it.* What we learn is about ourselves as part of what we observe and measure.

For most of anthropology, despite its avowed credo in the principle of the psychic unity of humankind, the encounter with peoples who are different from ourselves is not primarily one of reciprocal sharing, learning with and from each other, making ourselves vulnerable to experience, learning about the world and ourselves in a completely different way than the one we were taught, or finding these lessons enriching and even worthy of adopting. Any anthropologist who dares to learn from and emulate the people they study is criticized for "going native" and risks getting excommunicated from the discipline.[17]

As Swimme puts it so eloquently, when "we deprive ourselves and our children of direct contact with the numinous powers that fill the universe, we are choosing a diminished existence" (1997, 46). For me, the lack of spirituality in my academic life led me to a severely diminished existence. My experiences in India, most of which I went through during anthropological fieldwork, were what opened up for me those numinous powers of the universe. It changed me profoundly and permanently. But when I returned to academia and began my career as a professor, I had to perform the same split between my heart/soul and my head/rationality that Miller sees as so common in children who are raised and educated in modernity. I then passed many years in a sort of schizophrenic split existence between my academic self and my spiritual self. It eventually induced in me a pervasive form of depression. Having to stick things out at Smith at least until my youngest child had completed college, I decided to take action and, along with a physicist at Amherst College, created a Five Colleges Faculty Seminar we called "New Epistemologies and Contemplation," which was dedicated to healing that split.[18] My physicist colleague immediately let me know we could not recruit interested faculty over the internet, only by word of mouth. I realized that, by the year 2000, faculty could come out of any imaginable sexual closet—but not out of the spirituality closet. My colleague and I tried to change things through this Five Colleges forum, but we were unable to institutionalize this space, give it academic legitimacy, or ensure its continuity, even though we had the strong support of the then chancellor of the University of Massachusetts. It all came to nothing in the end, but at least this seminar enabled me to emerge from my depression and face academic life with more equanimity. When my youngest child graduated from college in 2006, I decided to retire the next year. I was determined to create an alternative space with an alternative pedagogy and an emphasis on integral ecology.

I found it thrilling to learn of Lisa Miller's work, coming from the heart of mainstream academia and supported by her massive scientific research, which demonstrates that lived spirituality is one of the most effective and powerful preventatives against drug addiction, substance abuse, and a host of mental and emotional disorders plaguing youth and young adults today in the United States. To me, it shows how much things have changed over the last ten years.

Although I was still deeply in love with Odisha in eastern India, where I did my first fieldwork, I had come to feel that it was unacceptable for me to be there pursuing my own agenda, uninvited and without anything to substantively reciprocate for the gift of teaching me. I decided that if I were to work anywhere in the global South, I would only come if invited by people who wanted something I could share. This happened some twenty-four years ago after I met a representative of a Peruvian organization of intellectual activists at an international conference in Canada. The end result of this meeting was an invitation to collaborate with them in their work in Peru, which I did on a part-time basis for ten years. After this period, I collaborated with the fair-trade coffee co-op in Lamas for an additional four years. Ten years ago, I experienced a severe emotional collapse due to happenings both in Peru and at home in Massachusetts. I did not want to ever come back to Peru; my zest for life seemed to be completely extinguished. Having had recourse to the healing powers of Amazonian sacred plants many times before, I went on a forest retreat organized by Center Takiwasi in Peru. During the retreat, I took the sacred plant ayahuasca in an initial ceremony and, for nine days thereafter, another master nonpsychoactive plant while living in my tiny hut in solitude. All was silence except for the trees and the birds and the wind all communicating with me, keeping me company. This experience not only extracted me from the deep well of despair I had sunk into, but, through precise images that vividly captured key episodes of my personal life, it also revealed my own role in this collapse. I emerged from this retreat a new person.

From the teachings and subtle guidance I received from those spirit plants, as well as the more literal suggestion from my "adopted son" there, Randy Chung Gonzales, I founded a nonprofit organization called the Sachamama Center for Biocultural Regeneration, located in the Peruvian Upper Amazon. At its field campus, where I am writing these lines, I have sought to create a space where the split between heart and intellect, mind and body, nature and culture can be healed and where my ethical quandaries concerning anthropology can be effectively addressed. Here I teach study-abroad courses and programs for students from the United States and Canada, attempting to integrate intellectual learning with hands-on work by recreating the perpetually fertile pre-Columbian soil in the Amazon known as *terra preta* ("black earth") and applying other techniques. The aim is to share with Indigenous farmers an alternative to their slash-and-burn agriculture, a practice that is no longer sustainable since it contributes to deforestation and climate warming (Tindall, Apffel-Marglin, and Shearer 2017). Reviving the ancient soil technology originally developed by these farmers' ancestors was a practical response to requests by some local Indigenous leaders.

Besides this activity, I also try to open up students' hearts to the numinosity of this earth and the cosmos. We participate in rituals when invited by Native communities with whom we collaborate. We hold weekly gatherings in the evening to share experiences and engage in enactments that feel appropriate to the group with the aim of awakening our natural spirituality. All of this and more I have called the practice of "integrated pedagogy." It involves personal healing, embodied hands-on work with the earth, and, of course, intellectual pursuits, many of them introducing students to postmaterial science and the new materialism. In tandem with ritual enactments, these activities bring about both healing and understanding in the students.

In this part of the Amazon, Indigenous spirituality represents a union between ancestral shamanism and Catholicism. Pope Francis's ecological encyclical *Laudato Si*, along with his apology to the Indigenous peoples of the Americas for the horrendous way in which Catholicism was introduced and brutally enforced through its history on the continent, have transformed the confrontation between the church and Native Amazonians here into a fecund union. By issuing the encyclical, Catholicism has reinvested the nonhuman world with not only intrinsic value but also with numinosity. For the Indigenous and mestizo *curanderos* in this region, Catholicism and ancestral Indigenous spirituality are not experienced as antagonistic or separate forms of religions. This is one of the blessings of living in an oral milieu where memories of such a history have been blunted and lost their sense of horror.

Of course, I am not suggesting to my students that they should adopt this kind of spirituality; this would defeat the principle that truly healing and protective spirituality has to be a lived, embodied kind of spirituality. The sort of experience I encourage in the programs at the Sachamama Center is one that opens up the young to their own heartfelt, lived, numinous experience. Many come troubled by disorders in themselves or in their immediate family members. Their own testimonies at the end of the programs attest to the healing powers of their experiences. The example of transforming a brutal personal history into a process of healing offered by local shamans inspires us to transform our own experiences of the ravages of Western modernity into a healing union in the new postmaterialist and new materialism science. It is through such a path that we may rediscover the numinous and enchanted nature of the world and of ourselves as part of this world.

NOTES

1. For a fuller treatment of this topic, see Apffel-Marlin (2011, chaps. 2–3).
2. The Zohar was written in thirteenth-century Spain in Aramaic (a language no longer spoken). In Spain in the thirteenth century, Ramon Lull wrote a Christianized version of the Kabbalah (called Cabala to distinguish it from the original Jewish version) in Arabic. Among other features he introduced in this version was the notion that the unpronounceable, four-letter name of God was a foreshadowing of Jesus's name.
3. On supersessionism as a doctrine of the church, see Carroll (2001). The anti-Semitic nature of most of Martin Luther's writings is well known.
4. The classic work on these processes remains Karl Polanyi (1944).
5. Personal communication from Harvard economist Stephen A. Marglin.
6. For a fuller discussion of this, see Tindall, Apffel-Marglin, and Shearer (2017), especially chaps. 2 and 3.
7. On Boyle's invention of the scientific experimental method, the two classic works are Shapin and Schaffer (1985) and Potter (2001).
8. On the institution of the nation state being central to the problems of modernity, see Duara (2015).
9. During and after the scientific revolution, natural philosophers, later named "scientists," took it for granted that our minds are derivative of the Godhead, which is the origin.
10. On toxic cities, there are numerous media reports of massive air pollution, particularly in New Delhi and Beijing. On the science backing the causal relation between agrochemicals and depression among agricultural workers, see Weisskopf, Moisan, Tzourio, Rathouz, and Elbaz (2013).
11. See also Nandy (2001) on the practice of many Indigenous peoples in the first British censuses to respond to the question of their religious affiliation by citing a multiplicity of traditions. Even today, in several famous pilgrimage centers, some people who each have multiple faiths attend without any sense of contradiction.
12. On this point, see especially Thomas Berry (2006, 25–27), who lists six types of transcendence, all of which are, in my view, related to the notion of a God transcending this world.
13. I would not be surprised at all if the publication of her widely read book did not put a serious dent in her reputation as a first-rate biochemist, especially since it contains a foreword by

Deepak Chopra, a physician who openly uses Eastern spirituality in his writings, anathema to most academics and an excuse to dismiss his work.

14. This is also what Neils Bohr's quantum physics showed. For a brilliant exposition and extension of Bohr's work, see Barad (2007).
15. For years, my nearby nonprofit center, the Sachamama Center for Biocultural Regeneration (SCBR), has had a collaborative relationship with Takiwasi Center.
16. For more on the new materialism, see references in the "Introduction" in this volume.
17. I recognize that things have changed greatly in the field lately; however, anthropology remains overall a discipline dedicated to "representing" other worlds, a posture, I have argued (Apffel-Marglin 2011) that is a direct descendant of the colonial anthropology that created the field.
18. The five colleges of western Massachusetts are: the University of Massachusetts at Amherst, Smith College, Amherst College, Hampshire College, and Mount Holyoke College. An organization called "The Five Colleges" administers a series of shared academic programs in which students from any of these institutions can take courses.

References

Apffel-Marglin, Frédérique. 2011. *Subversive Spiritualities: How Rituals Enact the World.* Oxford: Oxford University Press.

Barad, Karen. 2007. *Meeting the Universe Halfway.* Durham, NC: Duke University Press.

Bateson, Gregory. 1973. *Steps to an Ecology of Mind.* New York: Ballantine Books.

Berry, Thomas. 2006. *Evening Thoughts: Reflecting on Earth as Sacred Community.* Berkeley: University of California Press.

Boyle, Robert. (1685) 2012. *A Free Enquiry into the Vulgarly Received Notion of Nature,* edited by Edward B. Davis. Cambridge: Cambridge University Press.

Carroll, James. 2001. *Constantine's Sword: The Church and the Jews.* Boston, MA: Houghton Mifflin.

Curry, Patrick. 2011. *Ecological Ethics: An Introduction,* 2nd ed. Cambridge, MA: Polity Press.

Descartes, René. (1641) 1998. *Discourse on Method and the Meditations.* Translated by F. E. Sutcliffe. London: Penguin.

Duara, Prasenjit. 2015. *The Crisis of Modernity: Asian Traditions and a Sustainable Future.* Cambridge: Cambridge University Press.

Hillman, James. 1992. *The Thought of the Heart and the Soul of the World.* Dallas, TX: Spring Publications.

Miller, Lisa. 2015. *The Spiritual Child: The New Science on Parenting for Health and Lifelong Thriving.* New York: St. Martin's Press.

Nandy, Ashis. 2001. *Time Warps: The Insistent Politics of Silent and Evasive Pasts.* New Delhi, IND: Permanent Black/Ravi Dayal.

New York Times. 2017. "Inside a Killer Drug Epidemic: A Look at America's Opioid Crisis." *New York Times* (January 6). https://www.nytimes.com/2017/01/06/us/opioid-crisis-epidemic.html.

Panikkar, Raimon. 2018. *Cultures and Religions in Dialogue,* vol. 6, part 1: *Pluralism and Interculturality.* New York: Orbis Books.

Pert, Candace. 1997. *Molecules of Emotions: Why You Feel the Way You Feel.* New York: Scribner.

Pert Candace. 2007. "Allostatic load: when protection gives way to damage" *The Praeger Handbook on Stress and Coping*; eds Alan Monat, Richard Lazarus & Gretchen Reevy. Westport, Conn. Praeger.

Polanyi, Karl. 1944. *The Great Transformation*. New York: Rinehart.

Potter, Elizabeth. 2001. *Gender and Boyle's Law of Gases*. Bloomington: Indiana University Press.

Pouillon, Jean. 1982. "Remarks on the Verb 'To Believe.'" In *Between Belief and Transgression: Structuralist Essays in Religion, History, and Myth*, edited by Michel Izard and Pierre Smith, translated by John Leavitt, 1–8. Chicago: University of Chicago Press.

Scott, James. 1998. *Seeing Like a State: How Certain Schemes to Improve the Human Condition Have Failed*. New Haven, CT: Yale University Press.

Shapin, Steven, and Simon Schaffer. 1985. *Leviathan and the Air-Pump: Hobbes, Boyle, and the Experimental Life*. Princeton, NJ: Princeton University Press.

Stengers, Isabelle. 2012. "Reclaiming Animism." *e-flux* 36 (July). https://www.e-flux.com/journal/36/61245/reclaiming-animism/.

Swimme, Brian. 1996. *The Hidden Heart of the Cosmos: Humanity and the New Story*. New York: Orbis Books.

Tindall, Robert, Frédérique Apffel-Marglin, and David Shearer. 2017. *Sacred Soil: Biochar and the Regeneration of the Earth*. San Francisco, CA: North Atlantic Books.

Toulmin, Stephen E. 1990. *Cosmopolis: The Hidden Agenda of Modernity*. New York: Free Press.

Tucker, Mary Evelyn, and John Grim. 2016. "The Movement of Religion and Ecology." In *Routledge Handbook of Religion and Ecology*, edited by Mary Evelyn Tucker and John Grim, 3–12. London: Routledge.

Tucker, Mary Evelyn, and John Grim. 2017. "The Movement of Religion and Ecology." In *Routledge Handbook of Religion and Ecology*, edited by Willis Jenkins, Mary Evelyn Tucker, and John Grim, 3–12. London: Routledge.

Vigne, Jacques (Vigyânânanda). 2008. "Ce en quoi la sagesse peut aider à sortir de la souffrance psychotique." https://sd-g1.archive-host.com/membres/up/d66fa6346a6d63b34ce41ac9eb948df942066239/Article_JVigne_Infos-yoga.pdf (Accessed August 28, 2017).

Weisskopf Marc, G., F. Moisan, C. Tzourio, P. J. Rathouz, and A. Elbaz. 2013. "Pesticide Exposure and Depression Among Agricultural Workers in France." *American Journal of Epidemiology* 178 (7): 1051–58. https://doi.org/10.1093/aje/kwt089.

Yates, Frances. 1979. *The Occult Philosophy in the Elizabethan Age*. London: Walton and Maberly.

CHAPTER TWO

Lost and Found

Gifts, Dreams, and Sanity

D. AHMED

We are lived by powers we pretend to understand. (Auden 1991, 249)

This chapter is based on excerpts about an initial period of events forty years ago that are part of an ongoing story. Condensed for coherence, it is a mosaic of certain motifs or themes linked with Indigenous cultural knowledge systems in South Asian Islam and the problem of understanding embedded in the Western psychological "gaze." In the first part, I present an impressionist "case history" of events and characters. In the second, I construct a framework for analysis, leading to the third part, where I make personal reflections on the subject. For reasons of confidentiality, all names and some personal details have been changed.

An Event, a Dream, and a Gift

Approaching forty, in many ways I had "established" myself professionally and socially in one of Pakistan's largest cities. I had a successful private practice as a psychotherapist and was on the verge of becoming a professor at a leading art institution. Among my friends was Samantha Winter, with whom I shared a unique relationship. She was American and had come to Pakistan as a Christian missionary and teacher of English literature at the college I attended after high school. I was sixteen and she twenty-six. After college, our lives diverged. Samantha returned to the U.S., and I embarked on different journeys: travel, marriage, children, and almost a decade as a graduate student at an ivy league university. But we kept in touch through letters and became close friends.

The cornerstone of our friendship was not only a love of Western classical music, literature, and the arts but also the writings of C. G. Jung. By my mid-twenties, I had completed graduate studies in clinical psychology, training with Pakistan's foremost Jungian analyst. However, I felt I needed greater exposure to other disciplines and more pragmatic therapeutic methods. I left for the U.S., where Samantha and I crossed paths again. As always, I enjoyed her intellectually stimulating company. We shared intimate aspects of our lives and relationships and would analyze/understand/reflect upon them within primarily a psychospiritual Jungian frame. I was grateful for this friendship since no one else in my circle of friends at home or abroad at that time was interested in spirituality, religion, or Jung.

After almost a decade in the U.S., I returned to settle in my home city to pursue a dual career in teaching and the practice of psychotherapy. Within a few years, Samantha, who had always been keen to return, came back to teach in a school, and I welcomed her companionship. Beyond routine, the challenges of juggling career and children, and living (by then) as a single parent with my mother, life was generally good and full. The situation enabled me to also have an active social life, including frequently hosting big parties. In short, professionally, personally, and socially, there was an (ego-driven) sense of having successfully "arrived." Or so I thought. Until around this time, a woman came to me for psychotherapy.

Tania, who was about the same age as I, belonged to the Pakistani upper class, was married to a professional, and had two children. She was a scientist trained in biochemistry and a successful businesswoman. Like many educated in the sciences, she was skeptical about psychology. But a friend recommended she consult me about certain disturbing symptoms. For example, during business presentations or meetings, she would randomly experience states in which she knew she was speaking but could not hear her own voice or it would seem very distant. She felt disembodied, heard sounds, and was constantly uneasy for no reason. There were no major issues in her marriage or with her children.

Her parents had separated when she was around ten. Her mother decided to leave Pakistan and settle in Europe, leaving Tania in the care of her maternal grandparents. She recalled how unhappy and angry she had felt initially, but with time, she settled into a pleasant, "regular" childhood of school, friends, and the like.

Tania spoke of her grandparents with great affection. Her grandfather was a little remote but always kind and loving. She particularly loved and admired her grandmother, who was a well-known political activist for the educational rights of Muslim women. Because of this political work, there were always many people around their home. Additionally, as is still common in South Asia, there was the inevitable stream of house guests and extended families of relatives, some of whom would stay for months, even years. On the whole, it was a lively household with plenty of people, activities, and playmates for Tania. Thus, although the absence of parents was initially distressing, under the aegis of her grandmother's matriarchal umbrella, she never felt unloved or neglected. As such, she did not think that her childhood may have caused any psychological damage.

After finishing her university studies, she entered an arranged marriage. Her grandparents passed away some years later. Apart from the usual problems of living with in-laws, her marriage was stable, her children were doing well, and her technology business was thriving. In short, there were no major traumas in the past or any ongoing crises of relationships. She came across as an attractive, intelligent, sensitive personality. Self-contained, soft-spoken, and low-key, she could be classified as an introvert.

The Event

Within days of our first meeting, a strange event occurred. I had finished work for the day, including a session with Tania, when she telephoned after dinner. She sounded anxious and asked if she could come over. I agreed.

She looked tense and agitated. Entering my study, instead of sitting in the usual chair, she began pacing up and down along the bookshelves lining the walls. Occasionally she stopped, glance at them, resume pacing, then stopped in front of a different set of books. She said she did not know what she was looking for, but it was not a book. After a few minutes of this mysterious searching, she finally stood in front of a shelf and, in one sweep, pulled all the books on it and threw them to the floor. She then reached toward the rear of the shelf and picked up something that looked like a paper folded over and over into a small rectangular shape. By now, I was a little alarmed. Her face was extremely flushed with beads of sweat, and her breathing was so rapid that I could see her chest pounding. She stood there holding something in her fist and before I could say anything, she asked me where the restroom was, as she needed to wash her face. Given her agitated condition, I accompanied her. After splashing cold water on her face, she quickly regained her usual restrained composure. I asked if she was okay and what it was that she had picked up from the shelf. She answered in a tone that was calm but commanding: "Go and get your mother." She knew that my mother lived in the same house.

All this happened within no more than about five minutes. I was suspicious but felt it best to simply go along and see what would happen next. So I asked my mother to come and meet this woman. Tania greeted her and handed her the paper. She did not want me to touch it. My mother opened it fold by fold until its contents were revealed: an indecipherable grid of horizontal and vertical lines in ink, each square contained different letters of Urdu, Sanskrit, and Arabic script, numbers, and symbols. Below the grid were a few lines of Urdu that I could see. It was an invocation to the Hindu goddess Kali Mata (Mother) to bring death and destruction on me and loved ones. My mother calmly announced that this was a "black magic" amulet and should be immediately disposed of in running water. This was done by a friend later in a river on the outskirts of the city. By the end of all this late-night drama, Tania was exhausted and went home. I was at a complete loss, my mind full of questions and suspicions about her behavior, motivations, state of mind. Perhaps she was on drugs and had not told me. Had she staged it all? Had she placed the amulet there herself earlier and, if so, how? Was this an attempt at manipulating me? If so, why, to what end? My questions went on and on like this.

When we met again, she told me her symptoms were worse. She could only explain the event as a compulsion she *had* to carry out without knowing why. The compulsion was accompanied by a recurring thought that had started imperceptibly during our session earlier that day but had grown to a point that she felt compelled to call me. The thought was that I was in some sort of danger. Once in the study, she sensed the danger around her at some physical level. Her body was acting like a radar that eventually led her to the exact bookshelf containing the amulet. Like me, she had also heard of such magical amulets but had never seen one. As a trained scientist, she herself did not believe in such things. She was as baffled as I and exceedingly distressed. Both of us vaguely knew that people in our society believed in magic, especially the poor and uneducated masses, but both of us regarded notions such as the evil eye and black magic as superstitions based on ignorance and lack of modern education.

Neither my mother nor Tania knew that for at least a year I had become interested in Hindu mythology. Although I had long stopped using Jungian analysis in favor of shorter, more current clinical methods, some of Jung's ideas remained productively applicable to many aspects of religious art and culture. Thus, for example, Joseph Campbell's perennially popular *The Power of Myth* (1991), based on Jungian concepts, was required reading in one of my courses. My interest in Hindu myths and archetypal religious symbols was thus driven by secular, academic motives. When something seriously interests me intellectually, I tend to become passionate, almost obsessive about the topic, and this had also happened with my reading about Kali, the Great Mother Goddess, particularly her iconography.

Some days after Tania discovered the amulet, it suddenly struck me that the particular figure I had been incessantly engaged with mentally for many months was the very same figure evoked in the amulet. Was this a random coincidence or was it what Jung called "synchronicity," that is, a *meaningful* coincidence?

A Dream and a Gift

Following the amulet event, Tania's symptoms became more severe. Along with experiencing hypersensitivity to surroundings, she heard distant voices that she could not understand, sensed a growing unease, felt disembodied, and so on. She was also dreaming a lot about her grandmother, so I asked her to record these. One entry in her journal was particularly intriguing:

> I see a very bright and big shaft of light. I am on one side of the light and on the other is my grandmother. She asks me to come across toward her, but I am terrified of going through the light. Grandmother tells me not to be afraid, she keeps reassuring me to come across. Then I wake up frightened.

The dream led to various associations by Tania. Her grandmother was a devout practicing Muslim. She would wake up Tania as a teenager not only for the pre-dawn prayer but often also for the late-night prayer, which is not compulsory in Islam although ranked highly for its spiritual value. Because of her grandmother's influence, she had continued formal prayer regularly for many years. However, somewhere along the way of raising children, working, and so on, this routine had slipped away. She had not prayed in years. While I was still trying to figure out the amulet event and where to locate Tania on the spectrum of "psychotic disorders," given her fragile and anxious state of mind, I suggested she should consider bringing regular prayer back into her life. Perhaps the dream was a reminder and praying might help.

Another noteworthy association concerned the death of her grandmother. Tania was in her twenties, married, and living in another city. Informed that her grandmother had had a stroke and was unconscious in hospital, Tania went to see her. At some point, with Tania standing by the bedside, the elderly woman opened her eyes and gestured to Tania to place her head on her (grandmother's) chest. Tania complied. Placing her hand on Tania's head, the grandmother said "I gift it to you" and died shortly after. Tania felt confused, but there was simply no time to ask what exactly her grandmother had meant. Over time, she forgot this episode, but almost twenty years later, its memory was triggered by the associations raised in discussions of her dreams with me.

A Clue from My Mother

In the absence of a professional mentor in the city and while respecting confidentiality, I would occasionally broadly mention interesting client "material" with Samantha or with my mother (who had a master's degree in social work from the University of Chicago and was a trained counselor). When I asked my mother about her reactions to Tania and the amulet, she responded quite casually saying that yes, magic is common and sometimes real, but one should not get too involved and regular prayer was enough "protection." She felt that Tania was sincere and not "acting out" but was also intrigued as to how she had found the amulet. I mentioned how Tania would occasionally feel inspired to write down some phrases from the Quran in Arabic but she did not know what they meant and neither did I.[1] My mother's knowledge was extensive, and she would give the meanings and locate the verse. These phrases were mostly around God's Attributes of Compassion, Love, Friendship, Protection, and repeatedly about not being afraid.

One day my mother, in her usual casual way, remarked that between the amulet and dreams, perhaps this was "the work of Khizr," a legendary/mythical Islamic figure. The name seemed only vaguely familiar. I could not recall the precise context, but had a hunch it may have been in Jung's collected works, which I had read many years earlier.

Luckily, I owned the entire set of eighteen indexed volumes and reread what it had to say in different contexts about Khizr.[2] It was not much, mostly brief, in passing. But a few references caught my attention. The main one was an essay/analysis of certain passages of the eighteenth chapter of the Quranic story about Moses and an unnamed person who in Islamic folklore is called Khizr (Jung [1940] 1969, ¶5).[3] Other short references mentioned how Khizr and the prophet Elijah were friends and did the annual pilgrimage to Mecca together. Another concerned the legend of Khizr having discovered and drunk the mythical "fountain of life," hence—like Elijah—he did not die but lives forever, can take many forms, and so on. All of this was interesting but seemed to shed no light on Tania, her dreams and symptoms. It was a start but there were too many unanswered questions. A bit like hearing only the opening bars of Beethoven's Fifth, again and again, then—silence.

Samantha was the only other person I knew who was familiar with Jung, so occasionally I also shared some thoughts with her. She had never read Jung on Khizr and her extremely negative reactions came as a surprise. Interestingly, while both she and my mother had considerable knowledge of depth psychology, neither saw Tania's situation in psychological terms, rather, they both gave semi-*theological* explanations. That is, my mother spoke of certain obscure Islamic motifs, while Samantha, who knew about the amulet, saw Tania in the grip of Satan and became worried lest I too became "possessed." I had always respected Samantha's views, but her ideas about satanism seemed far-fetched. My mother's clues seemed equally remote/obscure but with no judgment either way about Tania. These medieval and gothic explanations simply fueled my curiosity even as the momentum of bizarre situations around Tania increased.

But I did make one decision. Based on my mother's remark which prompted me to find *some* fragmentary information from Jung, it was clear that Khizr was *some* sort of Islamic archetypal spiritual figure. Although I knew nothing more, suppose my mother was right about Tania and the "work of Khizr"? If so, then I had to stop charging her for the sessions since, one had been told that money and spirituality are incompatible. My inner swirling thoughts around Tania were still of suspicion, skepticism, at times seeing what are considered "classic" symptoms of certain types of psychoses. Simultaneously, I was registering that in every way, Tania was fundamentally normal and was *herself* skeptical and afraid of what was happening. Eventually I decided to get money out of the way.

Thus, within a few weeks of our first meeting, I told Tania that I had decided not to continue in this "doctor/patient" relationship, but I was aware she was going through a critical phase and something powerful was in process. Additionally, I did not think it was malevolent or crazy, but also had no explanation for what was happening with her. However, I strongly felt that instead of resisting, if she were to "let it be," something important would emerge and I would be there for her until this crisis was over. Thus, it was no longer a client/therapist relationship. Rather, both of us were together on a quest into the unknown for answers about what exactly was happening with her. And as events unfolded, somehow it seemed that I too was part of this "script-in-progress."

The Wounded Healer

Soon, my own life started "falling apart." Other than teaching and practice, my social life completely receded. All my spare time was spent reading or talking with Tania. Led by Samantha and her concern for my welfare, friends and family thought I had gone crazy, or as she had told them, I was in the grip of Satan. The therapist herself had gone nuts! I must confess, for a while I became insecure regarding my professional reputation. My mother did not say much, but listened and asked me to keep praying. I became a recluse. There was simply too much to research/read—and to talk about with Tania, who by then had returned to regular prayer.

I became an object of "concerned conversation." Many relationships cracked amidst an atmosphere of suspicion and hostility. There was an unbridgeable void between myself and my peers. I tried explaining that Tania was not evil or manipulative, that this situation may have to do with mysteries of religion/spirituality, yet, I myself could explain no further. Their reactions were typically "modern," of bemusement or alarm, because only the illiterate believe in magic. Since I was *not* illiterate—after all, I had multiple degrees from the West—in the face of such nonsense the logical conclusion was that I had gone mad. The one friend who did not think it was nonsense was Samantha. But she was convinced that Tania had tricked and dragged me into a dark, diabolical domain. All I could do was carry this inner/outer upheaval, continue praying and trying to understand what was happening. By now, not only with Tania, but to the extent that I myself was getting entangled with some bizarre, paranormal phenomenon, it was clear that this business also concerned my own being as much as it did Tania's.

I continued searching for answers through all available means. Pre-internet days, this was mostly books. Slowly, Tania started feeling that prayer was leading her somewhere strange, at times during the ritual and otherwise. She started reporting "symptoms" as related to a heightened intuition accompanied by greater awareness of her surroundings verging on a hypersensitivity which was not pleasant. That is, she felt that she was randomly picking up a person's thoughts or feelings and emotions and found this deeply disturbing. She would often ask me to help stop whatever "it" was, she simply "did not want to know." Like me, she was constantly skeptical about herself. She (and I) would question if her perceptions, were her own projections, or not. For example, there were situations when she had met some of my friends or extended family members for the first time, and subsequently would tell me of certain conflicts/issues they were "carrying" and which only I was privy to. Similarly, it seemed that her intuition regarding the presence of malignant magic "on" or "around" a person started to intensify.

During this period (about a year), she started "picking" up in my patients, what she saw as symptoms of malevolent spells but as filtered through me. I must stress that these "diagnoses" were made by her reluctantly, but since the idea was to share and understand, she would tell me. Simultaneously, because I was groping in the dark, and she herself could not rationally explain an intuition (which by definition is "irrational"), I also remained skeptical. Additionally, it was all very well to "see" the presence of magic and the mental suffering it could presumably inflict, but what was the point if one could not negate/cure/alleviate its consequences?

The answer came directly. Two years before I met Tania, I had developed a steadily increasing pain in my leg. I had consulted medical specialists at home and abroad, gone through all tests, and so on, to no avail. Eventually, it was diagnosed as pinched sciatic nerve which, barring surgery, can only be managed with strong pain killers. My mobility was restricted and I had resigned myself to using the support of a walking stick, especially when standing during lectures. One day Tania told me that she would like to try to do something about my leg, which she felt was linked to the amulet she had found. Without going into the details of the prayer ritual which transpired, suffice it to say that within days the pain receded, and since then has not recurred. Subsequently, we tentatively "tested" this with my mother in pain.[4] I gave due consideration to concepts of "placebo," "autosuggestion," and the like but did not find them satisfactory. Tania herself approached it all in a tentative manner, not sure of herself, unable to explain much.

It is difficult to convey the sense of those days in which everything I thought I knew and was certain about, was stood on its head. There was confusion and fear that I had become entangled in a huge mess of my own making, made even messier by what was becoming a friendship with a woman perceived by others as full of cunning and guile and out to destroy me; while I saw her as a sincere, soft-spoken, gentle person caught in the eye of a mysterious storm. And who now seemed to have a capacity to heal. She was totally aware of what my friends and some family were saying about her. It must have been deeply painful for her to be judged negatively with derision, but she took it all with grace and dignity. I still cringe when I recall what I put her through. In tandem, there was the feeling of being gripped by a mystery, of encouraging her not to be afraid, and both of us seeking refuge and guidance in prayer.

The Wise Old Man

Then, again via synchronicity, help came in the form of the teacher who had first introduced me to Jung, Dr. Jamal, and with whom I had subsequently trained as an analyst. He was a renowned scholar, not only of Jungian psychology but also of Sufism, the mystical dimension of Islam. I had last met him at least ten years earlier at his home in another city. He arrived out of the blue—now the archetypal wise-old-man—and I shared my turmoil with him: That, in my view, Tania's "symptoms" were not pathological and beyond a certain point Jung was no help. I was full of doubt, confusion, fear and anger since many were convinced I was either mad or bad (in the grip of Satan). Dr. Jamal met Tania a number of times and reassured me that she was quite normal. He suggested some books on Sufism. I had known of his long-standing interest in this, but had never really explored the subject.

It has been said that mysticism is the great river flowing under all religions. Slowly, I began navigating different streams of an enormous body of writing spanning many centuries and continents. In the process, Tania and I gained not only new insights into Islam, but also a deeper understanding of other major traditions. One started to glimpse light at the end of the tunnel and the steady culmination of what has been called the soul's "wandering in the wilderness" or the "night sea journey."

With time, confusion and fear gave way to a different sense of excitement/exhilaration, renewed intensity in prayer, and vast amounts of reading on subjects which had me, so to speak, "spellbound." At this stage, I started being inundated with patients at a scale which I had not experienced in many years of clinical practice—and after some time—never happened again. It was not a question of some sensational "fame" of mine, because much of the gossip around me (and Tania) was hardly laudatory. More important, at this stage, Tania's "gift" of healing had barely begun to emerge. My long social retreat ensured that only a few individuals, such as Dr. Jamal and my mother, knew about it.

But during that period, strangely, there was a seemingly endless stream of patients, from all walks of life, even different parts of the country. The initial wave seemed to be overwhelmingly victims of magic, at least according to Tania. Of course, none of these mostly middle- and upper-class people thought that their (different) symptoms were linked with magic. If anything, they had come because of my Western credentials. Suggesting the idea of magic would have been utterly counterproductive for them and professional suicide for me. In short, my interventions, as suggested by Tania, were entirely abstruse and indirect.[5]

By around the second/third year, Tania was settling into a reluctant acceptance of the steady unfolding of her grandmother's gift. Its "practice" was primarily on my patients but only with me and my persona of the therapist/healer. In this phase, her diagnostic abilities steadily progressed from having to physically touch the person, to being in the same room to eventually a couple of rooms away. One had to handle these situations carefully and devise different scenarios, such as Tania sitting in the waiting/ante room posing as a patient, or as my "assistant" who would check blood pressure. Eventually she could "diagnose" from a distance.

What was fascinating for me, was that as such, my working methods did not change. That is, a person's psychological "case history," personality and related "issues" remained *fully relevant* to the therapeutic process. But in different ways, I would indirectly add whatever Tania suggested, such as giving a prayed over glass of water to a patient. In each case, I had a good idea of what change/improvement was possible in a given time span with different therapeutic techniques I had trained in. In short, I was keenly aware of the limitations of my discipline/practice and had considerable knowledge of what is possible or not with different ailments within a certain span of time. And there were just far too many cases of improvement/recovery to be explained away on the basis of my personal skills or as "mere coincidence" or flukes.

It was like being trained all over again, a crash course in the "new" (old/traditional) perspectives I had started immersing myself in theoretically, and their validity was being repeatedly demonstrated practically, in the clinical situation. In hindsight, simultaneously, it was clear that actually it was Tania who was being taught/trained in the ABCs of the gift of healing and given an opportunity to practice her knowledge via "my" patients. I, in turn, was going through a parallel process of unlearning/learning, not just about my profession in "mental health," but more importantly myself—warts and all. In a way it was a sort of team work between Tania and I, in which she was, at least for a time, both student and teacher, while I was primarily a student (and still am). Whether it was my leg or psychological/spiritual/religious understanding, it was she who was the healer/teacher and I the patient/student.

Intellectually it was a very "rich" period. There were intense conversations with Tania and Dr. Jamal whenever he was in town. Still curious, and unsure of myself, whenever possible I would get him to sit in on some sessions providing a sort of second opinion. By this time, Dr. Jamal himself had developed a warm and respectful regard for Tania. As may happen in the Sufi tradition, at some point, he asked Tania to accept him as her "disciple student" that is, to be his spiritual guide/teacher/guru. By then, I did not care about what anyone was saying. There was simply too much to learn.

It is impossible to fully describe the swirling mixture of emotions and intellectual ferment of that period. I have barely sketched some highlights of the early stages of a process which unfolded over a period of about five to seven years. By then, Tania had assimilated what was required. Until today, when I think it can really help, she continues through concentrated prayer and other spiritual methods to share her gift. Like Khizr, she is mostly invisible to its recipients, since by now her physical presence is rarely required. Lest I create an impression of reclusive piety, let me add that she never stopped working and continues with her public persona as a wife, mother and businesswoman. For various reasons, including her low-key personality, we mutually decided to keep all this "below the radar" of personal and social lives. Until today, few know of this aspect of our lives. Our families have become close friends, but still don't know much, nor are they interested. They good-naturedly see us as two eccentric, close friends who, a long time ago, were constantly immersed in conversations about spirituality and religion and who now refer to God as "S/He" with "Her" sense of humor.

As for the rest, most of them have seen over time that the catastrophe implied in Samantha's view of my friendship with Tania, has not happened. As such, I am pretty much the same person they have always known. I think they simply do not want to deal with anything beyond that. Nor do I have the urge to explain, as I had so desperately wanted in my earlier clumsy attempts to prove that neither of us were crazy.

Making Sense: Decolonizing Subjectivity

> Today has meaning only if it stands between yesterday and tomorrow. (Jung [1928] 1970, ¶153)

The search for understanding led to concentrated research/study/ on, mysticism, Islam, contemporary post-Jungian psychology, and the interdisciplinary fields of what is broadly called "cultural studies" and the critique of modernity. Over time, a loosely structured explanatory framework emerged for understanding not just one's self but also the initial diffused sense of social trauma and reactions of others. This section will broadly sketch a general conceptual framework located within the critique of modernity. Since the encounter with Tania was in the context of psychology and part of her "problem" concerned religion, the focus will be on psychology, culture and religion though not always directly. Thirty-five years later, the world has dramatically changed. Extremist and violent ideologies, including religions, are ascendant globally. Indirectly reflecting many aspects of the encounter, the frame is a broad stroked alternative lens for understanding the past/present—and possibly the future.

Modernity, Psychology, and the Crisis of Meaning

It is easy to forget that current crises, from economics and the environment to different forms of violent extremism, have their source, and answers, in the human psyche. Originally a dense concept, today the psyche, is primarily understood as the "mind" within the brain in the skull, and which constitutes our sense of "self."[6]

The concept of the self has been a long-standing, ongoing concern in feminism, which has extensively critiqued its construction in Western moral and political philosophy in the twentieth and twenty-first centuries (Willet et al. 2016). Even outside of academic philosophy, the self is a crucial concept, since it ultimately concerns our notions of humans, who we are, you and I.

As key elements of the conceptual scaffolding of the Enlightenment and modernity, psychiatry/psychology are quintessentially modern disciplines. They are based on neo-Kantian-Cartesian notions of the human self as a freely choosing ethical subject functioning on the basis of pure reason leading to pure moral truth which transcends culture. Feminism has long contested this disembodied vision inspired by "dead white males," a vision that ignores gender issues in general and women in particular and valorizes the self as primarily *Homo rationalis* and *Homo economicus*. Cast as a lesser form of male, woman is the Other and thus, the non-person, nonagent, nonsubject. "In Western culture, the mind and reason are coded masculine, whereas the body and emotion are coded feminine . . . to identify the self with the rational mind is, then, to privilege a narrow idea of reason and to masculinize the self" (Willet, Anderson, and Meyers 2016). This split between an emotional, nonverbal, "feminine" body, on the one hand, and on the other, a rational "masculine" mind is simultaneously a critique of the Cartesian dictum about the self: "I think, therefore I am." This leads to what is called "the Cartesian masculinization of thought" (Bordo 1986) and what feminists in different disciplines call the "logocentric," "phallocratic," "disembodied" mind in Western cultural and intellectual consciousness.

Long before academic feminism, Cartesianism was absorbed into the ideas of Freud and Marx, and this combined vision of self and Other dominated the twentieth century. Not coincidentally, both saw religion as an "opiate," a psychological aberration, and an "infantile illusion" without a future (Freud 1927). Despite feminism and contemporary psychology, it is crucial to recognize how deeply Freudian-Cartesian ideals remain psychologically internalized at a global level. Invariably listed as one of the "greatest minds of the twentieth century" by mass publications such as *Time* (1999); as a laudatory *Newsweek* cover article put it, Freud's views are now "equated with universal common sense"(1998). Given that his main quarrel with Jung was over religion and the fact that religion is back with a vengeance, this "common sense" needs to be scrutinized.

Simply stated, for Freud, a mentally healthy person must have a well-developed rationality and the ability to control through logic and will power, all impulses which run contrary to reason. This "ego"-self must govern *all* that is "me" including the dark, chaotic "id" and the religiously inspired "superego." Any impulse/thought/emotion contrary to rationality is seen as "pathology" and must be controlled by the ego. Bringing a different, more culturally contextual view, Jung recognized that all ancient civilizations had mythologies which were actually religious *and* psychological in substance. The mythic pantheons were simultaneously a reflection *and* a projection of the psyche's inherent diversity and of the transcendent realm. Male and female divinities were uniquely embodied, symbolic representations of diverse *psychological capacities* within humans, *and* also different dimensions of the Divine. Thus, the psyche is diverse, polyvalent, having "feminine," "masculine" and numerous other dimensions. In view of our inner multiplicity, Freud's rational ego-self is modeled largely on just one youthful male "god" among many, Apollo. In contrast to Freud's categories of id/ego/superego, for Jung, mental illness was basically an imbalance, a one-sidedness in the presence of many other dimensions within us—including the spiritual/religious.

This inner multiplicity of the many is evident when we are alone or in initial stages of meditation. Even in company, there is nothing particularly rational about this inner stream of un/consciousness as thoughts, emotions and ideas come and go in no given order. Frequently they are beyond our control. Roving from sexual desire, to ambition, betrayal, anger, and more, the psyche as we *experience* it, is recursive rather than discursive, not consistent but insistent about its wide-ranging concerns.

Given the therapy supermarket today, it may seem passé to bring up Freud/Jung differences. But the fact remains that these differences nevertheless continue as distinct forms of epistemology, that is *worldviews* underlying the entire psychotherapeutic project until today, including the medical. If Jung sounds "unscientific," it simply confirms the dominance of the Cartesian-Freudian paradigm.

Mythos and Logos

Putting it another way: For millennia, humans inhabited two psychological "worlds" and "languages" in tandem: a world of facts and a world of meanings. Based on rationality, *logos* enables practical human functioning, it is a world of quantifiable facts best expressed by science. In contrast, *mythos* is about "making sense," giving *meaning* to the impossible complexity of a range of emotional experiences we call "life." Its language is symbolic.

Science/logos applies Cartesian logic to material facts. By definition, a fact can only have one meaning. Water will always be two parts hydrogen and one part oxygen, no two ways (or meanings) about it. But when drinking, swimming or walking in rain, we don't think of it as literal H_2O. Its meaning/s will depend on our range of experience of it, and will be qualitative, subjective, multiple. Similarly, when depressed we speak symbolically of feeling "blue" or "down," which convey a better idea of the experience than a biochemical label.

In post-Jungian psychology, the symbolic is synonymous with "archetypes" and is *not* about allegory, literary metaphor, or the substitutive shorthand signs of science. An archetype functions similarly to other axiomatic conceptual principles such as "matter," "health," "society," "art," none of which can be fully described, circumscribed or clearly identified, yet, they conceptually hold entire worlds together and function as psychological realities (Hillman 1983b).

In spite of technological progress, at heart life remains the same mixture of insecurity/love/defeat/despair/confusion/joy/ambition, and the like. The archetypal is about this universal "stuff" of life, experiences which can be considered stereotypical and banal—except when actually experienced by the individual. All of us have experienced being "in the grip" of emotions such as love, grief, fear, or "stirred," "moved" by an event/situation. When we are emotionally impacted by an idea or feeling, and the discourse of ordinary life is inadequate to express it—such as 9/11—it is an archetypal experience. In short, the most significant events of our lives, from love to despair, engage us at a different level than, for instance, the literal molecular structure of water (Ahmed 2013).

Faced with extreme emotions, from joy to horror and more, the archetypal enable us to cope with the inexpressible by giving it a name, image, ritual, place. Deeply linked to culture, religion and human creative imagination, the archetypes of *mythos* provide emotional *containers,* enabling us to "say" what simply *cannot* be expressed in any other way. Since its meanings are ambiguous, multiple and subjective, *mythos* always contains an element of mystery, which at times we refer to as a "sense" of the sacred. As such, *mythos* also points to a parallel, "higher" plane of existence or transcendence. Functioning as a sort of cultural DNA, the symbolic offers guidelines and cues through various individual life stages and has a transformative potential. Today even non-Jungian scholars note that " it is a mistake to regard myth as an inferior mode of thought which can be cast aside when humans have attained the age of reason" and "We must disabuse ourselves of the nineteenth century fallacy that myth is false" (Armstrong 2005, 4, 117).

The central question about religious stories was never "Are they factually true?" but rather, "Do we need to know this?" and if so, "What do they *mean*?" The meaning spectrum of science/l*ogos* is narrow, restricted, and for it, these stories must either be "proved" or are nonsense. Overwhelmed today by *logos* and the Freudian-Cartesian paradigm of modernity, the expansive, multiple and capacious spectrum of meanings provided by *mythos* has become almost extinct or reduced to literalisms. Today, East and West, billions of believers take scriptures literally but they *also* have advanced training/credentials in modern rational thinking. The top leadership of al-Qaeda, for example, were/are highly trained in sciences such as medicine, engineering and physics (Ahmed 2010).

As one of the most ancient human concerns, there is nothing new about religion. What is new is the way we *think* of it (and of ourselves) in the language of *logos*. For those who can't or won't believe, the fact remains that people kill as much in the name of God as they kill without it. In the twentieth century between the "death" of God, the two world wars, Hiroshima and Nagasaki, the Gulag, more than 300 million people died violent deaths for reasons that had nothing to do with religion. Until recently, there have been no non-Western equivalents to Stalinism, Fascism, the Holocaust, apartheid, or the inventions of horrific chemical, nuclear, and biological weapons. All this occurred either in Europe or originated in the modern "secular" West. Overall, the sheer scale of killing and genocide in the previous century (the Holocaust, the Indian partition, Rwanda, Cambodia, the Balkans), not to mention the Oklahoma bombings and periodic "mindless" massacres of schoolchildren in Europe and the U.S., indicate that there is something inherently self-destructive within the modern self.

Modernity and Religion

From Jung in the last century to Karen Armstrong, today, there is a long-standing and widespread academic recognition of a "crisis of meaning in modernity" (Berger and Luckmann 1995; Roberts 2007). Jung had anticipated this much earlier, linking it to the religious, cultural, and intellectual history of the West. A post-Jungian archetypal reading of Western civilization indicates how, starting over two thousand years, symbolic diversity in psychotheological consciousness has been steadily obliterated in the West, particularly with the loss of the "feminine." Tracing a symbolic upward trajectory from south to north, the surfeit of Mediterranean goddesses gave way to the gender-diverse Greek pantheon, which was steadily overwhelmed by the Age of Heroes. Eventually Zeus, who was just one among the Olympians, gave way to an abstract Judaic God of the Law, culminating with the God-Father of Christianity. Eventually, between the Inquisition and valorization of the heroic, materialist ethic of Protestant Christianity, the vestiges of the Divine Feminine of Marian Catholicism were erased (Hillman 1983b).

Following the Reformation and the emphasis on science, modern psychology emerged by the turn of the twentieth century. Viewed from the Freudian lens of rationality and willpower, the first two diseases to be "discovered" were hysteria ("the wandering uterus") in women, and schizophrenia or "multiple personality disorder" in men and women. In sum, the West's loss of symbolic consciousness is the loss *of psychological and spiritual diversity.* The heroic-ego of reason and will-power is just one important aspect of the self. Its dominance leads to not only a devaluation of the many feminine aspects of human un/consciousness, but also of other styles of masculinity different from the adolescent heroic model.

Given the intertwined roots of psychology and religion in Western cultural history, much of what we call psychopathology can be seen as a secularization of "heretical" tendencies, which in the Western public imagination is represented by Joan of Arc and the witch hunts of the Inquisition. In short, in both psychology and religion, women are either mad or bad.

Cartesian Christianism

Exceedingly critical of particularly Protestant Christianity, the post-Jungian perspective sees Cartesianism as psychologically reinforcing the dogmatic foundations of Christianity, turning it into "Christianism" (Hillman 1983a). For example, until recently, Christianity was the only religion which has tried to "prove" the truth of its claims through sciences such as archaeology, carbon dating, and the like. Christ has been literalized into an exclusivist idea of a male person-as-God. By simply accepting this literal "fact," your salvation is assured. Evil cannot touch you because of the Nicene Creed, which is essentially *a mental belief to be regarded as an incontrovertible fact.* Not to accept this fact means that you are damned and should be "saved." The splitting of faith and reason, good and evil, resonates with the Cartesian mind–body split. As Hillman points out, "The Christian mind cannot allow a destructive possibility copresent with love and goodness . . . it splits the negative and projects in onto the enemy . . . the heathens, Jews, Catholics, terrorists." Accordingly, "The Christian heritage is constantly at work invisibly inside our feelings, reactions, and ideas, preventing us from seeing ourselves and our world . . . you and me, too, we can't help but be Christian" (Hillman 1983a, 78–84).

Hillman is not referring simply to fundamentalist Christianity but, rather, from a post-Jungian perspective, to a psychological "double whammy" of "masculinized Cartesian thought" and its harmonious cognitive resonance with Christianist dogma. The resulting psychotheological ruptures in the modern self have created a moral malaise in the Western psyche, leading to a worldview that can be called Cartesian-Christianism. In this worldview, the tendency is to split phenomenon into literalized, warring dualisms and morally reduce them to irreconcilable opposites.

This tendency also unconsciously prevails in the "secular" modern individual. For Jung, the "white man in general" was utterly unaware of his unconscious and the existence of evil within the self (the "shadow") as a universal force in the human psyche (Jung [1945] 1970, ¶431). Frankly speaking, there is much about us that is neither rational, heroic, or free of evil. Christianist dogma makes what is basically one's own evil into a metaphysical principle of the devil, which is then endlessly mulled over philosophically by the rational problem-solving ego-mind as the "problem of evil." Denied psychologically within the self through a process of moral reductionism, evil is projected onto others. "Christianity has made the antinomy of good and evil into a world problem and by formulating the conflict dogmatically raised it into an absolute principle" (Jung [1944] 1968, ¶ 25).

Between an unconscious Cartesianism, which claims the "light" of scientific reason and insists on singularity of meaning, and a similarly unconscious Christianism, which is based on a literal and a theological exclusivism absolving one of the darkness within, evil is seen in others, outside of the self. This morally reductionist vision sees all issues in terms of good/bad, either/or. Its heroic masculinism is evident today in globally popular film series such as Harry Potter and the Lord of the Rings. Reflecting a literalist, reductionist, adolescent view of both divine and human, these narratives rely on technomagical weapons of destruction, not philosophical wisdom. The view of the world this involves is a decisive battle between good and evil. The goal of victory is destruction, extermination, annihilation (Warner 2001), even though Christianity speaks of the ideals of love and forgiveness, just as other religions speak of options such as deliberation, negotiation, redemption, and reconciliation.

A decade before 9/11, Hillman's (1983a, 81) observation on Christianism and the modern person remains relevant: "Terrorism and nihilism are inherent to the (modern) Western worldview and system of thinking," which is rooted in the West's "religious unconsciousness." As he says, the "world-conquering force of Christianity was not inspired by Christian love" but by successfully "mobilizing the will, which needs fundamentalism or it does not know what to do . . . there (has to be) only *one* meaning, *one* reading of the text, for instance, the *one* meaning of Christ's suffering" (1983a, 82).

In sum, the modern mind, irrespective of religion, is overwhelmingly Cartesian, masculine, literalist, positivistic, Apollonian, Protestant-Christianist, heroic, adolescent, and morally reductionist. It is unable to deal with, and reacts with violence to, that which it considers different from itself, including the feminine, other modes of masculinity, the intermediate, ambiguous, and the symbolic/metaphorical (Hillman 1983b). From a psychological perspective, fundamentalism may be expressed in a religious or nonreligious idiom but is imbued with Cartesianist-Christianist principles. To reiterate, this is not to contest Christianity as a religion, so it is best to let Jung speak: "I do not combat Christian truth. I am only arguing with the modern mind" (cited in Post 1978, 234).

For Jung, it was not that Westerners are more evil than their ancestors; rather, given the distortions and internalizing of Cartesian-Christianist ideals, human moral capacities have not evolved in proportion to the modern ability to destroy on an unprecedented scale. "That is the great problem before us today. *Reason alone no longer suffices*" (Jung [1957] 1970, ¶574). The issue, then, is not with any religion per se or with modernity; rather, it lies in the heroic, modern, male, adolescent vision distorting our construction of both knowledge and meanings regarding self, other(s), and religion.

Penetrations and S/Permutations

As Habermas (2001) has observed, "Despite its religious language, fundamentalism is, as we know, an exclusively modern phenomenon." Promoted and internalized via the universally "civilizing" projects of colonialism, modernization, and globalization, the nature of "secular" forms of fundamentalism today exhibit similar Cartesian-Christianist features in which the world of facts dominates at the expense of meaning(s): excessive literalism, an exclusivist attitude, singularity of meaning, and a reliance on history as "fact." This historicism bears directly on our individual and collective consciousness, which today believes that reductionist historical facts determine us to the exclusion of everything else. It ranges from evolutionary psychology to genetics, to the "case history" project of how psychotherapy and psychiatry have reduced the meaning/s of life to chemicals or what happened in childhood, to rigid ethnoreligious ideas of identity and the creation of nation states based on the politics of nationalism and literalized spiritual geographies. Similarly, it is evident that it is the literalist-historical view of religious and political identity that dominates, even beyond the Muslim world, where it is often ascribed. Hindu fundamentalists, for example, also relied on archaeology and geology to destroy the Ayodhya Mosque, and there is a similar psychology at work in Jerusalem (Ahmed 2010).

From this feminist, post-Jungian perspective, we can see that our thoughts and ideas ranging from politics and health to science, development, progress, and religion are permeated by a heroic, adolescent, confrontational, patriarchal, fundamentalist machismo underlying the logocentric worldview of modernity. In short, "Western civilization has changed the world. Nothing—not even religion—will ever be the same" (Armstrong 2000, xii). Whether we are religious or "secular" (Cartesian-Christianist), as participants in a global modernity we are all influenced by what Hillman (1983a, 43) calls "This extraordinary religion, the religion that we are all in no matter how hard we try to deny it or escape it." We are all in the same boat.

Today, Hinduism is increasingly a Cartesian-Christianist-Hinduism; similarly, we have Cartesian-Christianist-Judaism, Cartesian-Christianist-Buddhism, and, of course, Cartesian-Christianist-Islamism. Religious or secular, these forms of fundamentalism are misogynist, hegemonic, hypermasculine expressions that, through psychological and theological means, artificially negate psychospiritual diversities in favor of logocentric supremacist thought. As such, the terrorism of political Islam (or of political Hinduism, Judaism, etc.) is one side of the symmetrical psychodynamics underlying post/modernity where there is only *one* choice: "You are either with us or against us." In the absence of the Feminine, it is a mirror-image response to various masculinist "penetrations": psychologically, of Western Cartesianist-Christianist modernity, and literally, of the penetration of a homophobic, homoerotic dialogue/battle of masculinist violence into a language of war (Latifa 2017, 241–323). Ideas, like Prozac and other pills, are not gender- or culture-sensitive; they can be swallowed uncritically by both men and women. Similarly, we can all be (unconsciously) psychologically impacted by archetypal ideas. It should be clear by now that simply being female is not a prophylactic against these psychologically heroic (s)permutations, as shown, for example, by Condoleezza Rice, Hillary Clinton, Margaret Thatcher, Marine Le Pen, and the many women who join fundamentalist movements (Ahmed 2010).

An Environ-Mental Crisis

Generally speaking, Hindu, Islamic, Chinese, Buddhist, Judaic, and even the early Christian tradition all originally had the capacity to heal the psyche. Today, increasingly they cannot, since in a globalized modernity, by now vast numbers of adherents and leaders exhibit the core modern malaise of an inability to comprehend the symbolic. Reduced to a set of "mental beliefs," processed in Cartesian terms as "facts," religion has been reduced to literalized texts and rituals devoid of mystery and meanings. Stemming from the West's unique psychospiritual history, vast domains of the psyche have been rendered "pathological" by the heroic ideals of rationality. As Jung ([1929] 1968, ¶54) put it, "The gods have become diseases." As usual, women bear the brunt of this.

Since the advent of psychology and psychiatry in the twentieth century, the psyche's inherent diversity has been pathologized, drugged, erased, obliterated. In 1952, the Bible of psychiatry, *The Diagnostic and Statistical Manual of Mental Disorders* (DSM I) listed sixty disorders. By 1980, DSM III listed 265 conditions, and by 1994, DSM-IV listed 365. In 1966, pharmaceutical companies were offering 44 psychotropic drugs; today, there are 174 psychotropic drugs available; periodically more illnesses are added and drugs developed (Elias 2009).

Psychiatric drugs have made the global pharmaceutical industry the most profitable business sector in the world. The combined global sales of the psychopharmaceutical industry is more than $400 billion. By all accounts, there is a psychiatric drug crisis in the U.S. (Greenberg 2013; Reidbord 2015). Americans are less than 5% of the world's population, yet they consume 66% of the world's psychological drugs (Fraad 2011). In the U.S., 1 in 6 people, including children, take some sort of psychiatric drug (Miller 2016). Between insurance companies, global pharmaceuticals, and mental health practitioners, the human psyche is now big business.

While pharmaceutical corporations promote the Freudian-Cartesian masculine model of "mind," the World Health Organization (WHO) claims that gender disparities remain a crucial issue in mental health. Worldwide, *more* women suffer from depression than any other illness (WHO 2002). Large-scale global epidemiological studies indicate that overall, mental health conditions are more common in women than men by a factor of 20–40% (Ball 2013).

Males are also not immune to the onslaught on the psyche. It is notable that in just forty years, there has been a worldwide 60% increase in suicide. The WHO report on suicide estimates that in the near future, two million people will kill themselves annually. With the exception of China and parts of India, the majority of suicides worldwide are males (WHO 2002). Most alarming are the high rates of depression among adolescents in Europe and the U.S., where suicide is now the leading cause of death in adolescents, especially males (Norman and Jarosz 2016).

Globally rising rates of suicide and the pathologizing/erasing of inner (psychospiritual) diversity are outwardly reflected in the current murder/suicide of the planet and the steady worldwide extinction of different species, cultures, languages, and so on. The word "psyche" means "soul" in Greek, and it was frequently represented by the symbol of a beautiful woman or a butterfly. One wonders how many people today understand the meanings of these symbols. From time immemorial, Nature/Earth was seen as Feminine, a Mother full of mysterious spirits and power. But between *Homo rationalis* and *Homo economicus*, it came to be seen as lifeless (like women and the colonies), "mindless," and as something that must be tamed, owned, and controlled, as in mind over matter. Today, the destruction of Nature and the steady extinction of environmental and cultural diversity symbolically mirrors the ravaged psychospiritual and cultural diversity of the human psyche. Together, they once constituted the notion of Anima Mundi, the "soul" of the world. Still mysteriously connected by a mutual reflection-projection, both psyche and Nature point to what is truly a *man*made, the environ-Mental crisis.

Besides the negative impact on humans and other species of industrial waste, recent research suggests links between dramatic changes *within* species and the huge amounts of chemicals contained in antidepressants. Wastewater eventually carries these chemicals from human urine into ecosystems, leading to what is being termed the "feminizing" of different fish and birds (Ford 2018). The word "testimony" is derived from the testes. Today, it is confirmed that, in the past few decades, men have been facing a "fertility crisis": sperm counts are dropping. Particularly in Europe and the U.S., these counts are almost 50% less than they were a few decades ago (McKie 2017). Research indicates different environmental reasons, particularly the burning of plastic waste and changes in lifestyle (Fetters 2018). Meanwhile, after becoming the fastest-selling drug in history, sales of Viagra and other drugs for erectile dysfunction are anticipated to reach $3.4 billion by 2019 (Transparency Market Research 2015). In short, logocentrically inspired medical "solutions" for human psychological well-being are making the environment sick. Symbolically, such unfolding "revenge" of the Feminine on a *logos*-dominated masculinity can be seen as Nature's way of restoring its critical life-supporting principle of balance (Murphy 2016).

By now in the West, there are probably more than four hundred "schools" of psychotherapy (Health.com 2010), which is an indication of the psyche's inherent diversity—but now in suffering. Trapped between being reduced to chemicals and a bewildering array of treatments, the demeaned modern mind struggles for survival in a literally degraded environment.

Given the destruction of nature, one can only agree with Jung that "the white man in general is scarcely in a position to judge his own state of mind. He is too deeply involved" (Jung [1945] 1970, ¶431). Jung was referring to the moral crisis unleashed in the West by World War II and the valorizing of science, which has led to an arrogant attitude toward Nature. "Under the influence of scientific enlightenment ... for the first time, we are living in a lifeless nature" ([1945] 1970, ¶431).

In a five-thousand-year history, the present environmental imbalance is a relatively recent development. As Armstrong (2005, 103) notes, "Western modernity is a child of *logos* founded on the technological replication of resources and constant reinvestment of capital." Psychologically, it is the adolescent, Freudian, Cartesian-Christianist "mind" of *Homo economicus/rationalis* who has changed the world in a way that has had disastrous environmental consequences. The (Darwinian) adolescent hero myth of material "progress" aside, there is no evidence whatsoever that we are morally or ethically better people than our ancestors. The latest "solution" to the environmental crisis comes from the scientist Stephen Hawking in his last statement before dying. He said that because of humans, planetary destruction is inevitable; the only hope for our survival is to consider colonizing other places in the universe. It seems the situation is hopeless—but not serious.

To conclude, the above conceptual "re-viewing" of the psychological underpinnings of modernity illustrates how epistemological assumptions are not "just" philosophical abstractions. They function as worldviews with impacts and consequences. It took years of engaging with Western and non-Western sources to eventually see through the colonizing self's "common sense" mirrored in the postcolonial psyche, including the self. As Nandy (1983, ix) says, "the concept of the modern West is a psychological category .... The West is now everywhere, within the West and outside, in structures and in minds." Nandy was writing around the same period I encountered Tania, but I had no recourse to his book until much later.

Personal Reflections, Alternative Perspectives

> So go before the teachers of conventional knowledge, busy yourself with jurisprudence and become master of the science of "this is permitted and that is forbidden" . . . by the time intellect has found a camel for the hajj, love has circled the Kaaba. (Rumi, cited in Chittick 1983, 226)

What I will now refer to as the Experience can be compared to a peapod containing personally meaningful themes, prefiguring the past, present, and future. Foremost is the inner personal spiritual/religious journey. I will not dwell on this deeply private inner domain, for reasons that will become evident later. Instead, the reflections will be around some "outer" significant thematic "peas" within the pod of the Experience. Moving among the personal and the symbolic, I have selected themes to fill in some narrative gaps and give closure to the story. They also illustrate the meaning-making process within alternative worldviews.

Before proceeding, however, I want to clarify that I am not promoting Jungian therapy or suggesting that his work should be seen as a guide to mysticism or the like. As discussed elsewhere (Ahmed 2000), Jung's substantive understanding of Islam and its forms of mysticism is frequently ignorant, at best, and prejudiced, at worse. Nevertheless, relatively speaking, this foundational figure of Western psychology offers, particularly to the modern Western(ized) mind, alternative approaches and ideas for understanding the psychodynamics of religion and its place in the human psyche. Jung's entire opus, like the work of post-Jungians such as Hillman, principally address the "white person." More than Jung, Hillman's post-Jungian approach of archetypal psychology provides a conceptual and cultural bridge for mutually understanding both the modern colonial *and* postcolonized "white minds." Its bridging function is even more relevant if one keeps in mind that, according to Hillman, "the second immediate father" of archetypal psychology (after Jung) was the Sorbonne philosopher and scholar Henry Corbin, whose lifework was on Islamic mysticism (Hillman 1983b, 2).

Spiritual Emergency and Holotropic States

Tania's symptoms involved what the psychiatrist Stanislav Grof has termed a "spiritual emergency," when the spiritual contents of the unconscious "erupt" in a seeming psychosis. Researching certain clinical conditions as essentially spiritual emergencies within a psychospiritual frame, Grof discusses how a spiritual emergency can evoke fundamental and profound changes in consciousness with no serious pathological impairment in functioning. That is, even as a person remains basically sane, his or her field of consciousness is simultaneously "invaded" by contents from other dimensions of existence in intense and overwhelming ways. In short, "each foot is in a different world." In some ways similar to states induced by psychedelics such as LSD and Ayahuasca, holotropic experiences can be seen as a main source of cosmologies, mythologies, philosophies, and religious systems that describe the spiritual nature of the cosmos and of existence. "They are the key to understanding the ritual and spiritual life of humanity from shamanism and sacred ceremonies of aboriginal tribes to the great religions of humanity" (Grof 2000, 3). For Grof, the "most surprising discovery" was the different attitudes to these states as perceived by Western medicine in contrast to all other societies. That is, most non-Western cultures held holotropic states in great esteem, developing techniques and "mental" technologies around them for ritual and spiritual life, including the cultivation of intuition. Notably, Grof "found it difficult to believe that contemporary psychiatry does not recognize their specific features and does not have a special name for them" (2000, 2). Today, some education programs pay lip service to culture, but in practice, three decades later, the situation remains the same.

Life stories of shamans, mystics, saints, yogis, and so on are replete with spontaneous episodes of holotropic states, which, if guided by those who have gone through similar experiences, frequently lead to pragmatic wisdom and healing capacities. This is not to say that there is no genuine insanity or psychoses in these cultures. But when one respects and enters these domains on *their* terms, it is evident that non-Western cultures *do* have certain markers that distinguish episodes of "spiritual emergency" from psychopathology. These markers are also present in Islamic mysticism in South Asia, but Tania and I did not know about them until much later.

The Gift

I gradually learned how Tania's experience and, to some extent, mine had parallels in kundalini and some other forms of yoga, different types of shamanism and their psychosocial requirements, certain schools of Zen and Tantra, and the practices found in different forms of mysticism. Similarly, the powerful spiritual impulse, "seeded" and cryptically gifted to Tania by her grandmother, is not unusual, although not common, within certain mystical traditions in Islam and other religions. Such transmission or transfer of different types of psychospiritual knowledge is a strong part of the teacher–student, guru–disciple relationship in Hindu, Islamic, and other forms of South Asian religious mysticism and numerous Indigenous traditions.

Spanning different levels of spirituality from prophets to saints to "ordinary" individuals, the transfer of the "gift" has the same broad contours, for example, in the Biblical story in Genesis of Isaac and his two sons, Jacob and Esau. Family politics aside, with Isaac on his deathbed, there was conflict about which son should get the "blessing" as his rightful "inheritance," which eventually went to Jacob. In a similar spirit, when he was close to his own death, Jacob bypassed his twelve sons and put his hand on the head of one of his grandsons, thereby granting him the blessing. The interpretation of the blessing is, of course, religion-specific, in this case pertaining to the "covenant" between the Jews and God.

In the present context, what matters is the archetypal and universal nature of the *idea* of a covenant as a pact or bond between individuals or between humanity and a transcendent power, which entails a precious nonmaterial blessing that *must* be continued and passed on. In this universal context, what is overwhelmingly a nonmaterial gift, the "inheritance," concerns specific sorts of knowledge that, depending on individual capacities, may or may not be transferable to literal, biological next of kin. Only the carrier knows and decides accordingly.

In the mysticism of Islam (and other religions), the presence over the centuries of carriers of such gifts led to numerous Sufi "chains of transmission" across the Muslim world, known as *silsilahs* with their specific genealogies. Like the living phenomenon found in Nature, some have flourished, many have "died out," and new ones have emerged. Many are in a state of decay, primarily because of the steady literalizing of the notion of "inheritance" as the exclusive right of sons. Nevertheless, until today, the entire South Asian landscape remains heavily dotted with innumerable Muslim and Hindu shrines and graves of individuals who were regarded as wise, "gifted," spiritually advanced individuals.

The gift was and is not a matter of "one size fits all." Rather, it is invariably tailored to the personality and predisposition of the recipient carriers and their historical, and cultural contexts. Besides giving spiritual guidance, they serve certain needs of the communities they live in, such as overcoming infertility, diagnosing physical and mental ailments, knowing about healing plants and the future, and so on. As such, the status of "saints" comes from within the society in which they live, not from an institution or the specific religion to which the person belongs. A visit to many major Hindu and Muslim shrines in India will confirm that the masses who continue to come are from across the religious spectrum. The idea is that the blessing (*barakah*) carries an energy of which individuals can still partake to have their prayers answered.

The relationship between Tania and her grandmother recalls what Schimmel points out in her classic study of mysticism in Islam:

> It is remarkable that in modern times Sufi teaching is, to a large extent, carried out by women again. Not only does the interest in the mystical path—modernized as it may be—apparently appeal to women, but some of the most genuine representatives of mystical tradition, "directors of the soul," in Istanbul and Delhi (and probably other places as well) are women. (Schimmel 1975, 435)

The Ghettoizing of Mysticism

Jung's severe criticism of Christianity and its negative impact on the "white man in general" was centered around the loss and eventual extinction of its mystical dimension that had been active in it until the Reformation. In the context of the Experience, although I got some initial cues (about Khizr) from Jung, I went on to discover that Jung's theories were a sort of "pidgin English" compared to an original language, the "real thing," namely mysticism. That is, he was attempting to explain to a Western(ized) reader the psychological dynamics linked to mysticism, which he saw as indirectly related to mental illness and health. All major religions except Christianity still have a mystical tradition embedded within them to some extent, as do the Indigenous worldviews. Although there are many mystical texts in early Christianity by, for example Meister Eckhart or Teresa of Avila, what is missing is the presence of the requisite guides or gurus, without whom mysticism cannot "live," since its practice involves some form of verbal and physical initiation and the transmission of spiritual knowledge from the guide to the student/seeker. After all, one cannot learn swimming by reading a book. Though extremely limited in his understanding of Islam, Jung nevertheless recognized that mysticism was its intellectual and spiritual "secret backbone" (1984, 336). Ironically, one reason why he is marginalized in mainstream Western psychology is the criticism that he is vague, archaic, and "mystical."

The Body in Question. In the academic study of religion, mysticism is regarded as too "anecdotal" and "subjective," and the ideas and frameworks associated with it are generally viewed as academically, even religiously suspect. However, by now, as a contrapuntal response to the Cartesian mind/body split, numerous perspectives are converging on the body as a central category of analysis, particularly the connections among women, the body, knowledge, and religion (see, e.g., King 1995; Irigaray 1985, 1993; Cooey 1988, 1994; Jardine 1985). Collectively, this suggests that "the body as a central category has reinforced critiques and assumptions taken as normative since the Enlightenment, especially regarding claims about the nature of reality and the human condition, and claims regarding certainty and objectivity in respect to knowledge" (Cooey 1994, 5).

As a central category, the body is inextricably tied with sexuality. Along with a wide span of feelings around the lows and highs of "love," sexuality and spiritual experience have traditionally been linked in the language of mysticism. Terms such as "passion," "ecstasy," "rapture," "union," and being "ravished" occur frequently in mystical texts. Many non-Western psychologies that Westerners consider to be "esoteric," such as certain types of yoga, Tantra, and Sufism, would consider sexuality and spirituality as two sides of the one coin. As Hillman states, "Initiation as a transformation of consciousness about life, necessarily involves a transformation about sexuality," observing that "the absence of initiation and of mysteries in

Western culture is one main reason for its preoccupation with sexuality" (1972, 64). Like much else, sexuality keeps changing through life's phases and has numerous expressions even in the phallic world: Pan, Priapus, Hermes, Dionysus, Eros, are just some Western male images in this context. According to Hillman, the figure of Jesus is a rare textual and iconographic exception, wholly omitting this phallic dimension. Thus, individuals in Western culture are given no God-image as an example for the initiation and recognition of their sexual being as a spiritual being and vice versa (1972, 64).

Without understanding the nature of the symbolic, it is impossible to comprehend mysticism. For example, the Freudian view of seeing long objects such as a knife or a stick as nothing but a "symbol" for the penis basically substitutes one material object with another material one. It is more of a "sign," having a single meaning similar to male and female figures for public restrooms. In contrast with Freudian literalism, "The phallus is not just a sign that indicates the penis; it is a symbol because it has so many meanings" (Jung [1958] 1970, ¶637 note 3). As such, it is actually the penis which is "nothing but" a phallic symbol.

For millennia, people have used phallic symbols without ever conflating them with the penis, as in the example of the Hindu lingam. Similarly, in Indigenous and mystical worldviews, the phallus can be understood as indicative of potency, power, and the cognitive ability to "penetrate" and "see through" to deeper metaphysical levels, including the creative power of healing. As Hillman amplifies, "The Freudian error lies not so much in the importance given to sexuality; even graver is the delusion that sexuality is actual sexuality only, that phallus is always only penis" (1972, 63).

Although it is impossible to summarize the varied aspects of mysticism across most religions, Henry Corbin's work on Islamic mysticism deserves a brief mention. He is convinced of the ontological reality of what he called the *mundus imaginalis* (imaginal world) (Chittick 1994), which refers to an intermediate world that is neither concrete reality nor fantasy but in between the two. It is filled with forms and images, populated by incomplete or invisible bodies (Corbin 1977, 126). Moreover, this intermediate world is accessible by the "subtle imagination" or "agent imagination" (Qadir 2018). Corbin uses the phrase *mundus imaginalis* to distinguish it from the common associations of the word "imagination." His point is that, since the Enlightenment, the faculties of perception have been reduced to either the sensory (physical) or intellectual (conceptual). Post-Cartesian thought firmly conceives of imagination as a subset of the mind but only in the pejorative sense of "the imaginary, that is, the unreal, the mythic, the marvelous, the fictive, etc." (Corbin 1977, vii). In contrast, "This [subtle] Imagination does not construct something unreal, but unveils the hidden reality" (1977, 12). As all the Islamic sages and theosophists agree, the subtle imagination can be achieved by spiritual adepts or mystics who "polish the heart," but in the end it is a divine gift that cannot be demanded.

The central role of initiation and its deep links with formalized "mystery" rites of passage that include the body leads to the adage, "Those who speak do not know and those who know do not speak." Despite widespread presence of mysticism in different parts of the world, including similarly spiritual approaches in Indigenous traditions, its absence in the West has played a major role in the discrediting and marginalizing of mysticism in academic and theological discourse. This is unfortunate, since mysticism holds the potential for unifying discord among religions.

Body as Symbol. Drawing on contrasting metaphors of the male and female sexual organs, mysticism can be considered the hidden, inner, private dimension of religion. As the "feminine" counterpoint to more outer, public, moralistic, codified, "masculine" expressions of and about religion, the language and practice of mysticism as a whole tends to focus more on the "inner," experiential dimension of the transcendent. Whether as the "void" or the "divine," the process must be experienced and as such, is inseparable from the body. In contrast, the modern academic view reflects the Cartesian mind/body split of seeing the mystical as "soft" data compared with the "hard" (phallic) facts of patriarchal, historically legitimized texts and codified commentaries and categories (Latifa 2017).

In its theories, worldviews, and practices, mysticism aims to psychologically "de-masculinize" consciousness by "feminizing" it. That is, outer masculine heroic-ego attitudes, such as mastery, control, self-assertive action, and narrow material notions of rationality must take a back seat to more "inner," contemplative, and receptive attitudes. "In relation to God, we are all—men and women alike—basically feminine. Macho insights reveal nothing of God" (Ong 1981, 77). This religious "feminization" of the psyche exists across all cultures and religions.

In his study of shamanism, Eliade (1964) first looked at "primitive" initiation rituals into priesthood, which involve a symbolic change of sex, such as transvestism, homosexuality, or living as a wife to another man. Such a priest is often called a "soft man." By contrast, a "man's man" like Hercules completed his labors and became a servant to Queen Omphale. Similarly, in Homer's *Odyssey,* Ulysses surrenders to the goddess Circe. Contemplation, due deference, and worship of the feminine are integral to Hinduism. The Buddha is most often shown with feminine aspects: full-bellied and soft-breasted, with large ears indicating receptivity, and sitting in the lotus position, representing compassion. In Judaism, the Sabbath is feminine, welcoming Shekinah who brings peace and joy (Hillman 1998). In Islam, the most often repeated name or attribute of God is Ar-Rahman, having connotations of compassion, graciousness, mercy. Its etymological roots are directly related to those of "womb." The word "Islam" itself means "peace in surrender to God's intention," as, for example, when Mary accepted her role of bearing Christ.

Highlighting the "feminine" aspects of mysticism is not to suggest there are no problems around it in non-Western religions. There are many, but that is a different order of debate (Latifa 2017). The point here is to identify some reasons why mysticism is intellectually largely ghettoized in discussions on religion, which leads to a distorted understanding of phenomena such as Tania's gift and other events that she and I were part of.

Within worldviews and cosmologies that regard the psyche as inherently predisposed toward spirituality or religion, everything in Nature has a symbolic aspect, which is why "primitives" perceived gods in a thunderbolt or a spirit in a tree. In hindsight, we can see the natural wisdom at work behind Tania's grandmother's (forgotten) gift. It surfaced at the right age, when both of us were entering our forties, having the basic maturity required to understand the gift's meaning and purposes and generally how to live with and "handle" it. Its facets and depths were revealed organically, step by step over time, teaching both of us through direct experience what we *individually* needed to learn, including about the self. We did not recognize this organic process until years later because we had no other language apart from that of "modern" education and its excessive reliance on narrow notions of scientific rationality.

Our grandparents and my teacher, Dr. Jamal, who instructed me on Jung, were born in the early years of the last century. They were possibly the last of a generation for whom the traditional idioms of religion/culture/spirituality were still part of "higher" education in which the Sufi worldview of Islam was normative. This is why both Dr. Jamal and my mother had some idea of what was happening right away, such as my mother knowing about the amulet that Tania found in my study, or Dr. Jamal recognizing the links between mysticism and Tania's psychological condition. Both "elders" quickly recognized the "work of Khizr," but Tania and I were clueless for a long time.

In "outer" life, the Experience challenged my rational problem-solving ego-self at different stages. Surviving painful, difficult social and personal situations, internal confusion, and the like eventually led to (different) forms of knowledge and a certain empowerment for both Tania and me. In a way, the "wounds and scars" of the healer become part of a larger story. That is, we are challenged to expand the narrow individual sense of "self," leading to a more comprehensive awareness of the interconnected nature of life and the flowering of individual potential. Levy's description aptly sums up the Experience in terms of how the roles of patient and therapist can be flipped:

> We realize we are all sharing in and playing roles for each other in a deeper, mythic, archetypal process that is revealing itself to us as it acts itself out through us. We find ourselves instruments being moved by a greater, invisible hand, as if something vast, with more volume than our previously imagined selves is incarnating through us. To recognize this is to have a more open-ended and expansive sense of who we think we are, and who we imagine

> others are in relation to us. The wound is not only a personal experience, but rather, it is a doorway, a hyper-dimensional portal into the transpersonal/archetypal realm, which is a higher order (in terms of freedom) of our being. (Levy 2014)

What my mother called "Khizr's work" is the guiding of certain spiritually gifted individuals into a domain of knowledge that cannot simply be gained from books but requires a radically different context and approach. Tania's struggles were primarily around the absorption of spiritually mediated "higher" knowledge and profound wisdom.

Beyond a genuine fear that I was losing my (rational) mind, my own issues were more mundane, such as a painful leg, worries about potentially losing professional credibility, or concerns around social acceptability, friendships, and the like. For example, my egotistical sense of having "arrived" professionally was actually a trap I had fallen into, a trap of complacency based in material and social "success." The severe leg pain was my body's way of trying to slow me down, preventing me from moving in self-destructive directions. Between work and socializing, my children were neglected and inner life reduced to ritual prayer on autopilot. In short, I was spiritually stagnating, "crippled" at many levels. As Levy explains,

> Our wound introduces and connects us with the transpersonal dimension of our being, whose realization, amazingly enough, initiates the transformation and potential healing of our wound. Simultaneously containing both the pathology and its own medicine, our wound is a higher-dimensional event which has manifested in the flatland of our third dimensional life. Symbolically encoded in the wound, uniquely tailored to our exact sensibility and aesthetic, is both the seeming "problem" and its own re-solution co-joined in a state of open-ended and boundless, indwelling potentialities . . . . It is literally prodding and prompting us to evolve into a freer, more coherent, and higher order of ourselves. Hidden in our wound is its own re-solution . . . our wound connects us to life itself. We don't cure our wound. It cures us. (Levy 2014)

In accordance with an individual's specific personal history, inclinations, interests, and so on, and as catalyzed by the Experience, these "indwelling potentialities" have continued to outwardly unfold in my life over the decades, primarily through writing and teaching in different countries and institutions about issues linked to modernity, psychology, masculinity, women, culture, and Islam (Ahmed 1995, 2002a, 2006a, 2006b, 2010, 2013).

Religion and Culture

Although Tania and I belonged to different Islamic communities, our familial environments had broad similarities. My father was an ophthalmologist, and my mother a university professor. Both had received several higher education degrees in the West and were professionally and socially successful. Both were devout practicing Muslims. I was the youngest sibling, with four older brothers, and there was no discrimination or constraints as to what I could or could not do as a female. From horse riding to competitive swimming, my teens were full of adventure and discovery, including books and activities such as music and learning to play the piano. We were taught the simple basics of Islam: the short liturgical prayer, learning to recite the Quran in Arabic, and memorizing some brief verses in it. Apart from the short early morning prayer led by my father, we were not forced to pray at other times of the day. Nor were we forced to "believe" in God. If anything, no question was considered sacrilegious, and we engaged in endless lively discussions about the existence and idea of God.

In a way, the idea of an ever-present, all-encompassing divine presence was routine. Handel's *Messiah* was appreciated as much as the devotional music of Hinduism and Islam. Evoked in daily life for both major and minor situations, one of the ninety-nine names of God would be mentioned or some legendary story about the prophets would be told. A crisis would become a reminder of divine compassion and wisdom, the first mangoes of summer a manifestation of its generosity and abundance, and so on. These reminders were mentioned casually, frequently in passing. During mealtimes, both parents would often share their dreams, especially my mother, who had noted them in a journal. The interpretations were invariably symbolic and spiritual. Although ritual observance had a place, it was embedded in and secondary to a view in which the transcendent divine was the ever-present religiocultural backdrop to daily life, although not in an oppressive or stifling manner. In short, my parents were not saints nor was family life unmitigated bliss, and, of course, all had their foibles and flaws. The point here is that they were not exceptional. Rather, like Tania's and my grandparents, millions of Muslims in South Asia were able to appreciate Western modernity to varying degrees, depending on their socioeconomic status, without a wholesale rejection of their own religion or culture.

I started praying regularly in my late teens and it remains a part of my life. Ritual should not be conflated with piety. Apart from obvious un/conscious influences of my parents and culture, my reasons were also deeply private and personal. Beyond the basics, I never felt the need to explore Islam or any other religion in depth. If anything, because of my education in Christian missionary institutions, I knew substantively more about Christianity. Ever since childhood, I read widely but never about religion. Prayer was satisfying and sufficient. As such, the Experience did not make Tania and me born-again Muslims. Rather, it was like a blast of oxygen released on embers of a dormant spirituality marginalized by our education, training, and lack of consciousness about the nature of religion. This expansive, nondogmatic religiosity, lived by millions in South Asia, was part of the invisible cultural ground beneath the feet of our grandparents and Dr. Jamal. Despite the distortions brought about in our modern non/thinking about religion, we simply had the good fortune of additional exposure to that generation's knowledge and wisdom, giving us critical clues for making "sense" of the Experience.

Magic and Religion in South Asia

Given the postmodern "turn" in the social sciences and humanities regarding the social construction of meaning(s), it is increasingly problematic to define terms such as "magic" or, for that matter, "gender," "science," or "religion." Entire disciplines, such as history, anthropology, and art, have been similarly destabilized and deconstructed by exposing their West-centric biases. In the context of science, religion, and magic, as Knight points out,

> Presented as separate from religion, magic becomes simply an inferior version of religion, powerless at best and demonic at worst. The idea of religion and science as irreconcilably separate from each other positions magic as a scorned third space, marginalized by both. Religionists concerned with proving their rationalism and compatibility with science often make efforts to separate their religions from anything magical, while many pro-science atheist thinkers charge that any distinction between religion and magic is flimsy at best . . . . In both religious and scientific contexts, "magic" becomes a slur. (Knight 2016, 5)

Hitherto an unknown and degenerate domain to Tania's and my modern-trained minds, for the first time we started noticing the widespread belief in magic (not just in South Asia but in much of the rest of the world, except, as usual, in the West). We realized that the vernacular press had long carried classified ads by practitioners promising results for problems ranging from unrequited love to finding a job to ensuring protection against enemies or the evil eye and so on. So we conducted participant-observer "field work," visiting different practitioners for imaginary problems, eventually leading to conversational interviews. In brief, what we discovered was a cultural vortex in which it became impossible to discern where Hinduism ended and Islam began, and vice versa.

Most practitioners would initially speak an Islamic idiom, saying that they relied on verses from the Quran or had learned "white" magic from some holy man. Within the South Asian religio-spiritual universe, there is the distinction between "black" magic (sorcery), which is destructive, and "white" magic, called "knowledge of light," which has healing properties. In many ways, they share commonalities, such as ritual stages, discipline, and so on. Ultimately, the difference between the two depends on the intention of the individual using it and what purpose the knowledge will serve.

Many black-magic practitioners mentioned that they had learned their craft in Bengal, which is a major Indian center of devotion to the great goddess Kali. This reference to Bengal is frequently appended to the name of practitioners in advertisements as testimony of their caliber and credentials. Like any human endeavor from medicine to music, practitioners of black magic also represent a spectrum of competence, and we encountered quacks, amateurs, and grand masters. Genuine masters of "light" were, by definition, more difficult to locate and interview. They know that whether gifted or attained as a gift, this knowledge cannot be used for any sort of personal material gain, so advertising is out of the question.

What is striking in both these black or white domains is the seamless intermingling of Hinduism and Islam. Many of the inner, private "work" rooms of the practitioners were replete with photo posters of the Kaaba in Mecca alongside images of the Hindu great goddess and symbols such as the skull and swastika.[7] Although most self-identified as Muslim, they were simultaneously full of awe for Kali, who was evoked with great respect. What began as simple curiosity about magic led us into a deeper level of culture in which seemingly very different religions became indistinguishable. It also vividly illustrated the false modern tendency to separate religion from culture.

Much later, we came to understand that in the context of a religio-spiritual journey, the occult is often simply the first threshold encountered by the spiritual "traveler." That is, the individual is sensitized to the existence of an invisible dimension frequently evoking fear and amazement. Usually dramatic, the occult encounter becomes a sort of "test" of individual intentions, functioning as a sort of checkpoint along a path that may ultimately lead to the *mundus imaginalis*.

Many people get intrigued, dazzled when encountering the phenomena of magic. Frequently, they get trapped in this domain, seeing it primarily as an instrument of power. This prompted my mother's casual remarks about magic being "real" but not something to get involved with. Fortunately, and thanks particularly to Tania, once its introductory functions were served and lessons learned, both of us moved on. As an illustration of the magic pea in the pod of experience, meaning, and purpose, thirty years later, I served in a dual capacity as a psychologist and expert witness on Islam and culture in a law case in the U.S. It concerned a person of South Asian origin who had a psychiatric history and had committed murder. The singular defense of the individual was that it was totally unintentional and happened under the influence of magic spells. For various legal reasons, the psychiatric history was not enough to avoid the death penalty, which seemed inevitable (fortunately, it was commuted to life imprisonment without parole).

It is impossible to understand a person in depth apart from culture, which can serve as both a symbolic canvas and a diagnostic aid about individual and collective health or sickness. Today the psychotherapeutic project in the West has barely begun to acknowledge that culture shapes our mind and mental health (McRobbie 2018). The study of human behavior needs to be strongly anchored in culture, and not only as it appears in mythology and religion. The roots of culture and religion are inextricably interlinked. Poetry, music, drama, epic, dance, architecture, and other arts emerged originally from religion.

As part of the Indian subcontinent, Pakistani, Indian, and South Asian Islam is primarily Indo-Persian, subsuming the rich mythos of two ancient civilizations, not to mention the multilayered world of mythos that is the Quran. Along with Urdu and Punjabi literature, this intermeshing of culture and religion has been fully visible in the exquisite beauty of Mughal architecture and Indian classical music and performing arts to this day. As mentioned earlier, what today is called Sufi Islam was, in many ways, normative across much of the Muslim world. This is not to imply that every Muslim was or is practicing the sort of mysticism Corbin describes. Rather, it was the expansive, tolerant worldview of most Islamic mysticisms that was normative. As such, people did not see or call themselves Sufis but simply Muslims. I am both a product and witness to this normative aspect of South Asian Islam, popularly referred to today as Sufism.

Jung called Sufism "the secret backbone of Islam" (1984, 336) since, like many religions, Islam is indistinguishable from culture. Today, this backbone has been severely damaged. Between Cartesian-Christianist modernity and the Saudi-Salafi theocultural bulldozer, the psychospiritual (cultural) diversity of Islam is being steadily obliterated. This is a disaster for any natural phenomenon, and religion is no exception. All religions are languages of the soul and, like languages, they have numerous dialects and accents. Today, there is increasing pressure for all Muslims to "speak" with just one accent, in one dialect, namely Salafist/Wahabist Islam, even though 85% of more than a billion Muslims are not Arabic speakers. This sort of cultural terrorism can eventually lead to the extinction of the Islamic spiritual and cultural rainbow, replacing it with the monolithic, monochromatic bigotry of Salafis/Wahabis and other hypermasculine expressions of Islam.

Gender, Islam, and the Divine Feminine

In her path breaking *The Tao of Islam*, Murata (1992) identifies profound similarities between certain Islamic epistemological frameworks and Chinese-Taoist concepts of *yin* (feminine) and *yang* (masculine). *Yin* and *yang* are the two principles of existence, namely, the active and receptive. Conceptualized as a circular symbol divided by a sigmoid line, with each half containing the other, *yin/yang* constitute the Great Absolute or Ultimate. Most non-Western spiritual traditions use similar systems based on polarities.

The critical difference between Cartesianist dualism and these psychological/spiritual worldviews is that the former sees duality as irreconcilable, warring opposites in which, regardless of context, one must prevail over the other. "Contradiction," "paradox" "ambivalence" are dirty words for the logocentric mind. In contrast, the centrality of polar relationships in mysticisms is, from the outset, never about antagonistic opposites or a "problem" to be resolved. Rather, it is primarily about (gaining) knowledge of self/other/Divine. In a human-earth-life context, not to mention the universe itself, opposites are simply a given. Seen as such, in Islamic philosophy and theology, opposites function as a complementary, mutually defining, qualitative heuristic device. In contrast to antagonistic duality, opposites serve the purpose of ultimately establishing the idea of Unity (in Arabic, *Tawhid.*) Thus, Unity/*Tawhid* manifests itself principally through the dialectics of polarity. The mandalas of certain types of Hinduism and Buddhism graphically represent this unity. Functioning as a meditative device, mandalas are frequently based on the "opposite" forms of a square in a circle, symbolic of the principle of unity in Creation, including the unity of consciousness.

The *yin/yang* approach can also be applied to both the Islamic and Chinese traditions. The legalistic Islamic and Confucian view emphasizes the *yang* (masculine) principle, whereas the sapiential Sufi and Taoist stress the *yin* (feminine) principle. It is impossible to summarize these gendered ideals in Islam, Hinduism, and so on; suffice it to say that there is a highly sophisticated, vast body of metaphysical thought developed over millennia across different languages and cultures, and one can only offer a glimpse.

Unlike Christianity, God in Islam is beyond gender. The "He" is more of a linguistic constraint than a theological imperative, and there is no such notion as a divine "Father" (or Mother). However, for centuries, numerous commentaries on the (more than) 99 names or attributes of God have categorized them as names of majesty/power (*jalal*) and names of beauty/compassion (*jamal*). Like all religions, Islam generally urges balance. As mentioned earlier, the divine name Ar-Rahman, meaning mercy, graciousness, compassion, is derived from the same root as the word for womb/uterus. In the Quran, it is the only name made synonymous with Allah and is the most repeated among the names/attributes.

Corbin sums up the vision of the mystic quest in Islam: "The spirituality of Islamic mystics is led esoterically to the apparition of the Eternal Womanly as an Image of the Godhead … because in her, it contemplates the secret of the compassionate God, whose creative act is a liberation of human beings …. The Feminine is not opposed to the Masculine … but encompasses and combines the two aspects, receptive and active, whereas the Masculine possesses only one of the two" (Corbin 1997, 159–60).

It is important to recall that the genesis of violent Taliban extremism was the Cold War between the United States and the former Soviet Union. Decades of war led to tens of thousands of orphaned children, and many of the boys found refuge in orphanages run by extremist theologians. Educated in the prevailing zeitgeist of war, they were similarly indoctrinated into an Islam that emphasized only the "masculine" attributes/names of God. Most importantly, these boys had virtually no exposure to women. Starting with their mother, even in ultraconservative cultural and familial environments, young males experience interactions with the "womanly" in the form of mother, sister, aunt, grandmother, and so on. In short, given the absence of the feminine in their psychological experience and theology, the feminine was obliterated from the consciousness of Taliban members. This is an extreme example, and, of course, most males are not raised in such orphanages. But it illustrates the destructive distortions that can occur in the unbalanced psyche today when it comes to religion, which regards the feminine as wholly "other," something to be repressed, controlled, or denigrated.

Some have speculated that, for example, the more "feminine" a religious perspective—in idea, image, content, expression, interpretations, and meanings—the greater the chance of it being labeled as "heresy," followed by violent attempts to marginalize, stifle, and obliterate it (Ahmed 2006b). Apart from Joan of Arc as a symbol of heresy, many Westerners forget that the witch hunts were conducted with equal zeal by both Catholics and Protestants, the latter continuing to kill witches until the end of the eighteenth century (*Telegraph* 2008). Whereas religiously justified violence against women is common to all religions, there is nothing comparable in any other religion beyond Christianity to the Inquisition and witch hunts, not just in terms of the extended period but also in the sheer scale of killing women (and men) on religious grounds. It indicates yet again the extraordinarily negative relationship of (Western) Christianity with the feminine. In Greek, the word "heresy" actually means to "choose for oneself." As such, heresy as a "position" will potentially be present in any overly institutionalized domain with established power hierarchies. Within religion particularly, it is a position of rebellion, resisting centralized authority (Ahmed 2006b). To this extent, the issue of women and religion is, worldwide, the last "last frontier" in our understanding of religion (Ahmed 2002b).

Today, these various motifs of the goddess and women's spirituality, along with feminist revisionings of various religions, not to mention Jung's emphasis on the feminine aspects of the psyche, have all steadily converged on the idea of what Simmer-Brown (2002) refers to as the "Feminine Principle" in her study of Buddhism. At a more general level, the feminine is essentially an archetypal constellation of a different consciousness. After centuries of being eclipsed, suppressed by the ascent of a heroic, masculinist modernity, today it is slowly rising again as an archetype in multiple ways: in feminism, the empowerment of women and women's scholarship; in ecological awareness and the increasing sense of unity with the planet, in the widespread urge to reconnect with the body and emotions; in the broad popularity of the Gaia hypothesis; and in scientific ideas about chaos, dissipative structures, and holism. The list could go on and on. The main point is that it is not always an image but a psychospiritual consciousness and it is this consciousness that must be reclaimed.

Conclusion

To recap and conclude the story around Tania and the essentially symbolic nature of the Experience/journey: It began as "just" another abstract academic interest in the idea of a "goddess" well before I met Tania. Unknown to me, at some stage, "it" had directly entered my life in a room lined with books (none of which proved useful for fully understanding what followed). After Tania discovered the amulet hidden behind the books, which symbolically evoked the Great Mother Goddess Kali, we set out in search of her. The journey began "magically" and we had no idea where it might take us. In Hinduism, Kali presides over not only death and destruction but also re/birth, in short, life itself. Friends were skeptical. One older woman, a foreigner who had lived in my culture for a long time and whom I regarded as knowledgeable, tried hard to discourage me from traveling. She meant well, but we disagreed. In quest of the Great Mother, Tania and I were sustained by both mythos and logos, namely, faith/culture and the aid/support of three "elders," two women and a man: Tania's grandmother, who passed on the "gift"; my mother, who gave me the first clue and guided me into another "type" of Islam; and my teacher/mentor in modern psychology, who eventually became Tania's "pupil"/disciple.

Through a long, tumultuous, at times painful period, the journey toward the Feminine simultaneously enriched us in multiple ways. In the process, we discovered Kali's many names and multifaceted "Face" within humanity, in other religions, and in Indigenous traditions. In our own tradition, as the Quran puts it, "Wherever you turn, you will see the face of God" (2:116). So we returned "home." I was keen to tell my foreign friend about the journey with Tania, but she had left the country.

Then the world changed in ways that were never anticipated.

* * *

Almost two decades after 9/11, predictions are being made that a third-world war is imminent, in which religion will play a major role. Muslim societies are increasingly being held hostage to fundamentalism and violence from within, projecting their anger on the West and seeing it as hypocritical and hegemonic. In turn, the West is driven by ignorance, a paranoid addiction to power, and misguided notions of technocultural "superiority," which it projects on different "others," including Muslims, regarding them as unenlightened barbarians. From both sides, the clash is not between "civilizations" but between different types of ignorance and dominant masculinist machismo (Latifa 2017).

As mentioned in the first part of this chapter, beyond Samantha agreeing about the existence of magic, I simply could not get her to understand anything from the cultural perspective of Islam. As I started to learn, I would try and explain different concepts/ideas to her but to no avail. Despite my respect and regard for her intellect, it started becoming clear that, compared with my knowledge and comfort level with Christianity and Western culture, she brought a certain arrogance and willful ignorance toward learning even the basics about Islam.

She had taught me Western literature, introduced me to its mythos, and we both liked Jung. Although a missionary, she had never tried to "convert" me. But her self-evident interest in religion and Jung had made me assume that her many years of living in an overwhelmingly Muslim society had surely piqued her curiosity to study the Quran as literature, even perfunctorily, if only to understand "the natives" better. She had never done this nor had she expressed a desire to understand this or that cultural aspect of Islam at any point over the years we had known each other. Whereas she had no qualms about quickly declaring Tania to be demonic, Samantha remained adamant until she left the country that her understanding of Islam was not insufficient but simply unnecessary.

The "image" of Samantha, then, is a final example of another incipient thematic "pea" in the pod of the Experience/journey. In hindsight, I realize that at a personal level, I was being taught by a different type of woman teacher/friend, who became the catalyst for my understanding a different world of the "imagination." Unlike our normal binocular vision, its visionary span is triple, containing past/present/future. The tensions between Samantha and me symbolically prefigured the long process of questioning and deconstructing my personal intellectual relationship with the West and, more generally, viewing anew its education and culture, much of which I admire and will always love. That "West" is important and still comfortably within me. For example, I cannot imagine my life and self without the English language and the myriad literary, philosophical, and academic gifts it mediated, including feminism. Similarly, many aspects of Western cultures, from different types of music to poetry and art, continue to deeply engage and delight me. But by now, I also discern certain other attitudes in that same education and culture that can be exasperating in their ignorance and arrogance about the non-Western world.

Corbin has observed that in archetypal situations, life events that come through as autobiographical data can at the same time become "charged with transhistorical meaning" (Corbin 1997, 38). Jung referred to this as the "higher continuity of history" in the human psyche (Jung [1911] 1967, ¶1). To this extent, Samantha, Tania and I symbolically prefigured future tensions between Islam and the West. 9/11 happened a decade later. Thus, the Experience was also a microcosm of future, larger issues that, by now, have taken center stage in the global arena. None of us could have made these connections at that time.

With time, it became clear that all people in the story were dramatic personae and symbolic catalysts. Speaking for Tania and myself, it was all as it was meant to be, so there is no question of blame or recrimination. Personally speaking, much of the discord with Samantha or other friends happened due to my own ignorance, naiveté, and stupidity. The persona (mask) of "therapist" had to be ripped away and the rational-material ego had to be dealt with in order to accept its ignorance, not standing in the way of my struggle against social and professional conformity. Not feeling stung by peer reactions of extreme derision and disbelief would have meant sinking into complacent arrogance, learning nothing—in which case this story would have been different. Life carries on yet will never be the same. Ultimately, I am left imbued with an ever-present sense of wonder and gratitude, of how, as in nature, we too were and remain part of an exquisitely interconnected web, a microcosm of a larger living unity.

Notes

1. The Quran in Arabic has almost the same alphabet or script as Urdu. Eighty-five percent of Muslims are not Arabic speakers, but the vast majority learn how to "read" the Quran without comprehending it. The latter is accomplished through numerous widely available translations and interpretations in different languages, including Urdu.
2. Also spelled in English and German as Khider, Khader, or Khidr.
3. All citations for Jung in this chapter follow the convention of indicating the paragraph number (signaled by the symbol) instead of the page number. Since so many editions and translations of Jung's works exist, this convention is used by Jungian scholars to avoid confusion in citation practices. The first date indicates year when Jung first published the work; the second indicates the year when the work was published as part of the Princeton University Press Bollingen Series, The Collected Works of C. G. Jung.
4. Here, Tania's method was different: it took the form of a written "prescription" around a few lines from the Quran and three to five "names" (attributes) of God to be recited a specific number of times in a specific order.
5. These ranged from suggesting small prayers from the Islamic tradition to giving a glass of water over which Tania had prayed to those who did not believe. It was all very exhausting for her, since it required intense concentration that left her drained.
6. Parts of this section have been adapted from Ahmed (2010, 2013).
7. The Hindu Sanskrit symbol of the swastika means faith, sincerity, devotion, achievement. Its lines end in curves with a dot in each of the four corners. It is a highly respected image in Hinduism. The Nazi swastika, more accurately referred to as *Hakenkreuz* (signifying Aryan racial superiority), is a variant of the shape that uses diagonal straight lines and no dots.

References

Adams, Vannoy Michael. 1996. *The Multicultural Imagination: Race, Color, and the Unconscious*. London: Routledge.

Ahmed, D. 1995. *Masculinity, Rationality and Religion: A Feminist Perspective*. Lahore, PAK: ASR.

Ahmed, D. 2000. "Islam and the West: A Psychological Analysis." *Journal of the Henry Martyn Institute* 19 (1): 3–62.

Ahmed, D. 2002a. "Real Men, Naked Women and the Politics of Paradise: The Archetype of Lal Ded." In *Gendering the Spirit: Women, Religion and the Postcolonial Response*, edited by D. Ahmed, 155–83. London: ZED Books.

Ahmed, D. 2002b. "Introduction: The Last Frontier." In *Gendering the Spirit: Women, Religion and the Postcolonial Response*, edited by D. Ahmed, 3–34. London: ZED Books.

Ahmed, D., editor. 2002c. *Gendering the Spirit: Women, Religion and the Postcolonial Response*. London: ZED Books.

Ahmed, D. 2006a. "Gender and Islamic Spirituality: A Psychological View of 'Low' Fundamentalism." In *Islamic Masculinities*, edited by Lahoucine Ouzgan, 11–34. London: ZED Books.

Ahmed, D. 2006b. "The Zikris of Pakistan: A Case Study." *InterCultural: Listening to Other Islams* 151: 7–22.

Ahmed, D. 2010. "Penetrations: Modernity, Fundamentalisms and Islam." In *Islam West: Crises Are Challenges*, edited by Marie-Claire Foblets and Jean-Yves Carlier, 53–69. Leuven, BEL: Leuven University Press.

Ahmed, D. 2013. "Decolonising Muslim Subjectivities: A Psycho-cultural Perspective." *Journal for Islamic Studies* 33:185–94.

Armstrong, Karen. 2000. *The Battle for God*. New York: Knopf.

Armstrong, Karen. 2005. *A Short History of Myth*. London: Penguin.

Auden, W. H. 1991. "In Memory of Ernst Toller." In *Collected Poems*, edited by Edward Mendelson, 249. New York: Vintage.

Ball, James. 2013. "Women 40% More Likely than Men to Develop Mental Illness, Study Finds." *Guardian*, May 22. https://www.theguardian.com/society/2013/may/22/women-men-mental-illness-study.

Barone-Chapman, Maryann. 2014. "Gender Legacies of Jung and Freud as Epistemology in Emergent Feminist Research on Late Motherhood." *Behavioral Sciences (Basel)*. 4 (1): 14–30. https://www.ncbi.nlm.nih.gov/pmc/articles/PMC4219252/.

Berger, Peter, and Thomas Luckmann. 1995. *Modernity, Pluralism and the Crisis of Meaning: The Orientation of Modern Man*. Gutersloh, DEU: Bertelsmann Foundation.

Bordo, Susan. 1986. "The Cartesian Masculinization of Thought." *Signs: Journal of Women in Culture and Society* 11 (3): 439–56.

Campbell, Joseph. 1991. *The Power of Myth*. New York: Anchor.

Chittick, C. William. 1983. *The Sufi Path of Love: The Spiritual Teachings of Rumi*. Albany: State University of New York Press.

Chittick, C. William. 1994. *Imaginal Worlds: Ibn al-'Arabï and the Problem of Religious Diversity*. Albany: State University of New York Press.

Cooey, M. Paula. 1994. *Religious Imagination and the Body: A Feminist Analysis*. Oxford: Oxford University Press.

Cooey, Paula M., and Sharon A. Farmer. 1988. *Embodied Love: Sensuality and Relationship as Feminist Values*. New York: Harper Collins.

Corbin, Henry. 1964. "Mundus Imaginalis, or the Imaginary and the Imaginal." *Cahiers internationaux de symbolisme* 6: 3–26.

Corbin, Henry. 1977. *Spiritual Body and Celestial Earth: From Mazdean Iran to Shi'ite Iran*. Translated by Nancy Pearson. Princeton, NJ: Princeton University Press.

Corbin, Henry. 1994. *The Man of Light in Iranian Sufism*. New York: Omega Publications.

Corbin, Henry. 1997. *Alone with the Alone: Creative Imagination in the Sufism of Ibn 'Arabi*. Princeton, NJ: Princeton University Press.

Eliade, Mircea. 1964. *Shamanism: Archaic Techniques of Ecstasy.* Translated by Willard R. Trask. Princeton, NJ: Princeton University Press.
Elias, Marilyn. 2009. "Conflict of Interests Bedevil Psychiatric Drug Research." *USA Today*, June 2. https://usatoday30.usatoday.com/news/health/2009-06-02-psychiatry-drugs-conflicts_N.htm.
Fetters, Ashley. 2018. "Sperm Counts Continue to Fall." *Atlantic*, October. https://www.theatlantic.com/family/archive/2018/10/sperm-counts-continue-to-fall/572794.
Ford, Alex. 2018. "How the Antidepressants Polluting Our Waters Could Change Animal Behavior." *Independent*, August 13. https://www.independent.co.uk/environment/antidepressants-effect-animal-behaviour-pollution-fish-birds-technology-a8489156.html.
Fraad, Harriet. 2011. "Profiting from Mental Ill Health." *Guardian*, March 15. https://www.theguardian.com/commentisfree/cifamerica/2011/mar/15/psychology-healthcare.
Freud, Sigmund. (1927) 1989. *The Future of an Illusion.* Translated by James Strachey. London: Hogarth.
Greenberg, Gary. 2013. "The Psychiatric Drug Crisis." *New Yorker*, September 3. https://www.newyorker.com/tech/annals-of-technology/the-psychiatric-drug-crisis/amp.
Grof, Stanislav, and Christina Grof. 1989. *Spiritual Emergency: When Personal Transformation Becomes a Crisis.* New York: TarcherPerigee.
Grof, Stanislav, and Christina Grof. 2000. *Psychology of the Future.* SUNY Series in Transpersonal and Humanistic Psychology. Albany: State University of New York Press.
Habermas, Jurgen. 2001. "Faith and Knowledge." Acceptance Speech for the Peace Prize of the German Book Trade. https://www.friedenspreis-des-deutschen-buchhandels.de/sixcms/media.php/1290/2001%20Acceptance%20Speech%20Juergen%20Habermas.pdf (Accessed December 1, 2018).
Health.com. 2010. "7 Types of Therapy that Can Help Depression." September 9. https://www.health.com/health/gallery/0,,20310354,00.html?slide=119070.
Hillman, James. 1972. *The Myth of Analysis: Three Essays in Archetypal Psychology.* Evanston, IL: Northwestern University Press.
Hillman, James. 1983a. *Inter Views: Conversations with Laura Pozzo on Psychotherapy, Biography, Love, Soul, Dreams, Work, Imagination, and the State of the Culture.* New York: Harper and Row.
Hillman, James. 1983b. *Archetypal Psychology: A Brief Account.* Dallas, TX: Spring.
Hillman, James. 1998. *Insearch: Psychology and Religion.* Dallas, TX: Spring.
Irigaray, Luce. 1985. *This Sex Which Is Not One*, edited by Catherine Porter, translated by Carolyn Burke. Ithaca, NY: Cornell University Press.
Irigaray, Luce. 1993. *An Ethics of Sexual Differences,* edited by Catherine Porter, translated by Gillian C. Gil. Ithaca, NY: Cornell University Press.
Jardine, Alice. 1985. *Gynésis: Configurations of Woman and Modernity.* Ithaca, NY: Cornell University Press.
Jung, Gustav Carl. (1911) 1967. "Introduction." In *Symbols of Transformation*, edited by Gerhard Adler and R. F. C. Hull, translated by R. F. C. Hull. Collected Works of C. G. Jung, vol. 5. Princeton, NJ: Princeton University Press.
Jung, Gustav Carl. (1928) 1970. "The Spiritual Problem of Modern Man." In *Civilization in Transition*, edited by Gerhard Adler and R. F. C. Hull, translated by R. F. C. Hull. Collected Works of C. G. Jung, vol. 10. Princeton, NJ: Princeton University Press.
Jung, Gustav Carl. (1929) 1968. "Commentary on 'The Secret of the Golden Flower.'" In *Alchemical Studies*, edited by Gerhard Adler and R. F. C. Hull, translated by R. F. C. Hull. Collected Works of C. G. Jung, vol. 13. Princeton, NJ: Princeton University Press.

Jung, Gustav Carl. (1939) 1977. "The Symbolic Life." In *The Symbolic Life: Miscellaneous Writings*, edited by Gerhard Adler and R. F. C. Hull, translated by R. F. C. Hull. Collected Works of C. G. Jung, vol. 18. Princeton, NJ: Princeton University Press.
Jung, Gustav Carl. (1940) 1969. "Concerning Rebirth." In *Archetypes and the Collective Unconscious*, edited by Gerhard Adler and R. F. C. Hull, translated by R. F. C. Hull. Collected Works of C. G. Jung, vol. 9, part 1. Princeton, NJ: Princeton University Press.
Jung, Gustav Carl. (1944) 1968. "Psychology and Alchemy." In *Psychology and Alchemy*, edited by Gerhard Adler and R. F. C. Hull, translated by R. F. C. Hull. Collected Works of C. G. Jung, vol. 12. Princeton, NJ: Princeton University Press.
Jung, Gustav Carl. (1945) 1970. "After the Catastrophe." In *Civilization in Transition,* edited by Gerhard Adler and R. F. C. Hull, translated by R. F. C. Hull. Collected Works of C. G. Jung, vol. 10. Princeton, NJ: Princeton University Press.
Jung, Gustav Carl. (1957) 1970. "The Undiscovered Self (Present and Future)." In *Civilization in Transition*, edited by Gerhard Adler and R. F. C. Hull, translated by R. F. C. Hull. Collected Works of C. G. Jung, vol. 10. Princeton, NJ: Princeton University Press.
Jung, Gustav Carl. (1958) 1970. "Flying Saucers: A Modern Myth." In *Civilization in Transition*, edited by Gerhard Adler and R. F. C. Hull, translated by R. F. C. Hull. Collected Works of C. G. Jung, vol. 10. Princeton, NJ: Princeton University Press.
Jung, Gustav Carl. 1984. *Seminar on Dream Analysis*. Princeton, NJ: Princeton University Press.
King, Ursala. 1995. *Religion and Gender*. Oxford: Wiley-Blackwell
Knight, Muhammad Michal. 2016. *Magic in Islam*. New York: Tarcher/Perigree.
Latifa, D. 2017. *Penetration and (s)Permutations: A Psychological Exploration of Modernity, Islam and Fundamentalisms*. Ghent, BEL: Yunus.
Levy, Paul. 2014. "The Wounded Healer," parts 1–2. *Awaken in the Dream*. Online blog. http://www.awakeninthedream.com/the-wounded-healer-part-1/ (Accessed December 1, 2018).
Lewis, Bernard. 1993. *Islam and the West*. Oxford: Oxford University Press.
Lincoln, Bruce. 2003. *Holy Terrors: Thinking about Religion after September Eleven*. Chicago: University of Chicago Press.
McKie, Robin. 2017. "The Infertility Crisis Is beyond Doubt: Now Scientists Must Find the Cause." *Guardian*, July 29. https://www.theguardian.com/science/2017/jul/29/infertility-crisis-sperm-counts-halved (Accessed December 1, 2018).
McRobbie, Linda Rodriguez. 2018. "How Culture Shapes Your Mind." *Boston Globe*, November 11. https://www.bostonglobe.com/ideas/2018/11/28/how-culture-shapes-your-mind-and-your-mental-illness/sMlhWP5LGSOvQAFd83I3qN/story.html (Accessed December 1, 2018).
Miller, Sara G. 2016. "1 in 6 Americans Takes a Psychiatric Drug." *Scientific American*, December 13. https://www.scientificamerican.com/article/1-in-6-americans-takes-a-psychiatric-drug (Accessed December 1, 2018).
Murata, Sachiko. 1992. *The Tao of Islam*. Albany: State University of New York Press.
Murphy, Lauren. 2016. "How Medicine Makes the Environment Sick." *Earth 911*. Online blog. https://earth911.com/living-well-being/medicine-makes-environment-sick/ (Accessed December 1, 2018).
Nandy, Ashis. 1983. *The Intimate Enemy: Loss and Recovery of Self under Colonialism*. Oxford: Oxford University Press.
Nandy, Ashis. 1998. "A New Cosmopolitanism: Toward a Dialogue of Asian Civilizations." In *Trajectories: Inter-Asia Cultural Studies*, edited by Kuan-Hsing Chen, 142–49. London: Routledge.
Newsweek. 1998. "Freud's Enduring Legacy: How His Ideas Still Shape Psychotherapy." April 7.

Norman, Alicia, and Beth Jarosz. 2016. "Suicide Replaces Homicide as Second-Leading Cause of Death among U.S. Teenagers." *Population Reference Bureau*, June 9. https://www.prb.org/suicide-replaces-homicide-second-leading-cause-death-among-us-teens/. [Accessed November 10, 2018]

Ong, Walter. 1981. *Fighting for Life: Contest, Sexuality, and Consciousness*. Ithaca, NY: Cornell University Press.

Post, Laurens. 1978. *Jung and the Story of Our Times*. London: Penguin.

Qadir, Ali. 2018. "Doors to the Imaginal: Implications of Sunni Islam's Persecution of the Ahmadi 'Heresy.'" *Religions* 9 (4): 6–9.

Reidbord, Steven. 2015. "America's Top-Selling Drug." *Psychology Today*, March 11. https://www.psychologytoday.com/ca/blog/sacramento-street-psychiatry/201503/americas-top-selling-drug. [Accessed November 20, 2018]

Roberts, Marc. 2007. "Modernity, Mental Illness and the Crisis of Meaning." *Journal of Psychiatric Mental Health Nursing* 14 (3): 277–81.

Said, Edward. 2002. "Impossible Histories: Why the Many Islams Cannot Be Simplified." *Harpers*, July. https://harpers.org/archive/2002/07/impossible-histories/. [Accessed November 1, 2018]

Schimmel, Annemarie. 1975. *Mystical Dimensions of Islam*. Chapel Hill: University of North Carolina Press.

Simmer-Brown, Judith. 2002. *Dakini's Warm Breath: The Feminine Principle in Tibetan Buddhism*. Boulder, CO: Shambhala.

Telegraph. 2008. "Europe's last executed witch to be cleared." June 11. https://www.telegraph.co.uk/news/newstopics/howaboutthat/2110937/Europes-last-executed-witch-to-be-cleared.html. [Accessed October 22, 2018]

The Message of the Quran. 1980. Translated by Muhammad Asad. London: E. J. Brill.

Time Magazine. 1999. "The Century's Greatest Minds." March 29.

Transparency Market Research. 2015. "Erectile Dysfunction Drugs Market Is Expected to Reach an Estimated Value of US$3.4 Billion in 2019." April 16. https://globenewswire.com/news-release/2015/04/16/725113/10129251/en/Erectile-Dysfunction-Drugs-Market-is-expected-to-reach-an-estimated-value-of-US-3-4-billion-in-2019-Transparency-Market-Research.html (Accessed December 1, 2018).

Warner, Marina. 2001. "Fantasy's Power and Peril." Ideas and Trends. *New York Times*, December 16. https://www.nytimes.com/2001/12/16/weekinreview/ideas-trends-fantasy-s-power-and-peril.html (Accessed December 1, 2018).

Willett, Cynthia, Ellen Anderson, and Diana Meyers. 2016. "Feminist Perspectives on the Self," rev. ed. *Stanford Encyclopedia of Philosophy*, edited by Edward N. Zalta. Digital encyclopedia. https://plato.stanford.edu/archives/win2016/entries/feminism-self/ (Accessed December 1, 2018).

World Health Organization. 2002. "Gender and Women's Mental Health." *Gender and Health*, June. http://apps.who.int/iris/bitstream/handle/10665/68884/a85573.pdf?sequence=1 (Accessed December 1, 2018).

Ziai, Hossein. 1995. "Mundus Imaginalis." In *Oxford Companion to Philosophy*, edited by Ted Honderich, 599. Oxford: Oxford University Press.

CHAPTER THREE

Spirit Crossings

A Nimipu/Tejana Cultural Perspective on Mortality and Death

INÉS HERNÁNDEZ-ÁVILA

At death, they say
everything inside us opens,
mouth, heart, even the ear opens
and breath passes
through the memories
of loves and faces.
The embrace opens
and grandmothers pass,
wearing sunlight
and thin rain,
walking out of fire
as flame
and smoke
leaving the ashes.
That's when rain begins,
and when the mouth of the river sings,
water flows from it

back to the cellular sea
and along the way
earth sprouts and blooms, the grandmothers
keep following the creation
that opens before them
as they sing.

Linda Hogan, "The Grandmother Songs" (1993: 57–58)[1]

My Nimipu mother walked out of fire as flame and smoke, leaving only her ashes. My youngest son Tom learned how to mountain climb so that he could take her ashes up almost to the top of Mt. Rainier in Seattle, her beloved city. This was her wish, and she looked me straight in the eye and said, "I don't want anyone holding on to my ashes." My mom was a private person, a "loner," she often said. She did not want a wake or a funeral: she wanted us. I am her only child; I was with her when she died. So were my husband, my sons, my oldest son's wife, and their children—we were all with her in the hospital room where she finally succumbed, when her small, frail body gave out. We talked to her even though she could no longer speak because of the morphine drip. My youngest son sang softly in her ear the lyrics to Frank Sinatra's "My Way," one of her favorite songs (and favorite actors): "For what is a [wo]man, what has [s]he got? If not [her]self, then [s]he has not/To say the things [s]he truly feels and not the words of one who kneels" (Anka, François, and Revaux 1969). She truly "stood tall," facing life fully, loving, bravely taking risks, sometimes losing, always in her own way. On her birthday in January 2011, she told us she wanted to see my youngest grandson graduate from high school. His father, my oldest son, told her, "No, Grandma, you have to see him graduate from college." She answered, smiling, "Oh, I don't think I can last that long!" By January of 2012, she was slowing down; her birthday was still full of laughter and joy, but by February, she was weaker, then in March, worse, and by April 19, she left us. She was ninety-one.

We learn about dying from the dying of others. We learn about living with mortality from being with others who also contemplate this challenge, this life lesson of the highest order. Indigenous people believe in spirit. Everything and everyone in the universe has spirit. Indigenous belief systems acknowledge spirit in a myriad of complex ways. When we are born, we come into our bodies from the spirit world—some say we even choose our parents. When we die, we go back to the spirit world; we say, "she crossed over," "he crossed over." In our ceremonies, we pray for safe crossings, peaceful crossings, protected crossings, crossings blessed with light.[2] Spirit is what makes me who I am. I am spirit inhabiting a body for my time on this earth in this life. I am connected to the spirits of those who came before me and those who will come after.

As a Nez Perce woman, I have always felt Hinmaton Yalatkit close to me. This is the Nimipu name of the man called Chief Joseph in English. His Native name means Thunder Rolling Over the Mountains. Because of my mother and her brother, my Uncle Frank, I feel a strong bond with this ancestor. When I am walking on an unknown city street in another country, a street that could become dangerous, I think of him; I ask him to be with me, and I feel his presence. When I traveled with my husband to the Wallowa Valley (the original homelands of Hinmaton Yalatkit's band of the Nimipu) and we approached Joseph, Oregon (near where the elder Joseph, the father, is buried), the rains came down hard, and yes, thunder rolled over the mountains to greet us. Hinmaton Yalatkit, Chief Joseph, died in 1904—of a broken heart, the reservation doctor said. He had been tireless in negotiating on behalf of his people. He is known throughout Indian Country and celebrated and honored for who he was and *who he still is*.[3] Hinmaton Yalatkit has come to me in spirit, in paintings, I feel him close always. During a stay at the University of California-Irvine in the fall semester of 2015, I was participating in a residential research group on a theme entitled "The History of Mortality." One evening, I was in the "porch" area of my apartment, burning some sage and cedar as I offered prayers. When I went inside, I sat in front of a small canvas I had on my easel, and I asked Joseph to come to me, to show himself to me. Suddenly, my hand reached for my brush and paints, and I started painting fast, very fast. I realized that his image was appearing on the canvas, an energistic impression of his image that came like a lightning bolt. I finished the initial image in about twenty minutes (Figure 3.1). I was astounded and supremely happy: he had heard me and had come.

According to our oral traditions, Indigenous peoples affirm that we have been from here since the beginning of time. When the invasion of the Americas occurred, the imperial projects, the settler colonial projects, went voraciously after everything and anything material that was of value, principally the land (and, in some parts of the Americas more than others, also the labor), using any and all force necessary, even, in many cases, outright genocide. As Ojibwe scholar Gross says, our peoples, Indigenous peoples, lived the apocalypse, and over the centuries and the generations, we have been dealing with a postapocalyptic world that has violently destabilized our lives as individuals, as communities, and as cultures (Gross 2014, 33).[4] The missionization campaign went after our spirits, with the intention of breaking them irrevocably. In 1875, in the United States, Captain Richard Henry Pratt uttered his now famous quote, "Kill the Indian to save the man"; Pratt was one of the perpetrators of the forced assimilation policies of the federally established boarding schools like Carlisle.[5] The intent of missionizing was the complete eradication of our knowledge of ourselves as Indigenous peoples; in this sense, since invasion, we have lived with the palpable specter of death in every aspect of our beings. Yet memory runs deep, it is what we have as original peoples

Figure 3.1. Visit from Chief Joseph (Artist: Inés Hernández-Ávila, © 2015).

of this hemisphere now known as the Americas. *The earth holds us in her memory just as we hold her in ours.* While it may seem strange to Westerners, Indigenous peoples understand that the earth, a sentient being, not only has memory, but she also has been and is witness to all that happens on her body—the earth knows history, writ large, just as our own bodies hold and know memory and history.

As Judith Butler states, "Loss and vulnerability seem to follow from our being socially constituted bodies, attached to others, at risk of losing those attachments, exposed to others, at risk of violence by virtue of that exposure" (Butler 2014, 20). Indigenous peoples did indeed experience loss and vulnerability when their original socially constituted bodies were damaged and wrecked, sometimes beyond repair. Butler writes of the "inscrutability" that follows: "[when] we lose some of the ties by which we are constituted, we do not know who we are or what to do" (2014, 22),[6] which is what Gross terms the postapocalyptic world that Indigenous peoples face on a daily basis. Butler also addresses the "derealization of the 'Other'" that makes of the "Other" someone who is "neither alive nor dead, but interminably spectral" (2014, 33–34). For centuries, Indigenous peoples have lived this derealization, even though we have not called it such. Butler asks the questions, "Who counts as human? Whose lives count as lives? What *makes for a grievable life*?" (2014, 20). Elsewhere (2009, 38), she explains that "[a]n ungrievable life is one that cannot be mourned because it has never lived, that is, it has never counted as a life at all." Reminiscent of Butler's notion of derealization, Martínez Salazar notes how communities that are seen as subversive (as with Indigenous peoples in Guatemala who defend themselves against capitalist and state terror) experience an "evict[ion] . . . from humanity" (2014, 144).[7] Martínez Salazar goes on to say that "the process of vilifying, prohibiting, and silencing the memories of agents of decolonial and progressive change perpetuates the human hierarchy in which only certain lives and cultures are deemed deserving, while the remainder is construed as disposable and inferior" (2014, 145)—ungrievable, derealized, evicted from humanity to the end of silencing memory. Butler points out that grief "furnishes a sense of political community of a complex order, and it does this first of all by bringing to the fore the relational ties that have implications for theorizing fundamental dependency and ethical responsibility" (Butler 2014, 22). For Indigenous peoples, such dependency is an interdependency that is a core principle regarding their ways of knowing and being. The relational interdependent ties extend beyond the human species to all of the nonhuman world, indeed a political community of a complex order.

For Indigenous peoples, our lives count as grievable lives because we have lived, regardless of how we might be seen or not seen by anyone else; as much as we might have been excluded over the centuries, as original peoples we are from this hemisphere known as the Americas.[8] This is not to say that other lives do not matter to us; on the contrary, when we offer prayers, we offer them for all of life, all of humanity, with specificity upon specificity. But we are the ones who remember our histories, our people's suffering—we are otherwise invisible in contemporary society, individually and collectively, with few exceptions.[9] If we don't remember, then who will? If we don't grieve, then who will? In so many ways, Native peoples are born with this grief, and we live with it all of our lives, every day, every night, every moment, every breath. Do we succumb? Sometimes, thus the high incidence of alcoholism, illness, suicide. Butler asks, "Is there something to be gained in the political domain by maintaining grief as part of the framework within which we think our international ties?" (2014, 30). Grief is definitely a part of the framework for Indigenous peoples, "remaining exposed to its unbearability," at the same time allowing for "an apprehension of a common human vulnerability" (2014, 30). Memory serves either to trigger lifelong despair or to allow us to embody the courage and dignity of generations, to live in the face of constant and ongoing death, to (re)create ourselves *because of* our mortality, and for the future and past generations. Richard Ray Whitman, a contemporary Yuchi artist from Oklahoma, participated in an exhibition held in Canada in 2000, which was called *The Presence of our Absence*. Indigenous peoples throughout the hemisphere manifest the presence of our absence. Commenting on Whitman's work, curator Paul Chaat Smith notes that for the Yuchi, there is no word for history: "What comes closest is the word 'memory'" (Smith 2009, 138). Here again, memory surfaces as a most powerful weapon against the efforts of centuries to make Indigenous lives "unmarkable and ungrievable" (Butler 2014, 15).

"Whose lives count as lives? What *makes for a grievable life?*" (2014, 20). For Indigenous peoples, the phrase is relevant to nonhuman forms of life as well. The historical slaughter of the buffalo to destroy a food source for the Plains Indians and to appropriate the buffalo skins for profit—the lives of these relatives are still grievable—are deaths that live on in the memories of Native people in the United States. The ongoing struggle of the Caribou Nation in the Alaska National Wildlife Refuge, to survive against ecological disruption, like the struggles of so many other sentient beings we consider to be related to us, makes for grievable lives. Indigenous peoples have belief systems that bind them intimately to the land and ecological worlds from which they come. This intimacy is articulated often by Indigenous peoples claiming their relationality with the nonhuman entities that are most central to their way of life. Thus, there is the Buffalo Nation, the Caribou Nation, the Salmon Nation, and humans are members of the same family of relations that bind together the buffalo, the caribou, and the salmon in interdependency. Native peoples, who see themselves as closely related to these nonhuman beings, recognize their immense responsibility in assuring the sustainability of the ecosystems for their relatives.

In August 2002, a fish kill or "die-off" of more than 34,000 chinook and coho salmon had devastating effects on the Klamath, Modoc, Yahooskin, Karuk, Yurok, and Hupa people from the Klamath Basin in northern California. The fish kill was caused by the "high water temperatures, low water levels, and toxic algae levels caused by the overuse of water by agriculture . . . [in spite of] warnings . . . in spring 2001 by Native scientists and in reports . . . to the Environmental Protection Agency (EPA) by the National Marine Fisheries Service and other agencies" (May 2014, xi). The word "kill" is more appropriate than "die"; in the Klamath language "*[g]leega*, 'death' means literally to change into or become something else. This is a natural process, a change that transforms life into life" (May 2014, xiv). The word *č̓ooq'atk*, on the other hand, means "'dead ones,' or corpses . . . unnatural loss," a loss caused by human error (May 2014, xiv).[10] This human (ethical) error, the intentional diversion of water for the interests of agriculture, created a "wall of collective grief" for the Indian communities of the Klamath watershed; the horrifying images of the innumerable dead fish, scattered in (sometimes deep) piles along the shores of the basin, have remained engraved in their hearts.[11] The fish kill spurred Indian people from grief to action.

The other factor impacting the sustainability of the salmon is the fact that four hydroelectric dams on the Klamath River cause blockages to the salmon on their return journeys upstream to spawn. In April 2016, it was announced that these four dams (three in California, one in Oregon) will be removed, opening up 420 miles of historic salmon habitat on the Klamath River (Gilman 2016).[12] If this comes to pass, it will be a major victory for the salmon and the people who depend on the salmon for their sustenance and who recognize their responsibility for maintaining an ecosystem that, in turn, will sustain the salmon. In the previous year, a news article announced that chinook salmon were returning to the Columbia River in record numbers. "Scientists estimate 200,000 chinook are spawning in the [Hanford] Reach . . . the most since dams were constructed in the 1930s" (Geranios 2015). Native people have had a part in this resurgence and these dramatic changes.

As Suzanne Crawford O'Brien (2014) explains, the Coast Salish people have been devoted to carrying out their ceremonies, including prayer, drumming, song, dance, and a ritual of beauty and honor to welcome the first salmon (catch). Once the salmon has been received,

> [t]he fish is filleted, and the head, tail, and backbone are returned to the young woman in the canoe, who receives the remains. As the canoe paddles far out into Puget Sound, songs and drumming continue. The gathered crowd on the beach stands silent. Even children seem to know that this is a solemn moment, aware that the remains of Chief Salmon are being returned to his village under the sea where he will gain new life, ensuring that salmon will return in years to come. We stand on the beach, watching the canoe as it moves toward Squaxin Island, Mt. Rainier looming snow-capped above. (Crawford O'Brien 2014, 2)

This ceremony to ensure the survival of the salmon, Crawford O'Brien reminds us, is also the affirmation that for this culture, water and canoes are "the place where spirit worlds meet, passage and movement between this world and the world of the salmon people, the world of the dead, and the world of the spirit (2014, 3). Living in the face of the constant dangers posed to the salmon, the Coast Salish people, comprehending the impact on their own mortality and culture, carry out their cherished obligations to their relatives, the Salmon Nation, to secure their continuance.

In the History of Mortality residential research group (mentioned earlier), Daniel O'Neill, who is researching the Fukushima nuclear disaster, showed the film, *Nuclear Nation: Surviving Fukushima* (Funahashi 2012). I was struck, yet not surprised, by the resonances between the community of Futaba, where the nuclear plant was located, and the many removals of Native peoples in the United States. The Futaba residents were not only displaced; they were also dispossessed and rendered expendable. Their attachment to home was and still is more than a material attachment; it is about their way of life, their centeredness, their world as they knew it, a world that ended with the combined disasters of the earthquake, tsunami, and meltdown of the Fukushima Daichii nuclear power plant. The film presents us a man who stayed with his cattle to care for them, even though he and they were immensely vulnerable, explaining, "They are my destiny." His response marks the interrelatedness, the interdependence of life, his own indelibly entwined with theirs; his statement, for me, marks an ancient (Indigenous) voice sounding the relation to the land and all living beings. He could no more leave his cattle than take his own life. He pointed out other cows, mummified ones who were left in other people's barns (tied up and enclosed), the horror of their demise, the profundity of their grievable lives. He bears witness with us and for us. Like him, Indigenous people of the Klamath Basin and the Coast Salish could say, "The salmon are our destiny." What is our destiny as humans? Of what will we say, can we say, "This is my destiny"?

Death, Dying, and Mortality in Nimipu and Tejana Cultural Approaches

I have had wise, generous, exemplary teachers on my life's path toward the understanding of spirit, from within my family (like my father) and from others I am grateful to have had as my guides and mentors. I have also had false teachers, coming earlier in my path when I had not yet developed (or perhaps acknowledged) my intuitive contemplation regarding matters of life and death.

As a Nez Perce woman, I did not grow up on the reservation, so my knowledge of what this identity means comes to me from my mother and her parents and siblings, as well as from frequent renewal trips with my mom to the Colville Reservation in Nespelem, Washington. One experience stands out for me dramatically in terms of my consciousness of being Nez Perce. When my mother's youngest sister died, my mom and I went to the funeral ceremonies. The wake for my auntie was held in the Nez Perce longhouse (since we have a longhouse tradition). The wake was officiated by the Seven Drum Religion, in the *wa'lat'sa*, our original way. The singers sang into the night—seven men, seven drums, all in a row—before them, the simple pine coffin covered with a radiant Pendleton blanket in the center of the earthen floor up close to the singers. Family members took turns going out onto the floor to ecumenically offer prayers, songs, words, from Indian Shakers, Catholics, Pentecostal participants, Seven Drum people. At those moments, the drummers ceded the floor; we were told that this time, during the wake, was the time to say anything we had to say to or about the one who had departed; the following morning would be only for the burial of the body. When the drummers began to drum again, members of the family and community who were moved to dance danced a strenuous dance around the coffin, an up and down step dance, a jumping dance, and at the sound of each bell signal, they jumped-danced clockwise, continuing until the song ended. For me, the dancers' steps marked the place in the cosmos where my auntie's body rested. At the finale of the night, the Seven Drum singers, starting with the elder who was in charge of the ceremony, began to interweave the two sides of the community (women on one side, men on the other) with the dancers, beginning with the eldest of the women and the eldest of the men (although the most senior elders did not take part but were supported from the sidelines). I was the second eldest of my aunt's nieces, so I was placed in line from the very start. The circle grew and grew, spiraling until we were a multitude, circle upon circle, embracing with our bodies, our dance and song, my auntie who was about to take her journey to meet our relatives on the other side. This was one of the most joyous moments of my life, and it was this experience of death and dying, of crossing over into spirit, that brought me home to who I am as a Nez Perce woman, home to a wondrous place within my community. There was profound healing brought to all by this ceremony, a space created that is of this earth and beyond this earth.

On my father's side, I am Tejana, Texas Mexican. When I was growing up, my grandparents insisted to me that I was *mexicana*, while my dad simply called himself American and my mom always reminded me that we were the First Americans. I was raised Roman Catholic. As I became involved in the Chicana/o Movement, I was part of the community who searched for the roots of our indigeneity. The use of the term "Chicana/o" was an absolutely political act, one that not only put into relief our Mexicanness but also the Indigenousness embedded in the Mexicanness. Chicana/o consciousness allowed for a humorous retort to the damning phrase "Go back to where you came from." Chicanas/os could claim, along with their Indigenous relations throughout the Americas, "We are from here, you go back where you came from!"[13] At the same time, the exact details of Indigenous identity, for many of us, were elusive, fuzzy, just out of reach (although I am clearer now than I was in the 1970s about my Indigenous background on my father's side). Some of us found answers in the Conchero dance tradition, known popularly in the United States as Aztec dance. The story of my involvement in this tradition is complex, but regarding the issues of mortality and death, the tradition has been fulfilling for me. It was through this tradition that I came to learn about *Día de los Muertos*, the Day of the Dead, in a much more profound way than is more familiarly known in popular culture circles, both in Mexico and the United States. However, even at the level of popular culture, there is still much worth mentioning. The festivities that abound in the United States around Día de los Muertos are vibrant, joyous celebrations of memory in the communities that follow these traditions—the altars (both personal and communal), parades, Aztec dances, songs, the explosion of flowers, all, in effect, "call out" Death, recognizing with utmost respect this ultimate crossing that loved ones have taken and reminding those of us who are still alive that our time, too, will come.

The Aztec dance tradition recalls ancient Mesoamerica, specifically the Nahuatl peoples (of which the Mexicas, otherwise known as the Aztecs, are only one group), who believed in a Supreme Being of many names. In the highest sense, this Being was known as the Mystery; then numerous other names distinguished the various significant aspects of this Being. One name is Ometeotl, the Dual God,

He/She, She/He; the female and male principles were/are recognized not only in the Supreme Being but in everything that emanates from Her/Him. The divine couples that represent the attributes of the Creation are denoted in Spanish as *Señor y Señora* and in English by Lord and Lady.[14] For the ancient Nahuatl people, the Lord and Lady of Mictlan, the Land of the Dead, were el Señor Mictlantecuhtli and la Señora Mictlancihuatl. According to León-Portilla, Mictlanecuhtli and Mictlancihuatl were aspects of Ometéotl, and Mictlán was a region of shadows (León-Portilla 1990, 99). In the Codex Borgia, we are told that Quetzalcoatl and Mictlantecuhtli represent the opposing forces of the Wind (its priesthood) and Death (Anders, Jansen, and García 1993, 56–57). They are portrayed back to back, indicating that life and death are inextricably intertwined. In keeping with the emphasis on duality, each is the "other" of the "other," and each entity has both positive and negative aspects. The focus on duality allows for a fluidity rather than insisting on binaries that are absolute. In the well-known image of Quetzalcoatl and Mictlantecuhtli in the Borgia Codex, we see how the iconography shows that death is within life and life is within death.

The popular cultural performances on the Day of the Dead speak to this aspect of Mexican folk practices and beliefs, the understanding that *la Muerte*, death, is always with us and that it can come at any time, so it's best to simply call it out, recognize it, become familiar with it, laugh with it, drink with it, dance with it, and pray with it. It is a bold in-your-face and face-your-fears tradition, one that Mexican society knows intimately. The popular culture manifestations can be rowdy, raucous fiestas with la Muerte.[15] In another kind of space, communities in Mexico and the United States visit cemeteries on the evening of November 1, staying throughout the night of November 2, cleaning the graves, placing marigolds and other sunlit flowers, lighting candles, and singing songs. The preparations for this night begin weeks ahead of time, and as the day approaches, vendors get ready with their *pan de muerto* (bread of the dead, sometimes in the form of male or female human figures, sometimes decorated with crosses that resemble bones), candy skulls (with people's names available to add to them), and the *papel picado* (decorative paper banners with cut-outs showing death in various outfits and contexts). Different communities have developed other regional variations, so the graveyards in the state of Oaxaca in Mexico might look slightly different from the ones in the state of Michoacán. Although Día de los Muertos is a special night in Mexican culture, families often make regular visits throughout the year to the gravesites of their loved ones, bringing chairs, blankets, food, drinks, and flowers to have an afternoon picnic with the departed.

In the Conchero tradition, as I have learned through the oral tradition, a specific ceremony takes place for Día de los Muertos. In this tradition, the most important ceremony of the year is the *Velación de Ánimas*, the "Vigil for the Spirits," an all-night ceremony running from November 1 into November 2 that is filled with song. This is when a collective altar is constructed, adorned with religious images (Catholic and Indigenous), photos and mementos of the deceased, and flowers (especially the *cempasuchil,* marigolds, associated with this particular time of the year). This activity is imbued with ceremony. Beeswax candles are lit, representing the thirteen *tlatoani* (the leaders of the people up to Cuauhtemoc, the last ruler of the Mexica, who was tortured and killed by the Spaniards, still called by many "the young grandfather"). Other candles are lit for loved ones, family members, mentors. In the opening permission song of the ceremony, the spirits of the elders are called in by name to be with the community celebrating this day, honoring the dead, lighting up their paths in the spirit world to help them on their path to a place of peace and grace. This tradition is a solar tradition from ancient times, thus the marigolds represent the sun, light, the fierce and loving energy that the ancestors bestow on those who remember them. We give them light and they give us light. In the Conchero tradition I have learned as a dance captain since the early 1980s, this is the one time of the Conchero year when dance does not take place. All attention is given to the ancestor spirits who left us these traditions and to our circles of loved ones who have crossed over into spirit.

It is always amazing to me that even though the candles are all the same size and are all lit at the same time, some can burn the entire night into the next day while others burn out in a few hours. The candles represent the spirits who visit with us on this night. The ceremony is full of song, instrumentation, such as *conchas* (armadillo-backed guitars), mandolins, gourd rattles, percussion, and sometimes flutes. The room that becomes the oratory is transformed into a sacred space with flowers, paper banners, lights, offerings of fruit, liquor, tobacco, favorite foods, candies, and especially the images and mementos that mean the most to the participants. Participants take breaks at specific moments in the ceremony to partake of nourishment, food, drink, conversation, laughter, storytelling—in short, *comunidad,* a sense of community. The ceremony is a life-giving way of remembering and recognizing spirit, and yes, death and mortality, but above all, spirit. The night is a long one; it is easy to become tired. But the tiredness is felt most during the breaks; when we return to the ceremony, we return to the concentration of spirit, and the tiredness drops off our bodies. We are in the spirit world.

My Own Bouts with Death on My Spiritual Path in This Life on This Earth

When I was a youngster, perhaps twelve years old or so, I used to think I was going to die when I was thirty. I don't know if I said it to anyone, but I know I thought it. It turns out that when I was thirty-one, I did almost die. The story is long and complicated, but the short of it is that I got involved with someone who introduced me to cocaine. I had never taken it before I met him, and when I broke away from him, I never again used it. While I was taking it, I nearly quit eating (I would maybe eat a taco every once in a while), and I started dropping weight drastically. People went from telling me how good I looked to worrying about me. I withdrew so much into myself that I almost stopped talking and I began listening with excruciating attention to everything that was said and reading any words that I heard or saw (street signs, billboards, flyers tacked to poles, graffiti, newspapers, anything) as if they were messages meant precisely and only for me. I created a narrative (I won't say in my head—I'm not sure where it came from) that let me know my doom was approaching. Not only that, but in this narrative, God wanted me to die, and because I thought it was God's message, I didn't question it.

I remember one of my friends standing in front of me, asking me, "*Inés, qué te pasa?*" I wanted to answer, but I felt like my voice was too far away and too weak to get to my mouth in time to respond. I was actually teaching at this time as a visiting professor (my sons, thankfully, were with my parents) and somehow managed to carry out my responsibilities. Later, when I got better, I told a woman elder what had happened, and she explained that I had almost split in two. Another grandmother elder, who was close to me because I was close to her grandson and lived in the same city as I did, would come looking for me. She had her daughter drive her through the streets searching for me; when she found me, she would take me home and give me tea and food. When I left, she would press a few pecans into my palm. I would take them home and eat them one by one at different moments. When I slept at night, I slept as if I were in a coffin, with my hands folded across my chest. I was already sleeping with death.

What can I say? I was in a car one night with two people, one of whom was the man I was involved with, the other a man who was an immensely creative person. I had the feeling that this night would be the night when I would actually die, when I would be murdered. I could feel the moment coming. It arrived. It came and went. We drove along and reached a community center, where we stopped to visit. I understood that the moment had passed. I was alive and I would stay alive. I was so grateful that I took off my bracelets and gave them to two women at the center who were strangers to me, but it didn't matter to me. One was a bracelet I had inherited from my mother's mother, which I had received at the giveaway when she died. Even long after that night had passed, I never said a word to the two men about what had been going through my being.

What was my life lesson regarding mortality and death? There were several. I realized that it was not God that wanted me to die; the spirit of the cocaine had claimed me for its own. I realized that the man I was involved with wanted, above all else, to control me. I found out later that he had gone to someone to put a spell on me (I am not one to believe in this kind of trickery, or *hechicería*—this experience was the first time I realized just how desperate some individuals can get, leading them to resort to something like this). I cut him off and immediately quit using cocaine. I went to see a healer woman, who cleansed me and blessed me. The man continued to stalk me for some time, but I sought out all the spiritual help that I could. He finally realized he meant nothing to me any more. He no longer had a hold on me. He himself was death, and I had faced him and won.

The other life lesson that I am tremendously grateful for is that I was surrounded by people who loved me, beginning with my family who always sustained me in prayer. However, my family did not see me in the deteriorated conditions I was in (except for my father, who told me that nothing was going to happen to me and I wasn't going to die, because he said so). My friends saw me much more often, and they were tender with me, attentive to me, even when I couldn't talk to them. They didn't even think I was crazy, which has always left me with a sense of wonder; they could so easily have taken me to a hospital for having a breakdown. Instead, they weathered this turbulence with me, and I came back to myself. My spirit had been displaced for a while, much in the same way, I suppose, that sometimes women who are being raped leave their bodies to get through the violence.[16] After I recovered, I went to the reservation with my mom, my dad, and my sons and took part in a name-giving ceremony for me and several of my cousins (who are, in our way, my brothers and sisters). As part of the ceremony, four of us women were brought into the center of the circle to receive special healings and blessings. This was a turning point in my life. And to think that when I was a child I foresaw death coming to me, coming for me, at that period of my life.

II.

When my father's mother died, I learned of it by phone; I threw some things into a bag and rushed downstairs to my car. I was living in Austin at the time and I would have to drive to Galveston. My grandmother was ninety-three, and I had been very close to her all my life. I am named after her. When I got to the car, I opened the trunk to put my bag in, and suddenly, I felt an energy like a vacuum at my crown, pulling me out of myself. A friend who was with me saw what was happening. He grabbed me and started shaking me, yelling to me, "No, Inés, no!" I came to and was able to stand on my own and proceed with my trip. I feel I almost left with her; she was taking me with her.

After my father came home from his six-bypass heart surgery in 1991 and had started getting physical therapy, I was on one of my visits to Galveston to see him (which I did thirteen times during a ten-week quarter when I was teaching). I was in my bedroom asleep and dreamed that he and I were sitting on the sofa talking. I couldn't hear words, but we were communicating. In the dream, he let me know that he was afraid of dying and wanted me to go with him. He held out his hand to me, and I was about to give him mine, when I woke up sobbing so hard that both my parents came to the door of my room to see what was happening. My father and I exchanged a look, and I knew that he knew what I had dreamed.

My father was *in tlilli in tlalpalli*,[17] the red and black, a man of wisdom, a man of faith, a man of his word. When he first came out of the bypass surgery, he called me to his side and told me his last wishes. He had me read the Catholic prayer for a good death for him and told me to ask people to pray for him. He was ready to go; he felt it was his time. But by a cruel twist of fate, when people prayed for him, he stayed alive—I realized later that he might have wanted the prayers to help him on his path, but the prayers kept him here. For the next nine long years, he declined more and more, suffering seizures and ministrokes, becoming increasingly unhappy and dejected, the medications wearing him down until he finally took his last breath in my oldest son's arms. It was at the beginning of this downhill weary road when I had the dream about him suddenly being afraid. He had been cheated out of certainty and plunged into existential crisis—he was ready for his god to take him, but he was left to wait. It was more than he could bear.

III.

In the autumn of 2009, I was doing an artist's residency in New Zealand, at New Pacific Studios in Mt. Bruce. I noticed that I was bleeding every day, strange for a postmenopausal woman. Once I got home, I saw the doctor in early January; he did a biopsy in his office and said to me, all smiles, "Everything's fine, no cancer!" I kept bleeding off and on. I went to another doctor, a woman gynecologist, who sent me to a lab to get the biopsy; she subsequently confirmed that I did, indeed, have uterine cancer. I went for surgery in early July, but on the morning of the surgery, my blood sugar was 323 (my husband had taken me out the evening before for a meal that was obviously too rich). The surgeon wanted to operate; he didn't want his schedule diverted. He ordered an insulin shot for me and a nurse rechecked my blood; the reading was still high. The surgeon then asked the anesthesiologist, right in front of me, if he could just put me on an insulin drip during the surgery. The anesthesiologist said no. After the surgeon left, the anesthesiologist told me that if he had used the drip, I might have gone into a coma and they wouldn't have known.

IV.

I had to work on getting my blood sugar levels down. I went back in late July to have the surgery performed. In the meantime, many people were praying for me, among them my relatives on the reservation. They are powerful in their praying.

When I went for my follow-up visit with the surgeon, I found him perplexed. He told me that the postoperative biopsy showed "no evidence of cancer." He could not understand how that could be. I told him that many people had prayed for me.

I then told him the story of my oldest aunt, who had been diagnosed with a brain tumor in January of 2009. She had slipped into a coma and her doctor told the family to prepare for her fast-approaching death. My mother and I traveled to the reservation immediately. When we arrived at the hospital (one that understands the needs of Indian people), we found my auntie's room full of relatives and friends. The room was full of prayer at all hours. Our family had taken over the waiting room nearest her room as well. Indian Shakers came to pray (my auntie was a Shaker); the Seven Drum religion people came, and so did many others.[18] On Wednesday, our third day there, my auntie woke up and moved her lips. The next day she was eating broth and talking slightly. By Friday, after another round of tests, her doctor came in to tell her that there was no longer any evidence of anything wrong with her brain. He said to her, "Iva, I'm humbled."[19]

After hearing this story, my surgeon was quiet a slight moment, then he said, "That's all well and good, but I still need to understand this. I'm sending these to Stanford." I returned to see him again, and the message was the same: "No evidence of cancer."

My oldest son, Rudy, had a part in my oldest auntie's recovery. I wrote the following to commemorate what happened on the day my auntie woke up.

> For Rudy
>
> On January 21, 2009, my saint's day, the day of Santa Inés, your auntie woke up. On this day, shortly before, your other auntie was keeping count of all the prayers, the songs, the blessings arriving for her sister. The Indian Shakers had come to pray strong, to cleanse our auntie, to lift her up with song. The Seven Drum traditional people came to offer their words, their respect, their honoring, their songs. Younger auntie was happy, but she said, "Something is still missing. She's waiting for something." It came to me to call you, and you answered. You must have been waiting, too. I asked you to talk to her, I put the phone to her ear, I could hear some of your words, I heard when you said, "Okay, Auntie, I'm going to pray for you now." As your words moved, she began to move, to try ever so slightly to sit up, twice she tried to pull her body forward. She opened her left eye and closed it, opened it and closed it. When you were done, very shortly after you were done, she opened both eyes, and we all cried out to her. At this moment, she showed us she was coming back. At this moment. You are my son. We both know that what you did was part of the whole, but you are my son. When I called, you answered. Your auntie knows this. She knows you were waiting. She knows you were right where you needed to be.

After the cancer surgery, I went for yearly check-ups for two years, then I stopped, although I have gone again more recently. But I have to ask, why should I doubt the power of spirit? If I doubt spirit, if I become afraid of the cancer returning or another cancer coming, is this not calling the cancer to me? In my family's way, we are always reminded not to claim an illness as our own—we say, "the diabetes," not "my diabetes"; "the cancer," not "my cancer." The Western medical establishment wants us to live in fear, to make fear a home in our bodies.

V.

I have diabetes. I was diagnosed in 1998 (coincidentally, when I was Department Chair for the first time—academia itself is deadening). When I went to the doctor because my vision was blurring and I could not satiate my thirst, the doctor took one look at me and said, "You're American Indian, aren't you—you have diabetes!" That was it, no cordialities, no niceties. "You're this, and so you have that." I was infuriated. For the longest time, my attitude was "whatever." For the longest time, I did not feel that the medical establishment was serving my needs regarding this disease; it is only very recently that I finally feel like I have a doctor who is working with me. I came to hate the medications, so many of which have egregious side effects. However, my father's mother had diabetes and went blind from it; she was blind for the last thirty years of her ninety-three years. My dear friend, Gloria Anzaldúa, died at the age of fifty-five from diabetes. A few years ago, I learned about another friend, a Huichol brother, Miguel Carrillo, who died at the age of sixty-four from the disease. My oldest son has it. It is one of the major killers of Native/Indigenous and Latina/o people in this country and beyond. Precarity? Here is the protracted precarity facing these communities, and the disease must be historicized. The assault began with annihilation and then the violence of utter dispossession and forced assimilation. The related disease in Indian Country is alcoholism. When our worlds turned upside down, we lost many of our abilities to care for ourselves; recuperating them is part of our rebuilding today. Diabetes is my ongoing bout with death, my ongoing reality check regarding my own mortality.

What Do I Want When It Is My Turn to Cross over into Spirit?

I want to be ready. I want to have taken care of those things that pertain specifically to me so that no one has to wonder what to do. I am already thinking of these things; my husband and my sons would rather I didn't, but I told them it's good, I am not afraid. I hope I have the quiet time of reflection, if not exactly on the last day, then in the days leading up to the moment. I want to have time to remember the many lives I have lived. Many years ago, decades ago, a male friend who is a love of my life (an impossible love, but still a love) asked me earnestly, "Inés, will you think of me on the day you die?" I was stunned with the question, surprised that he had gone that far forward in time. A tenderness came over me, and I told him, "Yes, yes, I will." He was wise in presenting me with the question so early, because it has become engraved on my consciousness. He will figure in my recollections, for sure, as will many others.

My formation as an adult, from the time I was an undergraduate, was shaped by social movements. I was an activist and a cultural worker during the 1970s, and I have brought those social organizing skills, the critical and creative consciousness from that time, and the thirst and vision for social justice transformation throughout my life. When I read the letter written by the Russian revolutionary Natal'ia Klimova before her scheduled execution in 1908 (Morrissey 2006, 2015),[20] I found her spirit beautiful. I understood how she had made herself ready for death by realizing life in the way that she did, by coming to terms deeply with her own mortality. Susan Morrissey provides a complex analysis of Klimova as well as the revolutionary ethos that created the discourse of the time about whether "revolutionary self-sacrifice" was in fact suicide (Morrissey 2006, 294). At the same time, if we take Klimova's letter at its face value, it resonates with other expressions of revolutionary clarity from other places, other times. In the 1910 Mexican Revolution, Emiliano Zapata is famous for saying, "It is better to die on your feet than to live on your knees." In Chiapas, Mexico, in January 1994, the Zapatista Army of National Liberation (EZLN) took its name from Zapata. The (mostly) Mayan soldiers of the EZLN and their communities literally had nothing—Chiapas is an immensely poverty-stricken state of Mexico—and since they were dying anyway, they chose to fight, to give up their lives to create change for the future generations. Two of their many sayings are *Todo para todos* (everything for everyone) and *Un mundo en donde quepan muchos mundos* (a world in which many worlds fit).[21]

Decades ago, I recall seeing a Chicano movement silk-screen poster that had the phrase, "So that they will not have died in vain."[22] I have never forgotten that phrase. It meant particular communities to me when I first saw it; I quickly came to understand how broadly it could be read. Of course: it has deep meaning for any of us who regularly remember the ones who came before, the ones who did live, the ones whose lives still matter. To die in vain is to die ungrievable; to die in vain is to not have lived. How can this be? How can a person, how can people, how can our relations with the nonhuman world live yet not live? And yet the answer is simple: everywhere we see lives not being lived, lives thrown away, lives defeated, lives destroyed.

In Klimova's letter, she writes of the "feeling of inner freedom" (Morrissey 2015). I have a friend in Mexico City named Fabian, a street vendor and a storyteller; he told me about a short story he wrote, one about a caged bird. The bird belongs to a woman who adores it, cares for it, is tender with it, and is so grateful that the bird sings such beautiful songs for her. In Fabian's story, we are told that what the bird is singing is her lament to the woman who has kept her imprisoned; her songs are her cry for help, her cry for freedom.[23] *La reclama todos los días de su sufrimiento por estar enjaulado* [every day she laments her suffering from being caged]. Is the inner freedom that Klimova writes of knowing that soon her body will not hold her in check any more? Is this why she can look death in the face and say, "I am not afraid of you"?

My mother was not afraid. She suffered her share of illness; she had rheumatoid arthritis from childhood and it was the eventual cause of her death. She chose through her inner freedom what she wanted at the end of her body's journey. Her ashes mixed with the snows of Mount Rainier, the mountain that is the gateway to heaven for the Native peoples of that land. She went even beyond the radiance of that place. When I see Mount Rainier, I am filled with awe at the majesty and brilliance of the place she chose.

In perhaps the best of circumstances, what do humans want if not a way to make peace with death? I want to be at peace with death when my time comes. Intellectual inquisitiveness and readings help somewhat, especially readings of works like Klimova's, but only when the resonance happens, when we read something we have lived, when we confirm for ourselves what we have intimated in some way in our own lives, in our own hearts. The Guatemalan revolutionary poet, Castillo, tortured and burned to death in 1967 at the young age of thirty-one, wrote that being a visionary necessarily means suffering from the reality of the present. In "Before the Scales, Tomorrow," he affirms the hope and love of the vision, saying how "splendid [it is] to know yourself victorious/when all around you it's all still so cold so dark" (Castillo 1971, n.p.). For Castillo, this world that we live in is already death, a place that is cold and dark, without light. Klimova, herself a revolutionary, writes of the radiance of being, the immensity of universal love, the "limitless, all encompassing love . . . a tenderness to everything . . . the great unity of the entire world" (Morrissey 2006, 292). This radiance comes for her from her choice to "give [her]self completely to the struggle without return, without regret, and without hesitation before anything" (2006, 292). As Castillo, like Klimova, attests, "it's beautiful to love the world" so profoundly that death can be faced without fear. For those who yearn and work to see radical transformation in society, a stance that disavows the suffering of others is not possible.

The ancient Nahuatl people had a philosophy of education that focused on encouraging the "face-and-heart" (*in ixtli in yollotl*) of each learner to emerge in strength and wisdom. They saw poetry as *in xochitl in cuicatl*, the path to truth, to inner truth, perhaps what Klimova calls inner freedom. "Flower-and-song" (*in xochitl in cuicatl*) provides a way to manifest face-and-heart, and, I think, the singularity of being that Mari Ruti writes of in her interpretation of Jacques Lacan:

> [A]bsorbing moments of creativity . . . are characterized by a hyperfocused or elated state that temporarily makes us lose touch with the historical quality of human experience. No longer creatures situated on a personal continuum that extends from a past recognized as "ours" towards a hypothesized future—no longer creatures of either memory or hope—we fall into and embrace the immediacy of the present. Entirely immersed in the task at hand, we feel that we have been ushered to a place beyond time and self-reflexivity. The present takes up all of our experience, yet we do not feel in any way deprived or delimited, but are rather filled by an exhilarated sense of liberation and self-expansion; we feel vibrantly alive, connected to the deepest recesses of our being. (Ruti 2012, 26)

This is how I feel when I'm in the midst of creativity. The world falls off and I am alone, deeply embraced by the impulse that is moving me to paint or to write. In these moments, I am complete, reaching into myself and reaching toward the Mystery itself, surrendering to what is coming to me, from me, over me. Here is my response in painting to Ruti's passage (Figure 3.2).

It is possible that the ancient Nahuatl philosophers thought of poetry as the path to truth because the energy that is called upon to create is a call to dialogue with God (or the Mystery, the One of many names) in your heart (León-Portilla 1990,

Figure 3.2. Singularity of being: after Lacan (Artist: Inés Hernández-Ávila, © 2015).

75). In this ancient tradition, one of the names for the supreme being is *Moyocoyani*, the One Who Invents Herself/Himself (León-Portilla 1990, 95). The space that creativity brings forth is indeed one of elation, freedom, timelessness, the profundity of being alive and in intimate touch with oneself, with creation, with the creative force. This is why Nezahualcoyotl (Fasting Coyote) and the other poets of the time, facing their own mortality, wrote of leaving "at least flowers, at least songs," at least the evidence and witness of their singularity of being (León-Portilla 1990, 78). This is what I want to leave. This is why I create. As Judith Butler puts it,

> to be ec-static means, literally, to be outside oneself, and thus can have several meanings: to be transported beyond oneself by a passion, but also to be *beside oneself* with rage or grief . . . . I am speaking to those of us who are living in certain ways *beside ourselves*, whether in sexual passion, or emotional grief, or political rage. (Butler 2014, 24)

I venture to say that it is possible to feel the passion (the erotic energy of creation), the grief, and the rage all at once. The passion, the creativity, allows for a release, a transformation into something beautiful and worthy, of the centuries-old grief and rage that some of us feel on a daily basis. When I reach this *acercamiento*, this ecstasy of reaching the infinite, it does not matter that I must die, because I will have lived in fullness. As the ancient Nahuatl poet Nezahualcoyotl wrote:

> My flowers will not come to an end,
> my songs will not come to an end,
> I, the singer, raise them up;
> they are scattered, they are bestowed . . .
> Even though flowers on earth
> may wither and yellow,
> they will be carried there,
> to the interior of the house
> of the bird with the golden feathers. (León-Portilla 1992, 82)

I paint and write with the eyes of my heart. I hear and see with my spirit. My fingers bleed tears that are the ink of my dreams. The ancient Nahuatl philosophical concept called *nepantla* relates intimately to me. *Nepantla* is the space "in the middle, in the midst," an interstitial space, a place of liminalities, a neither here-nor-there, and yet an everywhere (Bierhorst 1985, 235). In her studied attention to Nahuatl philosophy and cosmology, Anzaldúa put emphasis on the currency of this concept for contemporary Chicanas. It is because of her work that *nepantla* is now a familiar term in circles of feminists, artists, and cultural workers. Anzaldúa writes,

> In *nepantla* you sense more keenly the overlap between the material and the spiritual worlds, you're in both places simultaneously—you glimpse *el espíritu*—see the body as inspirited.

> *Nepantla* is the point of contact where the "mundane" and the "numinous" converge, where you're in full awareness of the present moment. (Anzaldúa 2002, 549)

I am a Nepantlera, a daughter of Coyote, as a Nimipu woman, and a daughter of Revolution, as a Tejana. Because *nepantla* is that place in between, it holds all (im)possibilities—I am able to identify as a Nepantlera because, as a daughter of the trickster Coyote, I know deep in my being that within me I hold all the (im) possibilities as well. Within, without, I am whole. I had a mother and a father. I am an Indigenous woman of the Americas. When I am painting and sometimes when I am writing, I enter *nepantla*, not as an academic but as a creator being, a spirit being.

In the Nez Perce tradition, Coyote has an important role (which I have learned from the oral tradition). He/She is a creator spirit, trickster, teacher, savior, fool, the energy that teaches us that in life everything is possible and nothing is impossible. Coyote comes back from death; as long as there is one hair, one bit of him/her left, [s]he can be resuscitated. Native people are supposed to have vanished from the face of the earth, but we are still here, engaged and engaging with life and death. Coyote and all the tricksters have helped us resuscitate so many times. Coyote unites in me my mother's side and my father's side. Coyote might have even chosen them for me. In the ancient Nahuatl tradition, Huehuecoyotl, the First Coyote from the beginning of time, is said to have given the people the first dance, the first sound. In the Conchero tradition, there is a dance for Huehuecoyotl; it is one of my favorites. At each primary step (repeated twice between each changing step to reflect duality), the Ancient One marks a cross, a balanced cross, to remind us about balance, being centered, knowing our place in the universe, giving thanks always, even when we are being pursued, even when we face the utmost challenges, even when we face death. This is my interpretation, as a dance captain of the Conchero tradition, of the dance for Huehuecoyotl. I was born dancing, I know I was. When I cross over, I want to do so dancing my last dance.

I wonder whom I will visit when I die. My mother went to visit one of my colleagues and dear friends, a Chicano who has always been like a brother to me. She visited him in dreams the night right after the evening she died, as she was leaving this earth. He called me the next morning to tell me about the dream, how she took him back home to the reservation, to a beautiful home and a back room with a huge picture window looking out on to the mountains and the fields, how he received a blessing from her sisters. It was only after he told me the dream that I let him know she had died. She chose him—of all of my loved ones, she chose him to bless on her departure.

When my husband and I first got together, I went with him to visit a beloved Paiute elder, a true healer and wisdom keeper, Raymond Stone, called Grampa Raymond by many of us. I was excited about introducing my new partner to him.

Grampa Raymond was not impressed—he just kept shaking his head, because this man who became my husband was so much younger than me. I was hurt and stopped going to his sweat lodge ceremonies for some time, but he still was a major teacher to me and I still love(d) him dearly. One Saturday, my husband and I were on our way to a Native gathering when we ran into a friend who told me that Grampa Raymond was close to crossing over. My husband and I immediately decided to drive down to Lone Pine (by Bishop, California) from Davis. It was a long drive, but we drove with urgency. On arriving, we stopped at a grocery store to buy food to take to Grampa's home. When we arrived, his wife kindly let us in, and we saw that Grampa's bed had been moved to the living room (most likely because of the many visitors). He was lying there with his eyes closed. His wife told me, "Go ahead and talk to him, he will hear you." I grabbed his hand gently and told him softly who I was, I told him I was still with my husband, and that he was there with me, that we had been together all this time and that we were happy, that he was a good man. I thanked him for all that he had done for me (coming to me even in dreams to help me), and then I closed my eyes and prayed for him, for his safe crossing, for him to be blessed always on his path. I told him how much I loved him. When I opened my eyes, I suddenly noticed that there were tears flowing from his eyes. His wife came over and said, "Oh, he knows it's you." We thanked her and him, and then we left.

We got a room at a nearby motel, and on the next morning when we started our drive home, I looked out at the land and saw it was filled everywhere with radiance, a special blessed light. I learned later that he had crossed over between the time we left his home and the moment when we were beginning our drive. I am so grateful that he waited for me. I am so grateful that I knew to go, that my friend crossed my path when he did to tell me what was happening. This is the work of spirit. I trust in this work. This is why I am not afraid of death. This is what allows me to face my own mortality with calm, with my SpiritHeart:

> SpiritHeart be with me.
> SpiritHeart guide me.
> SpiritHeart I honor you.
> SpiritHeart I thank you.
> Always.
> Always. (Hernandez-Avila)

Acknowledgments

This essay emerged from a Faculty Residential Research Group on "The History of Mortality," which was supported by the University of California Humanities Research Institute (UCHRI) at University of California-Irvine, during the fall quarter of 2015. I am grateful for the support I received from the UCHRI through this research group. The paintings that are included in this chapter were done during this residency. I would also like to thank the University of Oklahoma Press for permission to quote a poem from Miguel León-Portilla's *Fifteen Poets of the Aztec World* (© 1992 by the University of Oklahoma Press), as well as Coffee House Press for permission to quote from Linda Hogan's poem "The Grandmother Songs," from her *Book of Medicines* (© 1993 by Coffee House Press).

Notes

1. Hogan is a distinguished mixed-heritage Choctaw writer (poet, novelist, essayist).
2. I am aware of Tweed's work (2006). He does note the different aspects of "crossing," including the idea of a final crossing, e.g., "religions mark and cross the *ultimate* horizon of human life" (emphasis in original) (2006, 76). The idea of "crossing over into the spirit world," from an Indigenous perspective, has had currency for a very long time, however. While Tweed's approach has some resonance with Indigenous cultural, epistemological, and theoretical frames, his approach falls short, ultimately, in his representation of Indigenous religions, in particular because he selects those aspects (e.g., of U.S. Native American religious traditions) that he considers examples that suit his theses (without providing the fuller picture of those traditions).
3. Hinmaton Yalatkit, Chief Joseph, is one of the chiefs with the most name recognition among Native peoples in the United States—the subject of songs, poems, sculptures, paintings, and commercial items. He is known for his wisdom, eloquence, steadfastness, diplomacy, and strength of spirit. "Indian Country" is the term used by Native people in the United States to designate everyone who is Indigenous/original to these lands now called by other names, and to designate the land itself, because no matter who has title or deed, the land is and always will be "Indian Country." Native peoples in the United States have made the term "Indian" their own, even though the term was a mistake made by Columbus and perpetuated over the centuries. Native peoples still use their own names for themselves, for their peoples, for their nations. One of the main news venues for Native peoples is *Indian Country Today*: http://indiancountrytodaymedianetwork.com/.
4. The devastation to individuals is manifested in substance abuse, suicide, domestic violence, mental illness, and despair, among other reactions. In terms of social institutions, Gross lists the collapse of family structures, religions or belief systems, governance, health care traditions, and educational systems.
5. Pratt is said to have first made this statement in 1875; he rearticulated it later in a speech read at a conference (Pratt [1892] 1978). See also the Carlisle Indian School Digital Resource Center, a "major site of memory for many Native peoples," according to the site's home page (http://carlisleindian.dickinson.edu).
6. Similarly, Tweed notes Charles Long's statement, "'For my purposes, religion will mean orientation—orientation in the ultimate sense, that is, how one comes to terms with the ultimate significance of one's place in the world'" (cited in Tweed 2006, 74). When the orientation is lost or fractured, it is difficult, sometimes impossible, to come to terms with this "significance."
7. I thank Margo Tamez, Nde scholar and poet, for introducing me to Martínez Salazar's work.

8. In Mexico, Indigenous peoples use the phrase "original peoples" to refer to themselves; Indigenous Canadians call themselves "First Nations." These expressions have been adopted by Indigenous peoples from other parts of the hemisphere.
9. The Cherokee writer Thomas King (2018, 54–59) argues that contemporary society puts Indians into three categories: "dead, live, and legal." "Dead Indians" comprise the most comfortable category for larger society; they are the clichés, the stereotypes that make for Hollywood fictions and sports mascots, the derealized Indians, the evicted Indians. "Live Indians," King says, are more annoying, while "legal Indians" are to be feared (largely because of treaty rights).
10. Various news sources say that the number of 34,000 is too low and that the actual number of deaths was possibly twice as much.
11. For heart-wrenching images of the fish kill, see https://www.google.com/search?q=Fish-kill,+2002,+Klamath+Watershed,+images&biw=1209&bih=685&tbm=isch&tbo=u&source=univ&sa=X&ved=0ahUKEwiuvqfKo6POAhWFOz4KHRFtCdwQsAQIGw#imgrc=.
12. The plan has been developed by federal officials, the states of California and Oregon, and PacifiCorp through the Klamath Hydroelectric Settlement Agreement, which created the Klamath River Renewal Corporation (Spencer 2017).
13. Chicana/o artists made silk-screen posters with an Aztec warrior looking furiously at the viewer, stating, "You go back where you came from!"
14. Miguel León-Portilla, the internationally distinguished scholar of ancient Mesoamerica, has a corpus of texts that address the ancient Nahuatl philosophical traditions. An excellent starting point for further study is his *Aztec Thought and Culture: A Study of the Ancient Nahuatl Mind* (León-Portilla 1990).
15. There are numerous festivals, parades, and community altars that appear throughout the United States in the period leading up to and including November 1–2 and sometimes a few days beyond. In California, two of the most well known are the festivals in San Francisco and Los Angeles.
16. Charon (2008, 8) writes about the "multiple subjectivities occasioned by the experience of illness." As she puts it,

    Illness intensifies the routine drives to recognize the self. . . . Illness occasions the telling of two tales of self at once, one told by the 'person' of the self and the other told by the body of the self . . . . The self depends on the body for its presence, its location . . . . The body is in the copulative position between the world and self. (2008, 87, 89)

    In the case of my "almost splitting," my body was telling one tale while my spirit was telling the other, one that I came to understand was the tale of real danger.
17. *In tlilli in tlalpalli* is an expression in the Nahuatl language that denotes the colors red and black, which signify wisdom and writing (León-Portilla 1990, 21, 23).
18. The Indian Shaker Church is a Christian movement in the northwestern United States, founded on the visions of John Slocum in the 1880s. Indian Shakers cure with prayer, a form of shaking, and dancing with singing.
19. My aunt went on to live another six, almost seven years, enjoying well-being even though she was more and more frail as she got older; she died at the age of 98 on October 10, 2015.
20. This letter was shared with those of us in the "History of Mortality" seminar by historian Susan Morrissey (2015), who translated it from Russian to English. The letter is a subject of her research, and she has written about Klimova in her work (Morrissey 2006, 291–93, 295, 306, 338–39, 346, 352).

21. For more information on the Zapatistas, see the website *Enlace Zapatista*: http://enlacezapatista.ezln.org.mx/.
22. This phrase could pertain to many communities worldwide. Thus, we have memorials, museums, commemorative statues, and the like. The poster is by an anonymous artist, which I saw in the 1970s.
23. I am also aware of Maya Angelou's *I Know Why the Caged Bird Sings* (1969).

References

Anders, Ferdinand, Maarten Jansen, and Luis Reyes García. 1993. *Los Templos del Cielo y de la Oscuridad: Oráculos y liturgia—Libro explicativo del llamado Codice Borgia*. Mexico City: Fondo de Cultura Económica.

Angelo, Maya. 1969. *I Know Why the Caged Bird Sings*. New York: Random House.

Anka, Paul, Claude François, and Jacques Revaux. 1969. *"My Way." Performed by Frank Sinatra*. Produced by Sonny Burke. Burbank, CA: Warner Bros. Records.

Anzaldúa, Gloria. 2002. " . . . now let us shift . . . the path of conocimiento . . . inner work, public acts." In *This Bridge We Call Home: Radical Visions for Transformation*, edited by Gloria E. Anzaldúa and Ana Louise Keating, 540–78. London: Routledge.

Bierhorst, John. 1985. *A Nahuatl-English Dictionary and Concordance to the Cantares Mexicanos, with an Analytical Transcription and Grammatical Notes*. Palo Alto, CA: Stanford University Press.

Butler, Judith. 2009. *Frames of War: When Is Life Grievable?* New York: Verso Books.

Butler, Judith. 2014. "Violence, Mourning, and Politics." In *Precarious Life: The Powers of Mourning and Violence*, 19–49. New York: Verso Books.

Castillo, Otto Rene. 1971. *Let's Go! Vamonos Patria a Caminar*. Translated by Margaret Randell. London: Cape Goliard Press.

Charon, Rita. 2008. *Narrative Medicine: Honoring the Stories of Illness*. Oxford: Oxford University Press.

Crawford O'Brien, Suzanne. 2014. "Salmon as Sacrament: First Salmon Ceremonies in the Pacific Northwest." In *Religion, Food, and Eating in North America*, edited by Benjamin Zeller, Marie Dallam, Reid Neilson, and Nora L. Rubel, 114–33. New York: Columbia University Press.

Encyclopedia Brittanica. n.d. "Indian Shaker Church: American Religious Sect." *Encyclopedia Brittanica* (online). https://www.britannica.com/topic/Indian-Shaker-Church (Accessed June 1, 2017).

Funahashi, Atsushi, dir. 2012. *Nuclear Nation: The Fukushima Refugees Story*, part 1. Big River Films. Tokyo: Documentary Japan. http://nuclearnation.jp/en/part1/

Geranios, Nicholas K. 2015. "Record Number of Salmon Return to Hanford Reach." *Seattle Times*, November 10. https://www.seattletimes.com/seattle-news/environment/record-number-of-salmon-return-to-hanford-reach/.

Gilman, Sarah. 2016. "This Will Be the Biggest Dam-Removal Project in History." *National Geographic Magazine*, April 11. http://news.nationalgeographic.com/2016/04/160411-klamath-glen-canyon-dam-removal-video-anniversary/.

Gross, Lawrence W. 2014. "Post-apocalyptic Stress Syndrome." In *Anishinaabe Ways of Knowing and Being*, 33–51. New York: Ashgate.

Hogan, Linda. 1993. "The Grandmother Songs." In *Book of Medicines: Poems by Linda Hogan*, 57–58. Minneapolis, MN: Coffee House Press.

King, Thomas. 2018. *The Inconvenient Indian: A Curious Account of Native People in North America*. Minneapolis: University of Minnesota Press.

Lee, James Kyung-Jin. 2015. "The Doctor, Undone: The Rise of Physician Chaplaincy in Asian American Medical Narratives." Paper presented to History of Mortality Residential Research Group, University of California Humanities Research Institute, Irving, November 2015.

León-Portilla, Miguel. 1990. *Aztec Thought and Culture: A Study of the Ancient Nahuatl Mind.* Translated by Jack Emory Davis. Norman: University of Oklahoma Press.

León-Portilla, Miguel. 1992. "Nezahualcoyotl of Tezcoco: Poet, Architect, Man of Divine Wisdom (1-Rabbit, 1402-6-Flint, 1472)." In *Fifteen Poets of the Aztec World*, 70–98. Norman: University of Oklahoma Press.

Martínez Salazar, Egla, 2014. *Global Coloniality of Power in Guatemala: Racism, Genocide, Citizenship.* Lanham, MD: Lexington Books.

May, Theresa. 2014. *Salmon Is Everything: Community-Based Theater in the Klamath Watershed.* Corvallis: Oregon State University Press.

Morrissey, Susan. 2006. *Suicide and the Body Politic in Imperial Russia.* Cambridge: Cambridge University Press.

Morrissey, Susan. 2015. "Letter by Natal'ia Klimova, 1908." Unpublished document presented to the Faculty Residential Research Group, "The History of Mortality," University of California Humanities Research Institute (UCHRI). Irving: University of California-Irvine.

Pratt, Richard H. (1892) 1973. "The Advantages of Mingling Indians with Whites." In *Americanizing the American Indians: Writings by the "Friends of the Indian" 1880–1900*, edited by Francis Paul Prucha, 260–71. Cambridge, MA: Harvard University Press.

Ruti, Mari. 2012. *The Singularity of Being: Lacan and the Immortal Within.* New York: Fordham University Press.

Smith, Paul Chaat. 2009. *Everything You Know about Indians Is Wrong.* Indigenous Americas Series. Minneapolis: University of Minnesota Press.

Spencer, Adam. 2017. "Meetings on Dams Removal Begin." *Del Norte Triplicate*, November 11. http://www.triplicate.com/news/localnews/5741606-151/meetings-on-dams-removal-begin.

Tweed, Thomas A. 2006. *Crossing and Dwelling: A Theory of Religion.* Cambridge, MA: Harvard University Press.

CHAPTER FOUR

The Sorcerer, the Madman and Grace

Are Archetypes Desacralized Spirits? Thoughts on Shamanism in the Amazon

JACQUES MABIT, M.D.

TRANSLATED BY GEORGE GOULD, MPHIL.

Can (Amazonian) shamanism hope to offer some insights on Jungian psychology? Can it answer questions plaguing contemporary Western culture? Because Amazonian shamanism is so much older than the Jungian school of thought, does shamanism have more to offer humanity in terms of self-knowledge? Does the concept of individuation make any sense when applied to the different ethnic groups of the Amazon region?

To begin with, is it even possible to establish a fruitful dialogue between these two approaches to understanding human nature? No, strictly speaking, a dialogue is not possible simply because, if Jung and his disciples are wordy and long-winded (no offense intended), shamanic traditions are never written down, and shamans themselves are silent guides whose teachings are passed on through practical experience, not verbal discourse.

I therefore find myself in the privileged yet uncomfortable position of speaking on behalf of a silent shaman, though I am not in any way an expert on Jung either. This being the case, I must beg your indulgence, as well as that of the members of the Amazonian tradition of which I speak, for any translation mistakes or any inadequate linguistic approximations to which I may have to resort. To avoid these inaccuracies, any verbal discussion of the subject would necessarily have to be accompanied by the living experience of the initiation process itself within a shamanic context, if only to allow the language of the shaman its full gamut of expression. And perhaps the true Jungian, as Jung himself said, is the person who ceases to be so by seeing the individuation process through to the end, thus distancing him- or herself entirely from the master, thus making Jung "vanish" altogether.

If I feel authorized to write on this subject, I owe it to a life-changing experience that took place at the very start of my work with Amazonian healers. During a session with ayahuasca, the vision-inducing plant of the central Amazon region, I found myself, in a vision, standing in the presence of a group of anthropomorphic beings who sat, like members of a jury, facing me in a semicircle and who introduced themselves as "guardians of the forest." I did not experience this as a remote vision but as an actual real-life occurrence in three dimensions, as a scene in which I was physically present. They asked me why I had taken ayahuasca, and I replied that I wished to learn more about this cure. After confabulating among themselves, the central figure turned to me, saying, "You are authorized to enter this territory," and added, "Your path leads you through here." To my amazement, he then showed me a scene in which I saw myself treating drug addicts. This is precisely what I have been doing to this day, although for three years I resisted the calling—which I considered an unattractive and difficult prospect, far removed from my initial goals.

This risky incursion into the shamanic practices of the upper Amazon involves experimenting with induced altered states of consciousness during which the "self" is displaced from its habitual position at the center of being. In such a state, other dimensions of the real emerge, both about an interior world and the world of invisible nature, accompanied by paranormal phenomena and extrasensory experiences in which the body is very much an active participant. The silence of the shaman about the emergence of this new cartography, this abundance of surprising new experiences and novel information, forced me to seek out personal adjustments to incorporate these new findings and re-establish some sense of order, helping to dispel the dangerous interior chaos I felt. Even if undeveloped, my two pillars of knowledge had until then been made up of the psychological propositions of Carl G. Jung, on the one hand, and my Christian upbringing, on the other.

At the risk of oversimplifying things or creating a caricature, let me now briefly describe my view of shamanism and its context. I will refer exclusively to the shamanic practices of the upper Amazon region of Peru, where I have been active for thirty years, without trying to include all the different strains of shamanism present in the Amazon as a whole, much less of those found in other parts of the world as well. I will trace different paths that, taken together, help illustrate how such a vast and complicated subject calls for a "complexity of thought" (Morin 1990) and cannot be encapsulated in a brief synthesis—which, at this stage of my research, would be pretentious, at best.

Objectives of Initiation and Individuation within a Tribal or Shamanic Context

Is it relevant to talk of individuation in a tribal or shamanic context? Individuation in such contexts cannot be taken separately from the personal objectives of each individual or from the aims of the survival and well-being of the group. Human beings do not aspire merely to personal and individual accomplishment but also to the fulfillment of a collective destiny so that each member contributes toward and shares in the benefits of the achievements of the entire group.

Initiation is traditionally the means whereby initiates can define themselves and find their rank within a community. Rites of passage, particularly those between childhood and adulthood, guide the individual and define qualities of a man or woman in a group and grant the non-fertile the chance to bear children, thus helping to increase the size of the community. Through it, individuals can find their true vocation, determining their role in the collective struggle for survival and development, whether as artisan, hunter, warrior, *cacique* (leader or chief), or shaman.

In a way, this collective dimension goes against the grain of personal individuation, which could lead to an individual becoming too different from the rest of the group. Indeed, from the perspective of a wise reciprocity at work within the community, individuals are not expected to distinguish themselves too much or rise too far above their peers. This sort of distinction or differentiation could be seen by other members as a kind of loss of identity in the community, entailing a rejection of the principle of reciprocity, possibly leading to imbalances. Gaining unwarranted status or power in the group can also give rise to envy among members, which is especially detrimental to harmony and cohesion within the group. Envy, manifested through aggression, can lead, in turn, to social disorder, making the community more vulnerable by menacing its stability. Violence, outwardly projected, would eventually revert to the group, threatening to destroy it by starting a cycle of vengeance or a vendetta.

As a result, the specific role of the shaman—potentially a dangerous being—must be strictly monitored and regulated by the community. He will have to account for his actions and, if he is found guilty of abusing his power or committing any misdemeanor, he will be punished and sometimes even sentenced to death. In this way, allowing for variations between different ethnic groups, several control mechanisms are set up to prevent any transgression on the part of these otherwise much feared and admired members of the community. For instance, the shaman must earn a living just like anyone else in the tribe and cannot dedicate his time fully to his main line of work; he must obtain permission from the *cacique* before conducting any activity of a shamanic nature; and he must abide by all applicable rules and prohibitions regarding diet, sexual activity, and so on. Also, any misfortune might be attributed to the shaman, since he is responsible for protecting the group and maintaining the general harmony, inasmuch as it is his duty to establish a connection between the community and the invisible world.

The shaman's journeys to the invisible world during his duties are thus not so much a personal process of individuation as they are a means of acquiring powers and allies, which will allow him to best serve the tribal society of which he is a valued member.

Nowadays, even in the Amerindian communities of the Amazon region, shamans have a hard time finding pupils or disciples. The young are put off by the dangers surrounding the shaman's activities, which can include attacks by evil spirits as well as by other shamans, and by the eternal pall of suspicion under which the shaman lives. He is, after all, a wizard, and any trouble within the community can potentially be traced back to him. Young people are also wary of the long and arduous period of apprenticeship required of the post, not to mention the restrictions that come with the job: strict diets, abstinence, and little in the way of material reward. The attractions of modern life have only aggravated the problem: young villagers dream of city lights far more than they do of the demands of the priesthood or of shamanic trances.

The Tribal Cultural Horizon

Shamanism belongs in the context of the tribe (and the component clans and extended families), where the main referent is the community itself. This means that the individual's needs and duties come second, the priority resting with the social group of which he or she is a member. If the sacrifice of an individual is deemed necessary for the well-being or survival of the community, then it is accepted[1] not only by other members of the community but often by the victims themselves, for whom dying for the greater good constitutes both a duty and an honor.

In this context, everything that lies outside the community is potentially a source of danger, threatening its integrity, whether from the visible or the invisible world. The union of the group is therefore reliant on the development of a system of alliances and defense against the outside world, whose perils can come in the shape of hostile neighboring tribes or entities from the invisible world.

The tribal structure is based on a foundation myth of justice.[2] The integrity and harmony of the community are maintained through fixed mechanisms that can re-establish the internal balance of the group following exchanges with the outside world. This balance can be upset in any number of different ways, whether through individual or collective misfortune: disease, accidents, famine, scarcity of game, and so forth. The spirits of the wild boar or deer must be appeased through ritual before tribesmen set off on a hunt in order to compensate the animal community for its imminent loss. If one tribe abducts young girls from another tribe, the latter must reciprocate, thus redressing the balance and making things fair and even again. If a present of great value is offered to someone who has no means of offering a gift of equal worth in return, this is seen as an act of humiliation, perhaps even as a declaration of war. Reciprocity in acts of exchange constitutes the basis on which harmony is maintained in these societies.

The same is true for the tribal shaman. In the eyes of the community, his job of keeping in check the attacks of shamans from enemy tribes constitutes a legitimate form of defense. His offensives in the invisible world are absolutely justified, answering the need for a just reciprocity, which forms the basis of all relations with "the other." Shamanic wars are thus as rife as intertribal wars. They involve essentially a bellicose form of shamanism in which the main goal is to obtain the greatest possible number of weapons in the fight against enemies, becoming more powerful than them in preparation for any act of aggression. Reciprocity can thus take on the guise of vengeance, with violence gauging its efficacy within a cyclic dynamic without beginning or end.

Within such a system, the projection of the individual or collective shadow toward "the other," if this other belongs to another tribal community or another family, is perfectly legitimate. Evil comes from outside the community, and, when seeking its source, it is to the outside that Amerindians will spontaneously turn, and it is there that they will try to vanquish it. This being the case, the individual is little inclined to look inwardly when trying to discover the source of his or her sufferings.

The result is that, for the shaman, any healing activity carried out on a member of his community inevitably entails an offensive act against the outsider or supposed adversary. Medicine and sorcery go hand in hand and are often indistinguishable.[3] The master healer can form alliances with both good and evil spirits, the former to heal, the latter to defend the members of his tribe.

In the current context of tribal disintegration in the face of globalization, modernization, and cultural interchange, magic has grown in importance because "the other" can now be anybody from the whole outside world and because tribal regulations can no longer provide clear-cut answers. Acts of vengeance in the invisible world are often surrogates for expressions of sorrow, since subjects find themselves unable to put their suffering into words,[4] especially in cases where the conflict is of an amorous nature.

The Western Cultural Horizon

While the cultural horizon of the tribal world is based on the myth of justice, that of the West rests on the myth of love. The lamb follows the ram, from whom he is descended and from whom he has inherited certain traits. Outside love is a universal society[5] where every "other" becomes my brother. Justice gives way to and is superseded by love. Tribal society, the clan, and the family have become a part of the larger family—the entire human race. Violence can no longer be projected toward the other with any sense of legitimacy, whatever or whoever the other may be. Deprived of this freedom to project their shadows outward, subjects are forced to find it within themselves by turning inward. The outside sources of suffering lead subjects back into themselves and to their manner of dealing inwardly with this projected aggression. This looking inward allows Westerners to identify the unconscious dimension hidden within themselves, which secretly stimulates them and sometimes manipulates them without their knowledge.

In this way, the Western individual has come to realize that his worst enemy is himself, or rather, a part of himself of which he is not conscious. This dynamic gives rise to the notion of the individual which has become so central that the acknowledgment of his dignity will take precedence over his duty to the community and its preservation. For political constitutions of countries in the West individual dignity and well-being represent the ultimate goals of society. A country is capable of spending a great deal and going to great lengths to save a single person.[6] This mutation means universal Western society is structured round a hierarchy of values which stand in almost diametrical opposition—at least apparently—to those of tribal societies, where the life of the group takes precedence over the lives of each of the individuals who comprise it. The shadow is, to some extent, individualized, privatized.

It thus follows that this process will lead to a decrease in the practice of magic among more Westernized tribes. Once they lose their legitimacy, shamanic practices become the focus of vehement—and sometimes violent—opposition by the Christian churches. The notion of pardon takes over, indicating that vengeance has been renounced and the cycle of violence is broken. It gains the upper hand precisely in answer to unforgivable faults,[7] and the meting out of justice is left to divine providence. Only then can pardon cease to be humanly unjust and give way to divine love, thereby re-establishing the balance of justice. Solomon's judgment in the Old Testament prefigures the emergence of this kind of wisdom: one which, in the name of love, appears at first to deny justice, only to then restore it entirely, thus revealing the transcendence of love over justice.[8] This episode illustrates, within the context of the Jewish tribe (which lies at the very root of Western culture), the shift from the myth of justice to that of love, of a tribal society toward a universal one.

Two Forms of the Unconscious for Two Cultural Horizons

The different cultural horizons of the tribal world and the West reveal equally different shadows, each consequently placed in quite distinct locations. The tribal shadow or unconscious is to be found in nature, in the cosmos, outside the individual, while for the Westerner, the invisible and the unconscious are found within himself.

Amazonian shamanism makes great use of vegetal preparations that can transport the subject to the world of invisible nature, there to explore its mechanisms and meet the beings who inhabit it. The drinking of brews thus prepared is governed by strict rules, particularly about diet and sexual activity. These sophisticated techniques serve to broaden the human perceptive spectrum, enabling the shaman to pass beyond natural sensory limits and glimpse things that would otherwise go unnoticed by ordinary perception. In this way, the subject can view never-seen colors and hear never-heard sounds. This perceptual expansion is not limited to the five ordinary senses alone; it extends to the proprioceptive senses (senses that recognize the body's schema from within, for instance), as well as psychic functions (memory, concentration) and those labeled paranormal (clairvoyance, clairaudience, psychokinesis, telepathy, etc.).[9] As the initiate develops these extrasensory faculties, his dream life becomes highly active and he begins to experience paranormal phenomena. At first, these experiences—which last throughout the initiation period—can be quite chaotic, until the initiate learns to master them and to impose some sense of coherence. The shaman's role in controlling this psychic production is of the utmost importance. He does this by ritualizing the induction process and intervening directly in the "energy body" of his patient or disciple. I will return to this subject later.

It is important to establish from the outset that these experiences of the external, invisible world reveal a living, inhabited universe. And it is interesting to note that this conclusion, arrived at empirically, from first-hand experience, is shared by Amerindians and Westerners alike, revealed to them in the same way as a microscope reveals to any observer the reality of a world that is invisible to the naked eye. In other words, these are objective realities, which will later, of course, be interpreted in varied ways according to the cognitive capacity, cultural background, and personal outlook of each individual subject. From this point of view, "animism," in the sense of the attribution of a living soul to natural phenomena, is not merely a belief but, rather, an acknowledged, verifiable fact, established by those who have taken the trouble to explore this world—the "otherworld." Every element in creation possesses a "mother" (*madre*) that represents its life-giving soul. This matrix has form, energy, and intelligence and can communicate with human beings through this shared dimension. This matrix or "spirit" is differentiated according to its evolutionary degree in nature as well as to the creation of specific qualities in each element. In this way, medicinal plants possess a more differentiated spirit than those of ordinary plants and, in the essence of their species, plants that can be used for initiations, the psychoactive plants, are distinguished by their higher level of differentiation. At the top of the Amazonian plant hierarchy are coca, tobacco, and ayahuasca, along with plants of the genus Datura. The spirits of these natural elements are collective, but they can also possess different degrees of differentiation. Thus, *pachamama*, or the earth spirit, can take on the specific form of a particular mountain, for instance, in terms of height or shape or exceptional placing, which will allow an extraordinary spirit *(apu)* to be recognized. Vegetal subjects are therefore relatively undifferentiated within their own family; each tobacco plant, for example, is not seen as an individual with a single spirit of its own but, rather, possessed of the collective spirit of the tobacco plant in general. These spirits present a differentiated personality and give rise to similar representations, even across different cultures.[10]

The universal nature of these figures from the invisible world has led some rationalist Westerners to interpret them as facets of the collective unconscious, projections of the inner world common to all human beings, denying them all objective external reality. This kind of logic is no more consistent than one that would deny the existence of trees, the sun, or the sea simply because these elements are often employed symbolically in dreams, in psychotherapeutic processes, or in artistic creations. Because a physical or psychical object serves a symbolic purpose is no reason to declare it nonexistent. Is this not rather an instance of a priori rationalism, through its obstinate reductionism, denying a vision readily available to anyone willing to take the trouble to seek it out?[11]

This is the same logic which classifies these vision-inducing brews as "hallucinogenic." To take the inexistence of the invisible world as a predicate, as a non-demonstrable preconception, entails the denial of any form of perception not prompted or stimulated by an object that is not readily detectable by one of the ordinary senses. According to this view, the initiate, during an altered state of consciousness, sees nothing real, nothing material, which leads to the conclusion that he must therefore be hallucinating. This kind of materialist reductionism is wrong for three reasons. First, it is wrong because psychic objects are just as real as those in the material world, and they can and do interact with the objective world perceived through the senses. Psychosomatic symptoms are proof of this: stress can provoke perforations in the stomach's mucous lining, trigger asthma attacks, cause high blood pressure, eczema, and so on. Second, it is wrong because this logic is based on the unfounded denial of the existence of the otherworld and its nonhuman inhabitants—although it not only fails to demonstrate this but also systematically ignores all evidence to the contrary.[12] And third, it is wrong because our organism itself secretes psychoactive substances very similar or sometimes identical to those found in the sacred plants of shamanic tradition.[13] The structural community that our organism shares with other species should be enough to call into question the purported toxicity of the first people's vegetal brews.[14]

That the Amerindians' observations are valid is substantiated in clinical practice, where we see that their model is a close fit with the real world. Their therapies are not only effective but also reach out beyond the Amerindians cultural frameworks.[15] Certain writers are willing to concede, at the most, that shamanic therapeutic practices may occasionally yield results within the limits of the local culture, meaning that we are thrown back on phenomena such as suggestibility and community identity to explain the surprising outcomes that can be achieved. This roundabout way of denying the global character of shamans' empirical knowledge is nevertheless disproved by clinical facts. The effectiveness of these practices is most clearly seen in cases when they have successfully treated pathologies that have proved unresponsive to the conventional allopathic approach or when the latter is very costly in either financial or human terms. In all probability, this denial points to a blind spot in the Western conceptual model.[16]

Although the reality of psychic evidence is generally recognized by psychotherapists, this recognition is not extended to nonhumans. Thus, the threshold of symbolic representation is never crossed, thus avoiding the need to verify the existence of nonhumans and their interaction with human beings.[17] According to this view, it is possible to ask whether nonhuman representations might belong, at most, to a collective subjective dimension but never to an objective one. Nonhumans do not benefit from a personal form of existence but merely from a degree of autonomy that varies according to the level of unconsciousness of the human being observing them. Beyond unconsciousness—and thus differentiation—these individual or collective psychic powers may have more freedom, but they nonetheless retain their fixed links with human activity. For Amerindians, by contrast, these powers are actual entities, creatures as real as human beings, inhabiting a spiritual dimension with no corporality or material existence. Nonetheless, these entities can be perceived by humans under anthropomorphic or zoomorphic guises. On a certain level, these beings form organized societies, such as the water spirits who live in underwater villages of sorts.[18] Sometimes a close bond is forged between these spirits and human beings, as in the case of the pink porpoises that inhabit the region's rivers and try to carry humans off to their watery world, sometimes to mate with them.[19] Another case is a story that was told to me of a man who became lost in the forest and was followed by a deer, which guided him to a beach on a lagoon where he was finally rescued. Sometimes children get lost in the forest, carried off by spirits living there; when they return weeks or months later, they possess special powers.[20] Such encounters between humans and nonhumans are generally considered dangerous, while any sexual contact between the two life forms is deemed a transgression. Formerly, at the highest levels of shamanism—nowadays almost extinct—was the *sumiruna*,[21] a shaman who was able to enter rivers and live among the water beings, only to re-emerge days later, completely dry.[22] Similar tales are still told daily, pointing to a cosmic vision that is not merely a symbolic one but, rather, is one that closely engages with the sensible world and can materially affect humans beyond the changes brought about by inducing altered states of consciousness.

We may thus draw a comparison between Western reductionism, for whom nonhumans are nothing but projections of our inner world, and the reductionism of the shaman, for whom psychic and emotional problems are caused by evil spirits and nothing else. Good and evil spirits coexist in this invisible world, the good helping humans in their affairs and the evil hindering them, assuming roles not unlike those of angels and devils. They can manifest themselves naturally or by means of possession: in the latter case, the preferred host is an animal through which the spirits can make themselves visible to their human interlocutor. The healer seeks to form alliances with good spirits, while the sorcerer seeks the collaboration of evil ones. In keeping with the myth of justice described earlier, more often the Amazonian medicine man will assume the roles of both healer and sorcerer in his attempt to bring about a balance between these opposing forces.

The invisible world is also peopled by the spirits of the dead. Close relatives or ancestors, the spirits of healers or well-meaning people, can project goodwill and kindness to provide help to those in need. Other spirits are dangerous and malevolent, from which people need protection. Journeys to the world of the dead are usually considered dangerous, and several rituals exist that are designed to stop the dead from interfering in the world of the living.

These different regions in the invisible world are ordered according to a specific, strict hierarchy that extends to the spirits themselves. Spirits, good or evil, are creatures and not divinities. The divine in the most general form of the Creator is deemed unknowable and unreachable unless mediated by the nonhuman or through sensory manifestation, and cults and rituals address these natural and supernatural intermediations.[23]

The Synthesis of Western and Tribal Unconsciousness

It seems to me that these two dimensions of the invisible world and the unconscious are not opposed: they complete one another while remaining distinct:

1. The personal unconscious is that of the inner depths, from which emerges subjective information relating to personal or transgenerational history.
2. The otherworld is extrinsic to the subject, transcendent, involving the world of the spirits and archetypes, providing forms to the visible world and information about what comes from above, which is autonomous in relation to the subject and is thus objective.

Both these views of the invisible are joined in the human heart (and body)—since human beings belong to both dimensions—and can form connections with the conscious mind by way of symbols, mainly through the human body's symbolic functions.[24] Ritual, through symbolic language, joins these two dimensions, subjective and objective, not by blending them but, rather, by differentiating between them. Madness would lie in confusing the two, as when the transcendental archetypes of the otherworld are appropriated by the subject, assuming them to have arisen from his or her own inner depths, or else when these transcendental archetypes take active possession of the ego. ("You have to conquer ayahuasca," say the shamans.) Through analogy, manifestations from the lower world or the subjective world echo those from the objective upper world, which means that they are similar but not identical.[25] To say that something "is like" something else is very different from saying it is "the same" as something else. In other words, things that are alike are not identical. Two twins may resemble one another in almost every particular, but nonetheless, they remain two distinct, separate beings. Analogy reveals a relation of meaning but does not invalidate the hierarchical position of the "high" in relation to the "low," nor does it obscure the primacy of the upper world—which is *princeps*—over the manifest world of the senses. This notion of the superiority of the invisible world over the tangible, created world is also found in ternary anthropology, with its concept of "forms" (Fromaget 1999), in Platonic thought, in the Christian concept of the Word or *logos* ("*In principio erat Verbum*" [John 1: 1–3]), as well as in numerous mythologies that situate the origins of culture in the far-off mythical times of the founding heroes.

In the same way that information from the upper world is conveyed through representations, in the sphere between the real (objective truth), which derives from the upper world, and its subjective apprehension in the lower world, there exists a gradient of interpretation and distortion that inevitably leads to a progressive approach to truth and so to its possible falsification. This is the same sphere that allows for the distinction between the immutability of God and the evolution of the *Imago Dei* as perceived by humans. Authentic mystical experiences help to reduce this sphere, this hiatus, this subjective distortion, and so representations can come closer to being "presentations," in the sense of evidence that lends itself neither to interpretation nor to distortion, since it has an objective existence: "It is thus." And so, God becomes the "I am" revealed to Moses (Exodus 3: 14).[26]

One can doubtless see the need, within the traditions of the initiation ritual, for a metalanguage that can "say things without saying them," like the parables of Jesus or the *ikaros* (sacred chants) of Amazonian shamans. In this way, analogy and ambivalence make it less likely that the information will be reduced to a single, clear-cut interpretation. For instance, no Christian takes the following statement by Jesus literally: "And if thine eye offend thee, pluck it out, and cast it from thee" (Matthew 5: 27–32). Nonetheless, a reductionist approach is often employed by those desperately seeking facile, clear-cut answers in the hope that they will serve as guides to life, replacing the need for the exercise of their own judgment and free will. By confusing the image with the thing signified—or, as Korzybski puts it in his course on general semantics (Korzybski 2009), mistaking the map for actual territory—literal interpretation of this kind can lead to sectarian fanaticism.

Support during Shamanic Initiation and the Risks Involved

The shaman is in effect a specialist who can navigate the invisible world and return from it unscathed in body and mind. His expertise rests on three pillars: the ritualistic function; corporal intervention; and a referential cosmic vision.

The Ritualistic Function

The shaman's journeys to the invisible world are supported by a ritualistic framework that serves as the interface between the two worlds and helps to guide him in his "voyage" to the "other side" and to find his way back afterward (Mabit 1999). In exercising the ritualistic function, the shaman determines his objectives in advance, makes sure he has permission to cross over to the otherworld, and invokes the aid of necessary protectors. The ritual reveals a doorway that serves as both an entrance and an exit. From a therapeutic point of view, the ritual constitutes a symbolic framework that both restricts and integrates the altered state experience.

Through its symbolic function, ritual establishes a coherent balance between this world and the otherworld, activating the order of the universe within the precise framework of its intervention, thus turning it into a reality. One must obey these strict albeit intangible laws, which make the ritual a sort of "technology of the sacred," laws without which it could not become operational (Apffel-Marglin 2011). These laws are not arbitrary, personal creations introduced for aesthetic reasons or merely for play but, rather, are passed down to the shaman from the otherworld. Breaking these laws can have serious consequences for the physical, psychic, and spiritual health of the shaman and his patients.

Corporal Intervention

The human body here assumes a symbolic function, representing the macrocosm, given that the body possesses three dimensions: the physical, the psycho-affective, and the spiritual (Mouret 1990). Through the principle of analogy, the subject's body relates back to the realities of the invisible world, while the shaman, acting upon the body, intervenes at the same time on the different levels or dimensions of the "I." Thus, the activity of the shaman includes intervening in the workings of the physical body (e.g., through massage), the psycho-affective or energy body (through exhaled smoke energized by the appropriate chants and applied to the different energy points of the patient),[27] and the spiritual body (through invocations, sung prayers, or *ikaros*). The psychic disturbances of a patient can sometimes be caused by problems relating to energy and, as a result, may require no psychotherapeutic verbalization during treatment but, rather, an intervention on the energy plane to alleviate the symptoms. These disturbances can also be spiritual in nature, caused by contact, infestation, intrusion, or even possession by an evil spirit, in which case liberation or exorcism techniques are called for. These three sources of disturbance can occur in combination, as in the case, for instance, of a sorcerer who makes a victim unknowingly drink a preparation of toxic plants activated by magic ritual.

In this context, it becomes essential to purge the body. Purgative plants are ritually activated to the level of the three bodies; for the patient, cleansing takes place simultaneously in all three planes: the physical, the psychic, and the spiritual. The patient's organism is purified of toxins, and at the same time, he vomits his "bad thoughts" and "bad feelings," expelling as he does so the evil spirit that has taken hold of him. Clinically speaking, the speed and efficiency of these methods, when properly applied, would surprise any Western doctor.

Referential Cosmic Vision

From infancy, a young Amerindian grows up hearing descriptions of an active, organized, operational otherworld. Legends, anecdotes, myths and tales make sense of this world and its manifestations. The otherworld soon becomes tangible through dreams, extrasensory perception, paranormal phenomena, and synchronicities that, unlike in the West, are not censored. The young initiate thus comes into possession of a conceptual framework that allows him to situate his experiences in the map of the collective invisible. On a journey to the otherworld under an altered state of consciousness, he finds a series of references and landmarks that in no way contradict ordinary waking reality. The gap between ordinary and extraordinary consciousness does not represent an insurmountable, senseless boundary for him. The coherence achieved brings peace and helps to re-establish a sense of harmony. When a patient's internal order is in tune with the transcendent order of the universe, this is a sign of healing or indeed good health.

On the other hand, for a contemporary Westerner, the disappearance of a coherent, tangible cosmic vision is promoted through modish relativism by the "masters of suspicion," who have been engaged since the Renaissance in the task of emptying the universe of all its meaning and sacrality.[28] The absence of collectively accepted points of social reference that individuals can claim as their own, which can guide and support them through their rites of passage, means that most Westerners have been cast adrift and left to their own devices. In this self-referential dynamic so typical of the current era, these devices run the risk of feeding on the individual's inner demons, his or her own unconscious impulses and projections. As for the body, it is reduced to its basic animal functions, dispossessed of its psychic and spiritual dimensions, while the mind, isolated and closed off, is surfeited with an overflow of data, a chaos of information it cannot possibly hope to assimilate without its original referential framework. The spiritual dimension, neglected because it has been labeled as nonexistent, lies fallow or is confused with the play of mental activity and introspection without end. Journeys to the otherworld undertaken in denial of the mind's transcendental nature, when embarked on for purposes of recreation or mere "aesthetic" appreciation, particularly when making use of drugs, dispense with ritualistic preparation and ignore the great powers involved. In this context, rituals become individual, subjective creations, doubtfully inspired,[29] and neglectful of what is really required of them. Never has the phrase "sorcerer's apprentice" described so many apprentices so aptly and all at the same time.

The individuation process involved in psychotherapeutic treatments often comes up against this *schize*, the split in the psyche, to a degree of dissociation that is more or less important,[30] encapsulated in endless mentalizations, disconnected from the body, and deprived of the spirit. According to this ancient clinical practice,[31] these psychic problems can result, at least partially, from physical and energy-related intoxication, from a spiritual infestation by one or more entities from the otherworld, or even from the patient's being when it is caught in a limbo between the dimensions of "this world" and "the other." The scant regard for the body in traditional psychoanalytical practice, which sometimes goes so far as to forbid its involvement in the treatment altogether, delays the healing process, significantly reduces the therapy's efficacy, and prevents it from attaining its full potential as a cure. As far as the spiritual dimension is concerned, the failure to detect a parasitic infestation by a toxic or malevolent entity altogether impedes the patient's liberation.[32] If this "evil" is judged to be a component of the patient's shadow, regarded as something that belongs to the self, this is tantamount to self-condemnation by the patient. There is no way this shadow can be assimilated other than to justify the pattern of a transgression authorized in the name of a relativism that allows the patient to exercise this exogenous evil guiltlessly, as something his own and unpardonable. This, in turn, entails the suppression of all morality and the distinction between good and evil. Clinical practice seems to demonstrate that the suppression of the moral conscience, which is innate and intrinsic to human nature, results only in increased aggression and violence, often manifested in harmful behavior toward others or in self-harm. Self-destructive behavior of this kind can take the shape of addiction, anorexia, bulimia, and suicidal impulses, and sometimes it can appear in the form of autoimmune or psychosomatic disorders.

In my own clinical experience, infestations[33] are remarkably frequent, but generally pass unidentified as such, even by religious practitioners, who often refuse to discuss the subject, as they find it extremely disturbing. Treatment of infestations in the West suffers from the double handicap of not being considered either by priests or by therapists. Members of the church, when they do not simply deny the existence of evil[34] altogether, tend to severely understate its influence, attribute it to psychopathological disorders, or, at best, deign to recognize it only in extreme cases of possession. Their usual reaction is to send the sufferer to a psychiatrist in order to rule out the possibility of mental illness, which hardly ever happens due to the general climate of ignorance about infestations under which psychiatry labors with relation tantamount, in practical terms, to a flat denial of its existence. The 1994 nosography set down by the psychiatric diagnostic manual DSM-IV (*Diagnostic and Statistical Manual of Mental Disorders*, American Psychiatric Association 1994) furnished a clear diagnostic framework which classified all "spiritual symptoms"[35] as forms of mental illness,[36] although the issue has begun to raise a lot of debate, stimulated in great part by ethnopsychiatric observations (van Duijl et al. 2012). Some recommend containment through prayer, while others suggest the use of neuroleptic or antipsychotic drugs. The chemical straightjacket does, in effect, serve to "contain" the problem but does little in the way of liberating the victim; the drugs tend to become a permanent presence in their lives, together with the many side effects they produce.

Sources of Infestations

Infestations are numerous because their sources are likewise numerous, and members of modern society are particularly vulnerable to them. The lack of spiritual protection plays a decisive role: people cannot defend themselves against an ignored danger. Ignorance or outright denial of their existence makes Westerners easy prey to malevolent entities.

We cannot hope to cover the entire subject here, but we can single out a few important aspects taken from clinical practice and from Amazonian practitioners' empirical observations, as well as from my own experience over the last thirty years. Shamanic ritual practice reveals that every infestation takes place inside the victim's body, which is a physical-psycho-spiritual unity, and that the most common kind of infestation affects the energy plane. By opening and activating the energy body, shamanic practices can stimulate the emergence of everything the energy body contains, right down to its most secret depths, and thereby make this content perceivable. Perception via the five more common senses, as well as the psychic functions, becomes heightened, in both the patient and the shaman. In other words, the energy content emerges and becomes momentarily amplified, enabling it to be identified. The links and intricacies of causality or resonance between the different spheres of the individual thus find an outlet and begin to make sense.[37] In any case, a kind of intervention on the energy plane has been undertaken following an act or situation that has in some way violated the integrity of the individual's body. This violation is not necessarily violent in nature; it can happen just as well through seduction or trickery. The individual may likewise have made him- or herself vulnerable to these intrusions through ignorance,[38] curiosity, or quite often as a result of reacting inadequately to deeply affective or existential problems. We often discover the origin of transgressive impulses and delusions in heartache. In a sense, these wounds have already dealt a blow to the individual's integrity, paving the way for an invasion by evil spirits. If the resulting pain is not eased by the balm of requited love itself or healed by the power of pardon, or even by cathartic expiation, it can easily create an opportunity for seduction by a vengeful spirit. The wounded individual desires to wound in return.

If we draw up a rough catalogue of the sources of infestations, we immediately see that they can be classified into consensual infestations, on the one hand, and nonconsensual ones, on the other. Among the latter, we find the inherited, transgenerational infestations, which deserve a separate discussion.

Infestations with the Explicit or Implicit Consent of the Individual

The most frequent kinds of infestation are the result of an active collaboration, often unconscious, between the individual and spirit entities.

Unprotected induction (without ritual or with badly conducted rituals) of altered states of consciousness through drug use leaves the patient's energy body exposed and permeable. The more frequently one is exposed to this "contaminated" environment through such experiences,[39] the greater the risk of contagion spreading to other participants. Spirits can move from one body to another or poison another body through contact with the diluted energy bodies left exposed by drug taking.

A similar process takes place during sexual intercourse when the energy bodies of the sexual partners, under the effect of intense vibration, experience intimacy at the level of fusion. The quality of the energy body of an individual will determine the possible outcomes of potentially contagious exchanges. Ritual and religious traditions promote the sacralization of unions in order to prevent mutual adulteration by the energy bodies involved, as well as by the inadequate overlapping of the physical, psycho-affective, and spiritual bodies (the latter corresponding to the energy body). Ideally, the spiritual body should govern sexual encounters.

Ritual transgression plays an important role in exposure to infestation by voluntary actions. Within the scope of this type of transgression, all voluntary actions converge toward the forced attempt to contact or penetrate the spirit world. The individual's motivations and objectives greatly influence ritual practice; if they are not clear or morally sound, or if they are deficient in any way, the ensuing experience may prove highly dangerous. Therefore different traditions rely on a precise liturgy and require thorough, painstaking preparation in order to direct ritual practices and to purify the intentions of the participants.[40]

It is important to point out the risks involved in joining so-called initiation groups that use different rites. The candidate is invited into the group through ritualized symbolic practices. The initiation rite creates a bond of spiritual dependence not only with the group as an institution but also with the spiritual entities that preside over its formation. The bond created by initiation in a sense connects the initiate with the master spiritual power of the *egregore* (collective group mind) through the chain of successive initiations.

By contrast, in trying to avoid the long and arduous work of individuation, contemporary Western thought reveals the magical, childish dimension upon which it is based, aggravated by a desacralized society that trades on illusion. We may ask whether the shamanic quest of certain Westerners does not stem from this postmodern social childishness and Rousseau's myth of the "noble savage," on the one hand, and from the supposed magical thinking of shamans, on the other. The shaman, considered a lunatic by the intelligentsia of the 1950s, has, just a few decades on, been upgraded to a sage, a master.[41]

Infestation without the Individual's Explicit Consent

There are numerous ways in which infestation can take place without the individual's consent or active participation, although this does not mean they are any less toxic. Infestations can occur passively through direct contact with locations, objects, or even animals that have been infected. These receptacles of evil spirits may have become infected when placed in the presence of malevolent entities or after being possessed by spirits of the dead. This can also happen when the victim is in a place where some form of spiritual transgression has taken place (where magic has been practiced, for instance, or where an act of sexual abuse has been committed), or when an act of violence has led to someone's sudden death (e.g., as in the case of murder). These acts may cause the victim's spirit to be cast adrift, so it latches onto the location.[42] In other cases, these receptacles have been "charged" intentionally through ritual practice with the express purpose of harming those with whom they come in contact. This type of evil practice has long been a staple of sorcery the world over.

Sexual abuse is not only a violation (sometimes nonviolent) of the physical body but also of the energy and spiritual bodies as well. The abuser, taken over by an entity (an unclean spirit), transmits this entity to his victim. This is why sexual abusers themselves were so often the victims of sexual abuse during childhood, something that cannot be explained by psychic mechanisms of confused identification with the modes of affective relations,[43] since subjects are aware of the transgression involved and refuse (or at least hesitate) to identify themselves with these mechanisms and put them into practice.[44] The a posteriori proof of real infestation—not just of the symptomatic manifestation of secondary processes stemming from psychic troubles—can be given once the therapist or the shaman is able to free the patient by acting directly upon his or her energy body without the need for the patient's conscious acceptance or participation in the process. Treatment can thus be done remotely without the patient's knowledge or presence through simple, unspectacular rituals and techniques, sometimes even in a language and culture that are foreign to the patient, where the possibility of suggestion is minimal or indeed nonexistent.

Transgenerational Inheritance

The most difficult sources of infestation to identify are those passed down to the patient via transgenerational inheritance. In these cases, patients are already infected from birth. In their eyes and those of the people around them, the disturbing manifestations seem to come not from outside but, rather, from the 'inner character' or nature of the victims themselves. In other words, their deviant behavior is attributed to inborn personality traits. Western medical practitioners will search for clues in the patient's genetic makeup, in disturbances during pregnancy or labor, while the psychotherapists will explore the possibility of psycho-affective disorders having been passed down to patients by their ancestors or in the psychic context of their birth. Even if these physical and psychical conditions do in fact exist, they do not in any way preclude the possibility of infestation, which not only takes advantage of these deficiencies that facilitated the process of intrusion but also, in their turn, can be the source of these problems, even on the physical plane. More frequently, anamnesis is unable to identify traumatic occurrences in a victim's experiences dating from the act of conception, which might explain abnormal behavior or impressions that fall outside the scope of what is considered normal (Mabit 2008).

Similarities between Jungian Individuation and Amazonian Shamanic Initiation

This brief description of the epistemological framework of shamanism in the upper Amazon of Peru and of its comprehensive concept of the universe, which contrasts so starkly with Western thought, allows us to detect certain similarities between the Jungian individuation process and shamanic initiation.

In both cases, it is a *process* that presupposes or determines a sequence of stages, slow and drawn out, comprising a succession of "little deaths" that gradually lead a person out of the shadows and into the light (Mabit 2000). An essential value is attributed to self-experimentation, and the results of this labor will correspond to the degree of personal dedication invested by the individual in the enterprise. In this context, real and relevant commitment is essential, while facile, effortless "magical" solutions are illusory. Individuals must therefore place themselves in the role of active protagonist in their own evolutionary process or apprenticeship, a process in which a master or teacher helps to channel the "energies" that have been mustered and to ensure their proper integration. It is taken for granted that such masters, in their role as guides, have undergone initiation themselves and can "lead the way," and that they will not take their pupils along paths that masters have not already explored beforehand. From one point of view, this pupil-teacher relation must consider mutual exchanges happening not only on the psychic plane but also on the energy plane, with the master taking responsibility to metabolize any of the student's or patient's processes that may have been triggered by the cure. The master is also expected to curtail the overeager and overabundant intentions of the pupil by establishing measures of containment and integration, which form the basis of both psychoanalytic cure and shamanic ritual. The master at the same time channels, guides, controls, and teaches, not in an academic sense but, rather, by working within a dynamic relationship involving full commitment to accompanying the pupil at every step of the latter's meandering, personal journey, a journey that is unique in every way.

Each process, too, is unique, and it demands that the master go in search of his or her own self together with the pupil, with the advances of the former matching those of the latter'. In fact, the pupil meets transcendental dimensions in a unique fashion through paths that are particular to each individual and often surpass those of the master. Therefore the master cannot intervene as a teacher in the *ex cathedra*, dogmatic sense but, rather, as a companion with his or her own particular connection with the otherworld. In other words, the master's role is to facilitate the pupil's entry into the "mystical dimension," understood as a first-hand consciousness or knowledge of the transcendental presence, which is neither intellectual nor speculative in nature but, instead, experiential.[45] The approach to this transpersonal dimension, in both the Jungian and shamanic contexts, is marked by reaching beyond the limits of the conventional space-time continuum, where the invisible world manifests itself through dreams, visions, paranormal phenomena, extrasensory experience, and synchronous events that bear a particular significance for the individual. Activation of the energy plane of the initiation process opens the doors to other dimensions, and the role of the master is to maintain a sense of coherence and to avoid any outside interference, any confusion in the process of integration, or any inflation of the ego, all of which could lead to chaos and dissociation. In effect, the phenomena of consciousness no longer have to do with physical laws on a molecular or even atomic level but, rather, with those on the subatomic or quantum level.[46] The master cannot rest contented with a materialistic reductionism or even with a certain reductive psychology but must instead possess the necessary instruments to enter the otherworld through actively symbolic, operational pathways. Indeed, the energies thus mobilized are endowed with a numinous quality, *tremendum et fascinans*, capable of robbing the individual of freedom through the sheer power of fascination, terrifying or alienating the self beyond recall. The danger of the powers invoked is recognized by both types of guide, the analyst of our psychological depths and the shaman. With no safety net, the occupation of the individual by forces from the invisible world risks becoming all-out expropriation or possession. It is at this point that the shaman could turn into a sorcerer and the psychoanalyst, in his turn, could play at being a "sorcerer's apprentice."

In relation to the individual, this unconscious or invisible world is peopled with autonomous nonhumans, Jungian archetypes, the "mothers" of plants, or spirits of nature. They intervene during the therapeutic process, whether as allies or adversaries, as protectors or guardians of the threshold. In any case, they take part in a possible structuring of the individual, who in turn is called upon to confront these powers and to adequately integrate or interiorize them. These powers of the invisible world play a fundamental role as teachers,[47] offering potential nourishment to help build up the individual. This invisible world is therefore clearly distinguishable from an unconscious serving merely as a region of repression and shadow in the classic Freudian sense.

The transpersonal dimension, in both cases, leads to the establishment of an inevitable bond with the past, a return to the legacy left by the individual's ancestors. This inherited capital, for better or worse, must be carefully managed in order to be recognized, accepted, and eventually purified. The individual's identity cannot be established outside this heritage, personal family history, and cultural and spiritual background. Managing this heritage entails, ideally, recognizing and paying homage to the spirits of one's ancestors (gratitude), giving thanks for the life one has been given, and, at the same time, freeing oneself from negative ties associated with behavior patterns or psychic configurations connected with traumatic events in the lives of these ancestors or of the community to which they belonged.[48] Ancestors and the dead in general can appropriate living people and interfere in their affairs, sometimes even to the point of feeding off experiences in their daily lives. The individuation process, like the shamanic ritual, must take this essential dimension into account and come up with solutions that can pacify these relations.

Both the Jungian and the shamanic approaches propose establishing a dialogue with the invisible world through a symbolic metalanguage that is essentially transcultural, a language that aspires to be universal. Metaphor and analogy are vital tools in the management of the therapeutic or initiation process. Both paths embrace the notion of transcendence, of spirituality, of the otherworld, and both accept the idea of a coherent reality beyond the manifest world. They also have in common the fact of not being religions but of serving as paths to knowledge, particularly self-knowledge, that are essentially therapeutic in outlook. In this respect, they are medicinal in the broader sense: neither is intent on founding a religious institution of any kind. The modern misappropriation of both psychoanalysis of the depths (Maxence 2004, 2012) and shamanism by churches or sectarian groups[49] seems, therefore, to go against the original intention of both these approaches.[50]

In this way, the Jungian individuation process and shamanic initiation converge on many issues. One might even be tempted to associate them more narrowly by describing them as two versions of a single or similar path: one Western, the other Amazonian. However, upon closer inspection, it will become clear that there are as many fundamental differences between the two as there are similarities, ruling out any such facile analogy.

Fundamental Differences between Jungian Individuation and Shamanic Initiation

When compared from the perspective of the Western and tribal cultural horizons that, respectively, gave birth to them, the Jungian and shamanic models present some glaring differences.

Individual and Community

Shamanic initiation cannot be understood separately from its ultimate purpose, which is to re-establish order and balance within the heart of the human community to which the shaman and his pupil belong. The principle of justice and reciprocity embraces the entire community, as well as creation as a whole, at least in its closest manifestations. In this sense, one cannot really speak of "individuation" because the individual does not fully exist outside the group. A member of such a group does not strive to become different from the others in the community but, rather, to blend in and become as intimate a part of the group as possible, assuming its function and rationale as his or her own. It would practically constitute a breach of the law for a member of the community to seek individuation by turning away from the extended family, and such a pursuit would not fail to provoke strong reactions among other members, since it would imperil the integrity and coherence of the group. By contrast, it is perfectly possible, from a Western, individualistic point of view—desirable, even—to strive for some form of personal, individual differentiation. Every Western individual feels free to choose this path irrespective of the expectations of the family, whether nuclear or extended, or indeed of professional, spiritual, and social milieus. The title of Jungian analyst confers a certain social status, and even if Jungian thought remains largely marginalized by dominant academic currents, this does not in any way threaten the integrity of its supporters and practitioners.

Time's Arrows

On a deeper level, however, these two processes are situated in diametrically opposed projections of time. In the tribal context, time's arrow points from the present back toward the past. Knowledge considered desirable to an Amerindian man or woman is that pertaining to the group's origins, the ancestors. They aspire to a reconnection with the ancestral lineage, trying to reach as far back as possible to a distant past where they may find the source of primal knowledge. The flow of time draws them further and further away from this primordial wisdom, as it does every successive generation, hence they 'respect' those placed chronologically closer to this source and ascribe fundamental importance to maintaining traditions and preserving the purity of the knowledge or science handed down from the dawn of humanity. From a Western viewpoint, the Amerindian moves backward. The Western man or woman, by contrast, faces forward, looking ahead at things to come; for them, time's arrow points from the present toward the future, and novelty excites great interest and fascination. Their expectations target a time ahead based on the principle of the evolutionary process, of progress, while with every passing moment the past loses its potential and becomes something of no consequence. Success, therefore, in the Amerindian case means rediscovering the past, while in Western case, success means discarding the past—or, put in a slightly different way, Amerindians try to retard the degradation caused by the flow of time, while for Westerners, its action is both welcome and wholesome.

Up to this point, one could still argue that these apparent differences or oppositions might be overlooked from a perspective that sees time as a spiral, its flow moving back toward the past in its horizontal rotations while simultaneously tracing a line forward to the future on the vertical plane. In the same way, a patient undergoing treatment may return to the past in order to assimilate the "knots" in his or her personal history, thus making it possible to forge ahead and progress toward a future cure. In the end, these regressive phases form part of the general forward thrust of personal progress into the future. We may equally well consider that in the furtherance of individuation, the patient's projections and phantoms weigh less on the group, and, in finding his or her true vocation, the patient also becomes of greater service to the collective well-being.

Word and Body

However, I find it more difficult to assess the comparative effectiveness of the instruments which, in Western and tribal therapy alike, enable therapy to proceed. On the one hand, psychoanalysis places too much importance on the spoken word, to a certain extent neglecting the body, while, on the other hand, shamanic initiation has as its starting point, and, above all, the bodies of the individual and of the shaman, with the use of the spoken word reduced to a minimum. Ever since I began my apprenticeship with Amazonian shamans, this point has seemed to me to be the cornerstone of initiation (Mabit, 1988). The Word of the Christian lamb, the offspring of the Jewish ram of justice, institutes the spoken word as the founding principle ("In the beginning . . ."), and this informs Jungian thought and practice, having their roots in Christianity. For the shaman, in contrast, it is the body that is the custodian of knowledge stretching backward in time, from the individual's personal, biographical memories to ancestral memories, and further back to human and even cosmic memories. The spoken word can be contaminated by the mind, by illusions, by phantoms, while the body cannot lie. The unconscious inhabits the whole of the body in its furthermost recesses, not only the brain, as Westerners generally believe. The body has become the laboratory in which we conduct experiments and verify the information received during altered states of consciousness, whether natural (dreams) or induced. The relative lack of importance attached to the body in analytical practice paves the way to psychic and even spiritual disorientation. The body needs to be purged of its various loads or interferences that inhibit the emergence into consciousness of truth-bearing memories. While it is true that psychoanalysis and even psychocorporal approaches can certainly enable awareness to develop, they cannot, however, reach the deepest levels of somatic memories that constitute the true encoded revelations of the Life we carry within us. The bodies of plants, judiciously energized through shamanic ritual and technique, can reach the human body directly at its deepest levels through the common physiological structure that we share with all created beings.

The body can also become subject to invasion. The practices of sorcery are founded in part on this ability of the body to be contaminated in its different dimensions: its physical dimension (through sorcery by contact); its energy (through sorcery at a distance by shooting magic darts or *virotes*); and its spiritual dimension (through magic spells, spiritualism, occult practices, etc.). These practices achieve effects that do not depend on belief or nonbelief in the phenomena but on the operation of universal laws of energy to which all human beings (and all created beings) are subject. Ignorance of the laws in no way protects people from their effects but, rather, makes them more vulnerable.

The potential of the corporal dimension is particularly visible in the case of the individual's psychic personality during sessions with ayahuasca, when it takes on an animal form. On these altered states of consciousness, the individual feels inhabited by a particular animal figure; he feels he has become that animal, even while retaining human awareness. He adopts the animal's movements, postures, reflexes, and instincts, even though the animal in question does not belong to his habitual environment. The experience is one of exaltation, in which the subject feels himself as strong as a bear, as agile as a feline, or as keen-sighted as an eagle. For the individual, the zoomorphic energy manifests itself as a constituent part of his deep and therefore unvarying psychic personality. Furthermore, these appearances often surprise the individual without corresponding to his personal phantoms or to his natural preference for one species over another. This experience is impossible to produce simply by imitation. Rather, it denotes unsuspected relationships between the human species and animals. Here we have reached a point that is very distant from abstract symbolism or imagination.

Integration and Cosmic Vision

This is an issue that has been mentioned above. Interiorizing the initiation process is made easier for Amerindians by their ability instantly to place their extraordinary experiences within a shared cosmic vision that provides them with landmarks and maps of the invisible world. For Westerners, partially divorced from their Judeo-Christian symbolic roots, the experiences of contact with the unconscious require the presence of a translator, a role that analysts may play if they have not become cut off from their roots. For a Westerner, it is at this point that the spoken word reasserts its importance. The blurred outlines of the average Westerner's spiritual landmarks tend to make this integration more difficult. While the Amerindian view of the world lends a global coherence to everyday life and to non-everyday experiences of awareness, the Westerner, living an everyday reality which is dissociated from the otherworld, will readily interpret a non-everyday experience in an irrational manner, whether as a sign of madness (denial) or as a sign of being divinely chosen, which in turn can lead to a dangerously inflated ego. In fact, Jung recognizes the danger of identifying the "I" with archetypical content, even though this inflated psyche may only be temporary in the course of assimilating the dimensions that arise from the unconscious. If the "I" is identified with the self, it loses its limits and fails to recognize the autonomy of unconscious forces.

Meanwhile, for the Amerindian, this map identifies independent spirits that can be harmful to the human soul and whose assimilation is thus unacceptable, unless it occurs for him to turn into a sorcerer or to become possessed. On the psychic level, the cause of the infestation or possession needs explaining in order to prevent it from recurring. It must certainly not be tolerated but calls for exorcism. The shaman will seek to discover sources external to the individual (spells, infested locations) and the means within the individual (improper behavior, transgressions) that have allowed the invasion to take place. While shamanic approaches seem to parallel those of Western psychotherapy, they involve ritual procedures that have the purpose of simply expelling the intrusive entity. The fact that most Westerners do not recognize the existence of malevolent spirits, or fail to distinguish them from unconscious psychic powers, may lead to a dangerous mix-up, closing the individual in upon him- or herself in a mental prison from which they can only escape if worse comes to worst either by dissociation (madness) or by identifying with the intruder (possession). Apart from these two extremes, the danger is still present, though more manageable, and less obvious. An instance of this, in my view, is the New Age spiritual narcissism, which is part and parcel of these seductive infestations shared by many people, but usually not identified as such, thus slyly poisoning the human soul. The question of detecting the material, tangible character of the spirits can clear the way for grace, a better option than the senseless extremes of insanity, on the one hand, and becoming a sorcerer, on the other.

Autonomous Powers

The fundamental difference between Amazonian shamanism and the Jungian approach undoubtedly lies in the definition and detection of fully autonomous invisible dynamic forces of the invisible unconscious, independent of the individual.

For Amerindians, the natural world is clearly peopled with entities that are created beings, generally invisible to anybody in an ordinary state of awareness, having substance, intelligence, a will, and the ability to communicate, which are fully autonomous in relation to human beings and which, though incorporeal, have substance. This otherworld is very much alive, active, and able both to communicate and interfere with human beings. These beings are differentiated by hierarchical levels: some of them, inferior to humans, may become either allies or adversaries, depending on how they are treated (the "mothers" of plants are a case in point), while others are superior to humans and are necessarily either benevolent (angels, good spirits) or malevolent (devils, evil spirits).

In the same way, the everyday perception of archetypes by Jungians confers upon them a degree of independence from humans, even though they do not clearly recognize them as spirits. Now, the very notion of the archetype in Jung seems to have arisen from Lévy-Bruhl's translation of what he calls "collective representations" or the symbolic content of "primitive" cosmic visions. As early as 1917, the Protestant theologian Rudolf Otto attributed a numinous character (*mysterium tremendum et fascinans*) to these "representations."

Overall, archetypes are defined as being contained within the psyche, which, in the final analysis, grants them neither full autonomy nor an existence of their own. Thus, according to Jung ([1927] 2001, 31), "the archetypes are inherited with brain structure," though he also maintains that they predate the psyche and may be assimilated to "psychoid entities." This means they can only be activated by psychotherapy, which removes the sacred in the full meaning of the term. In this context, shamanic rituals of curing or sorcery must either be inoperative or require "belief in them" by the subject. This brings us back to the issue of suggestion, though certainly in a more sophisticated form than in Bleuler's definition of hysteria but, in the end, equally lacking spiritual density. Western rationalist, positivistic thought is so deeply rooted that the very reality of the spirit world immediately smacks of brimstone. Not even the main current of Jungian thought can easily overcome this taboo. Without claiming to be familiar with the full extent of Jung's prolific output, it nevertheless seems to me that his approach seeks the approval of the scientific community, which necessarily prevents it from crossing the threshold into an area in which it would be possible to fully assert the reality of the existence of nonhumans as entities in their own right without being dismissed out of hand, inviting the supreme insult of being labeled as mysticism.[51]

And yet, Marilyn Nagy has pointed out that, in a letter to Fritz Kunkel, Jung specifies, "In each individual case I must of necessity be skeptical, but in the long run I have to admit that the spirit hypothesis yields better results in practice than any other" (cited in Nagy 1991, 86, n. 18). Nagy sees a parallel with Kant's statement: "I did not dare to deny completely the truth of the various ghost tales; on the contrary, I have always maintained a certain reserve and a sense of wonder toward them, doubting each story individually, but attributing some truthfulness to all of them put together" (cited in 1991, 86, n. 18).

Other statements by Jung show evidence of his feeling of hesitation about making a definitive decision, as, for example, when he says that "the appearance of the archetypes has a clearly numinous character which, if one does not wish to call it 'magical,' must be called spiritual" (Jung [1947] 1970, 185). He distinguishes between archetypes and archetypical representations, attaching to the former a character that cannot be known except by their indirect effects ([1942] 1962, 261). This once again denies a mystical dimension such as I defined it above, a dimension that presumes the possibility, by grace, of direct and immediate knowledge of transcendence. If spirits occupy a transcendent plane in relation to human beings, it is conceivable that they may be accessible by direct experiential knowledge, taking as a starting point the human spirit rather than human psychism, which can only reach, at best, as far as the representations. Jung's archetypes correspond to the notion of spirits found among the first peoples, since he recognizes their capability for inducing psychophysical phenomena of acausal coincidence (synchronicities) and triggering parapsychological abilities in the individual. The archetype may have a personal dimension, such as the self that directs our destiny or brings an "atmosphere for the I." It can equally represent a pattern of animistic transformation punctuating the major phases of a lifetime (birth, puberty, old age, death), thus drawing close to Stanislaf Grof's (1983) notion of complexes or perinatal matrices in the case of birth. The archetype can also represent the deep structure of the relational dimension established between two human beings (such as master–disciple, executioner–victim, lovers). In this last definition, the notion reappears of "mystical participation" so dear to Lévy-Bruhl, which once again sends us back to the spiritual dimension that is ever-present in an underlying form, although in the end it is neither fully explained nor clearly recognized. For Jung, however, the archetypes are not directly knowable other than by their effects. Consequently, they "do not belong to the order of direct experiential experimentation: they are of the nature of a postulate" (Biju-Duval 2001, 159).

What we see as Jung's hesitant see-sawing between a predominantly psychic notion of archetypes, having a scientific character, and another predominantly spiritual notion of a metaphysical order emerges very clearly in the question of the constellation of archetypes. Possession, observed in clinical cases that Jung did not neglect to mention in the service of psychiatry, is interpreted as the crushing invasion by an animistic power that originates when an archetypical psychic power buried in the depths suddenly rises to the surface of consciousness and emerges into the open. This power belongs, in a way, to the individual and comes "from below." For the world's great religions and for shamanic traditions alike, this invasion comes "from above" and is a manifestation of a human spirit being captured by another spirit. It is no longer a simple case of parasitism or of an infestation but of a capture that annexes the subject's will. He has no choice in the matter, and the invasion is effected without his consent. Exorcism is then called for, but Jung, in accordance with his conception, does not propose performing it: how is it possible, in fact, to expel a "force" that properly should belong to the individual, even though it is in his deep unconscious? Jung, in contrast, returns to the notion of possession seen from a different angle, this time more clearly psychic, when he considers that every human being is to a greater or lesser extent possessed by one or more archetypes, belonging to the order of "genies" (daemons), which may call the subject to certain vocations and to which he may declare his allegiance if he so wishes. For Jung, it is a matter of choosing, in a way, the "deity" to which the subject will submit, or, alternatively, the "dominant representation" he or she decides to subscribe to. This time, the subject's will power is called upon. Responding to an appeal, or following a vocation, has as its purpose the supreme welfare and liberation of the individual and, therefore, cannot be considered a form of alienating capture. This inspiration presupposes "inspirers" that elevate the subjects and "pull them upward," allowing them to overcome their limitations in a way that is completely distinct, and even opposed, to the lowering and crushing effect of possessive invasions, which "pull downward." This elevation represents a possible and desirable—or even necessary—outcome of the individuation process, while infestations of any kind, all the way up to full possession, degrade people by pushing them toward regression. The upward-pointing arrow tends toward a differentiated infinite, while the downward-pointing arrow takes them back toward an even greater nondifferentiation.

In his 1935 lecture, Jung evokes the "world of subtle bodies," where beings live who may possess a kind of corporality, and where immaterial beings may also be, the border between them being difficult to pinpoint (Jung [1935] 1989).

In the recently published *Red Book*, Jung (2009) records his conversations with spirits in the course of exercising his active imagination. In this surprising passage, the spirits demand—though without any great expectations—to be recognized as wholly existing beings:

> E [Elijah]: "You may call us symbols for the same reason you can also call your fellow human beings symbols, if you wish to. But we are just as real as your fellow men. You invalidate nothing and solve nothing by calling us symbols."
> I [Jung]: "You plunge me into a terrible confusion. Do you wish to be real?"
> E [Elijah]: "We are certainly what you call real. Here we are, and you have to accept us. The choice is yours." (Jung 2009, 187)

Despite these warnings, it seems clear to me that Jung never crossed this threshold. The proof, as I see it, seems to reside in the fact that if he had granted his recognition, it would necessarily have entailed putting in place a ritual formality on psychoanalytical consultations. This was not the case, as the initial therapeutic contract (detailing the length and frequency of each session, payment, couch or no couch, etc.) does not count as a ritual framework.

Jung does, however, recognize a natural disposition for religion in the human soul, "a current which deserves to be channeled and tamed through the symbolism of belief and ritual" (Jung [1934] 1970, 20), so as to avert the danger that this power might attack secondary, everyday objects (idols), which would then suddenly be raised to the category of the sacred. In this passage, he goes so far as to argue the necessity of religious institutions, justly stipulating that they must make possible, without obstruction, a true religious experience (Jung [1938] 1949). This accords with what I was saying earlier about mysticism as an individual experience being right and necessary in every process of individuation or personal evolution, and that it should therefore be "democratized."[52] Jung goes so far as to recognize a "pontifical" quality in dogma on account of its mediating ability to make the archetypical experience of the religious symbol situated in the human heart readily intelligible. This allusion to the magisterium of the Catholic Church should, however, not be seen as a concession to the revelation of which it claims to be the guardian (the deposit of faith), since he goes on immediately to specify that dogma, creed, and liturgy embody in a delicate and subtle manner the experience of many human souls who have traveled the paths of religious quest and, in this sense, have purified themselves individually in order to express an objective truth about the collective psyche (Jung [1938] 1949). God remains unreachable or else is reduced to this objective truth, in which case revelation is the fruit of individual or collective human effort aimed at uncovering the hidden depths of the psyche. The death sentence that "the masters of suspicion" have passed on God remains in force and has not been questioned other than in terms of the psychic consequences arising from the replacement of God with megalomaniacal science or totalitarian politics claiming to create an earthly paradise. Jung remains within the bounds of psychism and only concedes to the churches a role for them to play in mass psychism, but for him, God does not seem to be equally necessary for the elite of the initiated—to which, judging by the evidence, he himself belongs.

It is my understanding that ritual is a symbolic formulation that cannot be freely invented but must obey the rules of universal symbolism transmitted by the spiritual world, specifically through the proceedings and traditions of initiation. It is only on this condition that the approach to the spiritual world may become fully coherent and operational and may protect the person from numinous powers that dwell in him, such as regressive and incestuous tendencies toward nondifferentiation. In the absence of this condition, the doors to the spiritual world will not be fully opened, and grace will not abound.

The Jungian archetypes thus present themselves mainly as profane, desacralized spirits, flattened by an impoverishing psychic reductionism.

Discerning the Substantiality of Evil

These thoughts on a vast subject lead us to a few brief considerations on the question of the reality of evil in Jungian thought.[53]

The reality of spirits, whether good or evil, as autonomous material beings, extends to the recognition or denial of the material or substantial existence of God and the devil or Satan. In Jung, we find here once gain the same poorly defined borders mentioned above in connection with the psychic and spiritual dimensions. Jung certainly does not deny the reality of good and evil, but he confuses them on distinct and hierarchically segregated levels of reasoning, as Giegerich has shown (2010): the level of logical thought and judgment, that of empirical reality, and that of metaphysical substantiality. As principles governing our ethical judgment, good and evil are seen as being complementary and opposite in the same way as light and dark, or high and low (Jung 1951 [1986], 273). He then interprets the Christian principle of *privatio boni* as a way of denying the opposing poles of good and evil and asserts that the Christian cosmic vision is described as monistic, but in reality, it conceals a de facto dualism ([1951] 1986, 278). He interprets this definition as an attempt by the Catholic Church (specifically through St. Basil) to defend itself from the dualism of the Gnostics and Manicheists. He displays a strong aversion to the concept of *privatio boni*, which he criticizes for being an indirect way of emptying evil of all reality, although it can be empirically observed and its influence on the human soul is plainly visible. "Only the unconscious is oblivious to good and evil," he adds (1951, 65). *Privatio boni* harks back to the Christian concept of the *summum bonum* in which God is defined as the supreme good, and this, Jung says, seems to leave the human being as the only one guilty of sin, of the fall, and of evil. Toward the end of his life, however, Jung realized that these categories of principles of ethical judgment, "pushed to their furthest ontological roots, deal with divine aspects, with the names of God" (1959, 425), and recognized that his earlier "critique of the doctrine of *privatio boni* is valid only within the bounds of psychological experience" ([1951] 1986, 65). Furthermore, he finally stated that, "if one wishes to see the principle of evil as real, one must also call it the Devil" (1959, 432). While it is true that the principles of *privatio boni* and the *summum bonum* become absurd when reduced to the psychic dimension, they nevertheless retain their consistency in the spiritual sphere. The mystery of iniquity, which determines that Satan is a creature that rejects its own nature in a sort of ontological self-contradiction (Giegerich 2010), can only be conceived after free and sincere attachment to love, not as an object but as the very essence of the divine. As a result, the devil (or devils) ceases to be a "person,"[54] and this insensate choice no longer rests on logical reason.[55] Jung similarly takes an empirical attitude toward the question of evil and makes a point of specifying that "this does not mean that I am relativizing good and evil in themselves; I can clearly see that this or that is evil" ([1959] 2001, 426).

When Jung turns to the question of metaphysical substantiality, he appeals above all to Old Testament sources when discussing God—or at least the *imago dei*—while, by contrast, he too often approaches the figure of Christ from a modern standpoint. The animistic figure of Satan evolves step by step from the Old Testament and seems to cast doubt on its substantiality unless one admits while one may not confuse the immutable deity with the *imago dei* in constant movement, which is the understanding or apprehension of God in accordance with the current state of spiritual evolution. Similarly, it would be wrong to confuse Satan with the *imago diaboli*. Satan holds untold powers, and Jung rebukes present-day Christianity for presenting an overly benign picture of evil, reducing it to "a little prank-playing trickster." If the Christian message, under pressure, has in a sense flattened itself in order to get under the door of modernity, as some criticize it for doing, it still remains true that the substantial existence of evil has been constantly reasserted by the magisterium of the Catholic Church.[56] The combat in which exorcists are engaged remains fully relevant today (see, e.g., Amorth 2010). Elsewhere, Jung, once again referring to modernity, says that "the symbol of Christ is incomplete as a totality in the modern sense, because it does not include the dark side of things, but expressly excludes it as a Luciferian counterpart" ([1951] 1986, 54). Jung is indeed intrigued by the difficulty of discovering the shadow in Jesus. Considering that he is Christ and has a divine nature according to Christian tradition, this ought to reinforce the concept of the *summum bonum* if Jung had not already rejected it and reduced Christ's real substance to a mere "symbol." As Giegerich (2010) rightly points out, the theory of *privatio boni* presupposes an evolutionary leap in thought and understanding in which good and evil are not in opposition to one another but, rather, are now seen as two among the many points on a single, integrated continuum. Jung's discrepancy on the question of evil seems to stem from the fact he established a direct connection between, on the one hand, the unreality of evil as a metaphysical substance in its own right in *privatio boni* and the *summum bonum*, and, on the other, the automatic elimination of evil from the domain of logical judgments and empirical reality. In fact, this connection is not necessary. Jung, once again, has stepped outside his role as a clinical psychologist by risking metaphysical propositions that he had hoped to elude. To apprehend the real, one should not attempt to turn it into an abstraction.

The question of the opposition between the symbolic and the real looks like a wrongly formulated question: "The question therefore is not one of symbol or reality but, rather, of symbol and reality, since invisible reality can only be expressed in symbolic terms" (Laurentin 1995, 13).

Discernment therefore remains a central question for identifying the dimension of evil to which one wishes to refer; to draw a distinction between, on the one hand, the phenomenological level of evil and its representations on the psychic (and even physical) plane, and, on the other, its ontological status on the spiritual plane as a spiritual entity, without denying its status as a creature, and not a deity. Jung implicitly recognizes this when he describes human discernment as limited, since the certainty of judgment is an attribute of the deity alone. The apodictic certainty of judgment in humans is a symptom of madness. This is more reason to appeal to the need for revelation as a source of discernment. Christian tradition has contrived to produce a list of criteria for spiritual discernment, which would doubtless advance this necessary differentiation.[57] The latter may prove complementary to the process of discernment between "dissociative crisis" and "spiritual emergency," as proposed by Stanislas Grof (1983), or in other words, the differentiation between pre-personal experience (prior to the constitution of the self) and transpersonal experience (a mystical event situated beyond the everyday awareness of the self), as suggested by Ken Wilber (2007).

The horizon of modernity converges with Jung in the sense that the Judeo-Christian myth, when it concerns the reality of Satan and the devils, corresponds to the evolutionary state of primitive, rudimentary, outdated thought patterns that can now be discarded. The present-day collective psychic state is said to transcend this mythical thought pattern, which has now become obsolete, corresponding to the psychic and cultural needs of a past era.[58] Any practitioner who dares to affirm the reality of devils comes across as an animist at best or, at worst, a slave to medieval superstitions and magic. At the same time, whether Christian doctrine is true or not, Jungian theory does not fail to point out the psychic risks that arise when Satan is cut out of the Western animistic world view as an archetypical image having the capability to symbolize the dark side of the collective psyche. Without Satan, the psychic energy of this archetype of the dark side of the self is said to invade the individual, who may be plunged into negative hubris. One may question whether the massive collective emergence of satanic elements in Western society is a sign of the return of the repressed.

In the face of this double denial of the divine and the ontological reality of evil and its manifestations in the form of devils and evil spirits, altered states of awareness described in Eastern traditions of spiritual contemplation (Wilber 2007) parallel those proposed both by Christian tradition, on the one hand, and by shamanic practices, on the other. In some of these heightened states, awareness ceases to be limited by the self, and the individual can then consent to being overcome or inhabited by a oneness that is luminous, benevolent, absolutely good, and just. This nondual experience, in its verticality, goes beyond that which is seen as the horizontality of the *coincidentia oppositorum* in which good and evil are placed in juxtaposition without ever giving way to a transcendent third term. These mystical experiences, as I have had occasion to stress earlier, do not arise from a process of conjecture but operate beyond the reach of the techniques used, starting with the incarnation of the individual and as a grace bestowed. It is therefore, to a certain extent, the divine spirit descending into the manifest world and into the subject's body. Far from a case of the addict fleeing from the harsh realities and sufferings of this world below, or a "mysticism that arranges the subject's departure from his own body," it is rather "a mysticism of divine indwelling in the suffering body of the witness" (Marguerat 2012, 486).

In this context, the experience of the divine is not a response to a theoretical, intellectual, or abstract postulate, nor is it an attempt to prove a metaphysical truth; it is, rather, an offer to live it. If we postulate the existence of a wholly good God without the slightest trace of evil, it follows that each and every individual has the chance to experiment and verify its validity by adopting a methodology and precise conditions, among which, by necessity, must be sincerity of heart. In this place, in the presence of the deity, struggles and conflicts die down and even disappear altogether. The fact remains, however, that although this immersion in the "kingdom of heaven" requires welcoming and submission on the part of the individual invited to forget him- or herself and demands a certain degree of effort, even suffering, it is not a "thing" to be obtained or a goal to be achieved but a free gift of divine grace.

This grace proves to be perceptible and tangible, entering the individuals' body as a gratuitous and merciful inflow.

Conclusion

In the face of everyday spiritualism disguised as materialism, Amazonian shamanism has the potential to make a significant contribution to facilitate a much-needed return to the body and to the earth—in a word, to incarnation. The West is thus called upon to renounce its own modern magical thought and to consent instead, through pragmatism and the practical wisdom of ancestral traditions, to a true re-enchantment of the world that is neither factitious nor dangerous, as John Paul II, states.[59] In the context of the end of an era and of the human age,[60] this desacralization can result in two "heresies" which are a mirror image of one another, angelism (removal from the body) or animalism (profanation of the idolized body).[61] Both give way to spiritual infestations while denying the reality of spirits and the possibility of their intervention in human life. Amazonian shamanism reminds us of that reality and offers to provide treatment of this parasitism to rid oneself of them and to bring the sacred back into the core of matter and the manifest world. Individuation in the Jungian sense shows a lack in this regard, arising from its denial of the reality of spirits, which it reduces to mere symbols of the psyche, even if collective. The expertise of the shamans in the controlled inducement of altered states of awareness, within the constraints of just and established ritual practice, can equally facilitate the process of individuation by opening the way to mystical and transcendental experiences. This ritual opening broadens the concept of the environment to include the otherworld in its psychic and spiritual dimensions by, once again, forging the links of vertical integration with our ancestors and horizontal integration with our contemporaries.

The two earlier foundation myths of Aries-justice followed by Pisces-love give way to the tentative emergence of the new foundation myth of Aquarius-freedom, destined to absorb and thereby replace these earlier myths in a coherent manner. The body (incarnation) and the word (logos) are called upon to fertilize and explain one another, particularly within a ritual (liturgy), where the Word incarnate announces the coming of the Spirit, which is the source and breath of freedom.

True individuation, which allows itself to be inspired, reintroduces the dimension of the mystery of grace to the heart of the human being so that it can "lay itself open to something greater than itself,"[62] leading to a humbled and active fellow feeling with differentiated nonhumans and humans.[63] This can greatly benefit from the contributions of shamanism to save us from the daunting choice between becoming either mad or sorcerers by delivering us to freedom.

NOTES

1. In the context of the tribally based Jewish world, this thought pattern gives rise to the argument put forward by the High Priest, Caiaphas, when he demands the death penalty for Jesus: "It is expedient for us, that one man should die for the people, and that the whole nation perish not" (John 11: 50).
2. Once again, this is seen in the Old Testament, where the Jewish tribes worship a god of justice, even a god of law enforcement or a god of vengeance.
3. Describing the Candoshi, Alexandre Surrallés (2013, 129) gives the following explanation of "thinking with their hearts": "The Candoshi have the ability to think/love the strategy of an act of war because, when they prepare an attack on an enemy, they do so in terms of vengeance, and 'to love' means 'to avenge the loved one unceasingly.'"
4. I have mentioned elsewhere the urge for liberation of and through the spoken word as a means of overcoming the collective depression prevailing on our "Sad Planet." Several studies have confirmed this (e.g., Pepper and Cunningham 2004) by demonstrating that in non-Western cultures verbalization therapies produce excellent results (see Mabit and González Mariscal 2013).
5. Or "catholic" in the etymological sense of spatial and temporal plenitude, of cosmic universality, in which the term "other" includes not only other human beings but every element of creation.
6. Instances include the U.S. Embassy hostage crisis in Iran in 1979 and, more recently and more eloquently, Israel's extraordinary mobilization to secure the release of a single soldier, Gilad Shalit, held captive since 2006.
7. For excusable misdeeds, an apology is sufficient.
8. When two women appeal to Solomon, each claiming the baby as her own, he recognizes which of them is the child's true mother because she is moved by compassion, or, in the words of the King James Bible, "for her bowels yearned upon her son" (1 Kings 3:26). The Hebrew word here translated as "bowels" is *rachamim*, a form of the word *rechem*, whose literal meaning is "womb, uterus." The plural form seen here is the so-called plural of plenitude, signifying a mother's feeling of tenderness, pity, or mercy toward her child. The same word is used elsewhere to designate the entrails of Yahweh, the entrails of the Lord, the very source of all mercy.
9. Harmine, the active principle in *Banisteriopsis caapi*, the plant better known under the name of ayahuasca, was initially named telepathine when it was first isolated in the laboratory, until scientists realized that it is in fact the same active principle already known from *Peganum harmala*, or Syrian rue, a plant occurring in the Mediterranean.
10. One example in the Amazon region is the sprite with one leg shorter than the other, known by the Quechua name Chullachaki (literally, "unequal foot"). The sprite plays tricks on human beings, like the Dahu of rural areas in France. The sirens of the Amazon sing as seductively as their African sisters or as the ones whom Ulysses had to deal with in the *Odyssey*.
11. Thus, for example, the anthropologist Jean-Loup Amselle criticizes the use of ayahuasca and speaks mockingly of "shamanic faith" while loudly proclaiming his own "atheistic faith" and boasting that he never once took part in a session of that kind during his four months of field research in the area (spread out over a four-year period) (Amselle 2013). This is an echo of the senior ecclesiastics who mocked Galileo's Copernican model of the solar system and who, when urged to observe the planets for themselves, declined his invitation to put their eyes to his telescope.
12. The energy body has been dismissed as nonexistent (and is still dismissed by conventional medicine), but it can be visualized and measured by the Kirlian method or by Konstantin Korotkov's

bioelectro-photography. The extraordinary quantity of data on paranormal phenomena is simply disregarded by most members of the scientific community who, not knowing how to handle it, arbitrarily put it aside.

13. DMT (dimethyltryptamine) is one such substance. It is found both in the ayahuasca beverage and in the blood and cephalorachidian liquid of mammals, secreted by the pineal gland, known in Eastern traditions as "the third eye."
14. This toxicological nonsense of "intrinsic toxicity" did not prevent the toxicologist Laurent Rivier, for example, from saying, in the course of a presentation on the hallucinogenic plants of the Amazon region, that "certain among the so-called depurative plants are known to be highly toxic" (Rivier 2002).
15. This is what we seek to demonstrate at the Takiwasi Center when treating drug users from all cultural backgrounds, drawing on resources from traditional Amazonian medicine, and when dealing with a pathology that has defeated Western medical practice (see Giove 2002; Mabit 2007).
16. In particular in the case of degenerative diseases, autoimmune diseases, and mental health issues such as drug addiction (see Mabit 2012).
17. Significantly, anthropology has recourse to a subterfuge of the same kind when it draws a distinction between an "emic" approach (from the Indians' viewpoint) and an "etic" approach (a neutral description of the facts as observed by the anthropologist), thus dodging the question of the real existence of the nonhuman independently of the human. Under the "emic" approach, the interpretation of the Natives is presented (spirits exist objectively) while the anthropologist refrains from giving an opinion or evaluation of the truth of that interpretation, which is held to be "symbolic." Under the "etic" approach, the anthropologist describes the outward observable facts, lacking the ability or the will to verify the dimension of the phenomena that is not directly observable. This methodology introduces a blind spot, an unexplored and unverified area, convenient for rationalist thought, which thus justifies its refusal to reconsider its conceptual framework and consequently permits itself to engage in interpretation without confronting reality.
18. The discovery of this kind of relationship between humans and nonhumans tends to shift the borders between nature and culture as proposed by Lévi-Strauss, while giving rise to new philosophical propositions within the field of anthropology, such as that of Philippe Descola (2005) and, above all, the "perspectivism" of the Brazilian anthropologist Eduardo Viveiros de Castro (2009), who goes so far as to propose, by adopting the viewpoint of the shamans, a conception of the world in which the humans see animals as humans while the animals see humans as animals. This relationship has been replicated by David Dupuis in the context of the treatment for drug addiction offered at Takiwasi (see Dupuis 2009).
19. There is a remarkable physiological similarity between the sexual organs of this aquatic mammal and those of human beings.
20. My wife, who is a medical doctor, was consulted in the case of a child from the town of Tarapoto who had undergone an experience of this kind.
21. Literally, "underwater man."
22. These facts were related to me by elderly healers who had been personally acquainted with such cases, although they had not themselves attained the same level. I also heard a similar account in Gabon in 2001.
23. This discretion or lack of a direct invocation or words addressed to the deity has led some observers to believe that the pantheon of many early peoples was limited to these mediators whom they deified, a practice classified as animism or, in a restrictive religious view, idolatry.

But the Amajun, for instance, practice this form of "animism" while according recognition to a higher being, a creator named Arutam. This "primitive monotheism" is observed worldwide, even though in many cases it has subsequently changed into polytheism.

24. According to Michel Mouret (1990, 37), "The human body fulfils a psychic function of integration with the world order. The abdomen corresponds to symbolic integration with the vital forces, the thorax to that with the rational forces, the face to that with the spiritual forces, and the cranium to that with the archetypes contained in the very shapes observed in the world."
25. The analogy principle is already found in the Wisdom of Solomon, which asserts that the visible realities enable the observer to discover the invisible realities: "For from the greatness and the beauty of created things their original author, by analogy, is seen" (Wisdom 13: 5).
26. See also Revelation 1: 8: "I am the Alpha and the Omega," says the Lord God, "who is, and who was, and who is to come, the Almighty."
27. The energy centers correspond to the famous chakras of the energy body in Eastern traditions.
28. The sixteenth and seventeenth centuries were also the beginning of something new, as the philosophy of Descartes shows, besides inaugurating the scientific revolution. The world was entering a new era, as though the world spirit had vanished, and the ancient world ontologies had foundered. Modernity was emptying the world of magic. Modern man, despite his wealth and his vast culture, began to suffer from a spiritual emptiness. Modernity focuses only on economic and political problems (see esp. Shayegan 2001 and Apffel-Marglin 2011 on these issues).
29. Who are the "inspirers" behind the proliferation of the rituals of spiritism, occultism, and outright satanism?
30. The concept of dissociation is seen in the diagnostic classifications of borderline, bipolarity, psychosis, and schizophrenia.
31. This has been verified by our practical experience over the course of three decades.
32. "The awareness of continuity between the subtle world of the spirits and their incarnation in living beings constitutes the basis of the shamans' knowledge .... Modern materialism fails to grasp the reality of the subtle world, and that is the reason why it has fallen victim to it" (Daniélou 1992, 26, 29).
33. For the sake of simplicity, I shall here use this term alone, in accordance with Church practice on the subject, although there are several degrees of differentiation and gravity among the forms of parasitism on the part of malefic entities such as obsession, oppression, and possession.
34. This is in conflict, however, with the Gospels and the Magisterium.
35. This is illustrated by the saying, "If you speak to God, you may well be a holy man, but if he gives you an answer, you are surely a madman."
36. I might add that spiritual infestations affect mental health and, conversely, psycho-affective disorders provide a doorway by which such infestations can enter. In consequence, it is altogether theoretical and illusory to split up these disturbances, assigning them to mutually exclusive categories. One may ask to what extent this medical *schize*, projected onto infestation patients, may contribute to the proliferation of schizophrenia diagnoses. It should be mentioned, however, that the DSM-IV recognizes the existence of "unspecified dissociative disturbances" such as Dissociative Trance Disorder (DTD), defined as the replacement of the subject's personal identity by a new identity that is attributed to the influence of a spirit, a divine entity, or a deceased ancestor, often in association with episodes of amnesia. In Asia, it is known by a variety of names: in Indonesia, as *amok*, a term designating a state of trance marked by aggressive outbursts; in Malaysia, as *latah*; among the Inuit, as *pibloktog*; and in India, simply by the English word *possession* (see Canonne 2013).

37. For example, a physical symptom may be found to have a psychological origin, a mental pathology may have a spiritual cause, a physical malady may involve psychological and/or spiritual causality, and so on. Thus, a young girl who, for several years, has been undergoing treatment for rectal bleeding discovers that her chronic condition is connected to the activity of a grandmother who is an abortionist, or a man obsessed with thoughts of death finds a link with spiritist games that he played as a teenager, which he had forgotten or believed to be innocent.
38. The idea that a person's innocence will automatically afford him protection is a widespread error nowadays, not least as the outcome of the New Age doctrine that suggests that good intentions and strength of character are all that is needed. However, we have known for a long time that the road to hell is paved with good intentions, just as we know that a child who unwittingly drinks poison or who finds a hand grenade lying around and pulls out the pin enjoys no immunity from the consequences of his action.
39. For instance, at "raves" where drugs, chaotic sexuality, distorted sounds, and satanic symbols combine to leave the individual vulnerable.
40. Any attempt to identify this ideally non-infested human being, meaning one who is perfectly free from any deterministic influence external to his human nature, unavoidably begs the question of how this "human nature" is to be defined.
41. "With Kroeber, Linton and La Barre I assert, therefore, that the shaman is psychologically ill" (Devereux 1976, 16).
42. I have known the case of a student who in his free time was employed as a night watchman in a Peruvian university, and who became infested by the spirit of a deceased child. During the period of terrorism in Peru, a family that included a young girl had been held captive and subsequently murdered in the place where he spent each night.
43. A classic explanation which considers that the subject, having established a connection between amorousness and violence, and not having developed any other mode of amorous expression, reproduces the sole mechanism known to him for expressing his feelings.
44. One of my patients, an adult male, presented recurring obsessions with sexual abuse of his sister. The obsessions, which had begun at a young age, caused him great distress and he resisted them strongly. Conventional psychotherapy had produced no effect. The therapeutic process incorporating Amazonian practices brought to light an incestuous family context, which was resolved by a spiritual liberation approach. It emerged that a "spirit of incest" had been infesting the family over a period of several generations.
45. "Since Thomas Aquinas, mysticism has been described as *cognito affectiva seu experimentalis* (affective or experiential knowledge) of God. This knowledge of the deity is neither intellectual nor speculative, but experiential. A mystic experience is an immediate experience of divine transcendence" (Marguerat 2012: 478).
46. The theory of "agential realism" is relevant in this context. Developed by Karen Barad, a physicist, the theory is discussed by Frédérique Apffel-Marglin (2011, 55–63).
47. It should be recalled that the shamans' term for psychoactive plants is "master plants" on account of this teaching function, and that ayahuasca has been awarded the title "master of masters."
48. It is interesting at this point to establish a connection with the concept of the "structure of sin," boldly proposed by John Paul II, which "draws together contributions from (spiritual) history and sociology, on the one hand, and, on the other, social ethics," since "demonic action cannot be unaware that it also operates on the collective level and over the long term" (Ide 2011).

49. Especially in Brazil with the churches of São Daime, União do Vegetal, Barquinha, and others, which now have numerous offshoots in the northern hemisphere, including Japan, and in Australia.
50. Richard Noll (1997) rebukes Jung for believing himself to be the "Aryan Christ": "It is in the spirit community that most meet Jung, and there are those who see in the personal myth that provides the framework for *My Life* a present-day gospel laying the foundations for a new religion," while Denis Biju-Duval (2001) writes that "since the 1950s, Jung has been unafraid to set himself up as the prophet of the Age of Aquarius . . . . That explains how easily Jung is made use of nowadays by the New Age movement."
51. Biju-Duval (2001, 156) cites Jung on the epistemological limits necessary to science, and adds, "While it undoubtedly has access to the mental image of God in man, it can have nothing to say about God in himself, about God as a hypostasis outside the mind."
52. Daniel Marguerat, an academic at the University of Lausanne, uses the term "the democratization of mysticism" to convey the idea that, "in Paul, these formulae of indwelling are not restricted to an elite of charismatic performers but characterize the condition of every believer" (Marguerat 2012, 490–91).
53. For the writing of this section, I am deeply indebted to the ideas put forward by my friend, the psychologist Camilo Bautista Barrionuevo Durán, both in private conversation and in his master's thesis (2012).
54. "Some theologians use the expression 'non-person,' which may have given rise to a certain misunderstanding, enabling the notion to gain ground that the devil is *something*, not *someone*, that he is no more than an abstraction corresponding to something like the extinction of the self or that Evil, unlike the way in which it is usually represented, is not *something* but *nothing*. It is not an evil being, which is the Manichaean view: that would be a contradiction in terms, since being is good, inasmuch as it is being. Evil is the lack of being. It cannot be spoken of except in imagery and metaphor" (Laurentin 1995, 94).
55. "How was it possible, how did it happen? This remains obscure. Evil is not logical. Only God and good are logical, are light. Evil remains mysterious" (Benedict XVI 2008, n.p.).
56. "According to Sacred Scripture, particularly the New Testament, the dominion and influence of Satan and other evil spirits extend worldwide" (John Paul II 1986, n.p.).
57. Father René Chenesseau (2007), an experienced exorcist, stigmatizes equally those who see devils everywhere and those who see them nowhere, while admitting that every case needs investigating.
58. Even an André Gide, however, is not deceived: the devil, he says, "knows he has no better hiding place than behind such rational explanations, which relegate him to the plane of the gratuitous hypothesis. Satan, or the Gratuitous Hypothesis: probably the alias he prefers" (Gide [1927] 1959, 466).
59. John Paul II has said that, in the human dimension of the mystery of the Redemption,

man finds again the greatness, dignity and value that belong to his humanity . . . . [Man] becomes "newly expressed" and, in a way, is newly created. He is newly created! . . . The man who wishes to understand himself thoroughly—and not just in accordance with immediate, partial, often superficial, and even illusory standards and measures of his being—must with his unrest, uncertainty and even his weakness and sinfulness, with his life and death, draw near to Christ. He must, so to speak, enter into him with all his own self, he must "appropriate" and assimilate the whole of the reality of the Incarnation and Redemption in order to find himself. If this profound process takes place within him, he then bears fruit not only of adoration of God but also of deep wonder at himself. (John Paul II 1979, n.p.)

60. According to the Shivaist tradition of cycles governing the evolution of the world, the close of the present cycle of humanity is described as the twilight of the Kali Yugä (the age of Kali or the Iron Age), a period of conflicts, wars, genocide, malpractice, aberrant philosophies and social systems, and malignant developments of knowledge falling into irresponsible hands. Races and castes become mixed. Everything tends to be leveled out, and such a leveling in every field is the prelude to death. According to the Lingä Purânä, "The men of the Kali Yugä are governed by the lowest instincts. They prefer to choose false ideas. They do not hesitate to persecute the wise" (cited in Daniélou 1985, 211–12, who further presents a compilation of this tradition). On these points, see also Shayegan (2001).
61. On the three temptations of Jesus, Hadjadj comments,

To state the case briefly and thus, for the moment, oversimplifying, the issue here is flesh and Spirit. The bread is flesh without Spirit. The angels are Spirit without flesh. The kingdoms of this world are flesh virtualised by a worldly spirit, combining to produce an alluring contradiction: prayer as spectacle, faith as entertainment. (Hadjadj 2009, 46)

62. Fabrice Hadjadj restates the necessity of this tearing apart:

> Whoever elevates self-mastery to the rank of the supreme moral imperative falls into "the greatest of all vices": lacking all knowledge of that love which takes us out of ourselves and makes us weep "despite ourselves," he rejects the morality of mercy (Hadjadj 2009, 216).

He further comments,

> Not having the last word, allowing oneself to be torn apart by transcendence, to be drawn out of one's narrow clarity into a blinding light—that is the meaning of the first commandment, "Thou shalt have no other gods before me." . . . Leaving idolatry behind involves leaving pride behind and, consequently, a departure for which the initiative is not our own: an act of grace, the flash of a revelation (2009, 265, 269).

63. I mean that they differentiate, given that devils and angels are both nonhuman but the former, unlike the latter, lead to nondifferentiation.

References

Note: For this paper, I consulted works by C. G. Jung and some of his commentators almost exclusively in their Spanish translations, hence my listing them here under their Spanish titles.

American Psychiatric Association. 1994. *Diagnostic and Statistical Manual of Mental Disorders,* 4th ed. Arlington, VA: American Psychiatric Association.

Amorth, P. Gabriele. 2010. *Confessions: Mémoires de l'exorciste officiel du Vatican.* Neuilly-sur-Seine , FRA: Éditions Michel Lafon.

Amselle, Jean-Loup. 2013. *Psychotropiques: La fièvre de l'ayahuasca en forêt amazonienne.* Paris: Ed. Albin Michel.

Apffel-Marglin, Frédérique. 2011. *Subversive Spiritualities: How Ritual Enacts the World.* Oxford: Oxford University Press.

Barrionuevo Durán, Camilo Bautista. 2012. "El Demonio y la sombra: Sobre el problema del Mal desde la psicología analítica de Carl Gustav Jung." Master's thesis, University of Chile, Santiago.

Benedict XVI. 2008. "Saint Paul (15): The Apostle's Teaching on the Relation between Adam and Christ." *General Audience*, December 3. Vatican City: The Holy See. http://w2.vatican.va/content/benedict-xvi/en/audiences/2008/documents/hf_ben-xvi_aud_20081203.html.

Biju-Duval, Denis. 2001. *Le psychique et le spiritual.* Paris: Ed. de l'Emmanuel.

Canonne, Justine. 2013. "Le délire de possession en débat." *Le Cercle Psy*, January 30. https://le-cercle-psy.scienceshumaines.com/le-delire-de-possession-en-debat_sh_30186.

Chenesseau, René. 2007. *Journal d'un exorciste.* Saint-Benoît-du-Sault, FRA: Ed. Bénédictines.

Daniélou, Alain. 1985. *Nature et destin du monde dans la tradition shivaïte.* Paris: Ed. du Rocher.

Daniélou, Alain. 1992. "Las divinidades alucinógenas." *Revista Takiwasi* 1. https://takiwasi.com/docs/revista01/Rev_01_art_03.pdf.

Descola, Philippe. 2005. *Par-delà nature et culture.* Paris: Gallimard.

Devereux, Georges. 1976. "Normal et anormal." In *Essais d'ethnopsychiatrie générale.* Paris: Editions Gallimard.

Dupuis, David. 2009. "Une ethnographie de la Clinique Takiwasi: Soigner la toxicomanie avec l'aide des non-humains." Master's thesis in Social Anthropology, École des Hautes Études en Sciences Sociales, Paris.

Fromaget, Michel. 1999. *Corps, âme, esprit: Introduction à l'anthropologie ternaire.* Brussels: Edifie L. L. N.

Gide, André. (1927) 1959. *The Counterfeiters: With Journal of the Counterfeiters.* Reprint ed. Translated by Dorothy Bussy and Justin O'Brien. New York: Alfred Knopf.

Giegerich, Wolfgang. 2010. "¿La realidad del mal? Un análisis del argumento de Jung." Translated by Enrique Eskenazi and Alejandro Bica. *Ale Bica Blog*, September 8. http://alebica.blogspot.com/2011/08/la-realidad-del-mal.html?m=1.

Giove, Rosa. 2002. *La liana de los muertos al rescate de la vida: 7 años de experiencia del Centro Takiwasi.* Lima: Ed. Takiwasi-Devida.

Grof, Stanislav. 1983. *Royaumes de l'inconscient humain: La psychologie des profondeurs dévoilée par l'expérience LSD.* Translated by Paul Couturiau and Christel Rollinat. Monaco: Éditions du Rocher.

Hadjadj, Fabrice. 2009. *La foi des démons ou l'athéisme dépassé.* Paris: Ed. Salvator.

Ide, Pascal. 2011. *Combattre le démon: Histoire, théologie et pratique.* Editions de l'Emmanuel.

John Paul II. 1979. *Encyclical Redemptor Hominis* 10, March 4. Vatican City: The Holy See. http://w2.vatican.va/content/john-paul-ii/en/encyclicals/documents/hf_jp-ii_enc_04031979_redemptor-hominis.html.

John Paul II. 1986. "The Fall of the Rebellious Angels." *General Audience*, August 13. Vatican City: The Holy See. https://www.ewtn.com/library/PAPALDOC/JP2ANGEL.HTM.

Jung, Carl Gustav. (1927) 2001. "Alma y tierra." In *Civilización en transición*, translated by Carlos Martín Ramírez, 29–48. *Obras completas*, vol. 10. Madrid: Ed. Trotta.

Jung, Carl Gustav. (1934) 1970. "Sobre los arquetipos y lo inconsciente colectivo." In *Arquetipos e inconsciente colectivo*, translated by Miguel Murmis, 9–56. Buenos Aires: Paidós.

Jung, Carl Gustav. (1935) 1989. *Nietzsche's Zarathustra: Notes of the Seminar Given in 1934–1939*, vol. 1. Edited by James L. Jarrett. Jung Seminars 99. Princeton, NJ: Princeton University Press.

Jung, Carl Gustav. (1938) 1949. *Psicología y religión.* Translated by Ilse T. M. de Brugger. Buenos Aires: Paidós.

Jung, Carl Gustav. (1942) 1962. "Ensayo de interpretación psicológica del dogma de la trinidad." In *Simbología del espíritu: Estudios sobre fenomenología psíquica*, translated by Matilde Rodríguez, 229–310. Mexico City: Fondo de la Cultura Económica.

Jung, Carl Gustav (1947) 1970. "Consideraciones teóricas sobre la naturaleza de lo psíquico." In *Arquetipos e inconsciente colectivo*, translated by Miguel Murmis, 125–212. Buenos Aires: Ed. Paidós.

Jung, Carl Gustav. (1951) 1986. *Aion: Contribuciones a los simbolismos del sí mismo.* Translated by Julio Balderrama. Buenos Aires: Ed. Paidós.

Jung, Carl Gustav. (1959) 2001. "El bien y el mal en la psicología analítica." In *Civilización en transición*, translated by Carlos Martín Ramírez, 423–36. *Obras completas*, vol. 10. Madrid: Ed. Trotta.

Jung, Carl Gustav. 2009. *The Red Book: Liber Novus*. Edited with an introduction and notes by Sonu Shamdasani. New York: Norton.

Korzybski, Alfred. 2009. *Séminaire de sémantique générale, 1937.* Translated by Isabelle Aubert-Baudron. Paris: Interzone Editions.

Laurentin, René. 1995. *Le démon, mythe ou réalité?* Paris: Fayard.

Mabit, Jacques. 1988. "El cuerpo como instrumento de la iniciación shamanica." Paper presented at the Congreso Internacional de Medicinas Tradicionales, Lima, Peru, July 1988. https://static1.squarespace.com/static/59669c60a803bb304147f72a/t/597bba536f-4ca3948e272142/1501280852444/El+cuerpo+como+instrumento+de+la+iniciacion+shamanica.pdf.

Mabit, Jacques. 1999. "Ir y volver: El ritual como puerta entre los mundos, ejemplos en el shamanismo amazónico." *Amazonía Peruana* 13 (26): 143–65.

Mabit, Jacques. 2000. "Approches de la mort ou apprentissage de la Vie." *Cahiers de l'IANDS* 3. https://www.takiwasi.com/docs/arti_fra/approches_de_la_mort_ou_apprentissage_de_la_vie.pdf.

Mabit, Jacques. 2007. "Ayahuasca in the Treatment of Addictions." In *Hallucinogens and Health: New Evidence for Psychedelic Substances as Treatment*, vol. 2, edited by Michael J. Winkelman and Thomas B. Roberts, 87–103. Westport, CT: Greenwood.

Mabit, Jacques. 2008. "Héritages transgénérationnels, de la servitude à la filiation" [Transgenerational inheritance, from servitude to filiation]. Lecture presented at IdéePsy, Paris. https://www.youtube.com/watch?v=CIGIAp1d-LU.

Mabit, Jacques. 2012. "Introduction." In *Medicinas tradicionales, interculturalidad y salud mental.* Memorias del Congreso Internacional, June 7–10, 2009, Tarapoto, Peru. Edited by Renaud D'Abaddie, Christine Mikaelian, Patrick Díaz, and Jacques Mabit, iii–iv. Lima: Ed. Takiwasi.

Mabit, Jacques, and Jesús M. González Mariscal. 2013. "Hacia una medicina transcultural: Reflexiones y propuestas a partir de la experiencia en Takiwasi." *Journal of Transpersonal Research* 5 (2): 49–76.

Marguerat, Daniel. 2012. "Paul le mystique." *Revue théologique de Louvain* 43: 473–93.

Maxence, Jean-Luc. 2004. *Jung et l'avenir de la franc-maçonnerie*. Paris: Dervy.

Maxence, Jean-Luc. 2012. *Dictionnaire comparatif: C. G. Jung et la franc-maçonnerie*. Paris: Dervy.

Morin, Edgard. 1990. *Introduction à la pensée complexe*. Paris: Editions du Seuil.

Mouret, Michel. 1990. "Le temple du corps." *Actualités Psychiatriques* 20 (4): 37–43.

Nagy, Marilyn. 1991. *Philosophical Issues in the Psychology of C. G. Jung*. Albany: State University of New York Press.

Noll, Richard. 1997. *The Aryan Christ: The Secret Life of C. G. Jung*. London: Macmillan.

Pepper, T., and J. Cunningham. 2004. "Sad Planet: Depression Has Become a Global Disease; Breathing New Life into Talk Theory." *Newsweek*, June 21, 40–45.

Rivier, Laurent. 2002. "Ethnotoxicologie de l'ayahuasca." Paper presented at the Toxicology Congress, Martigny, Switzerland, June 19–20. http://www.psyvig.com/index.php?menu=34&page=5.

Shayegan, Daryush. 2001. *La lumière vient de l'Occident: Le réenchantement du monde et la pensée nomade.* Paris: Les Editions del'Aube.

Surrallés, Alexandre. 2013. "Destino, muerte y regeneración entre los Candoshi." In *Selva vida: De la destrucción de la Amazonía al paradigma de la regeneración*, edited by Stefano Varese, Frédérique Apffel-Marglin, and Roger Rumrill, 117–34. Copenhagen: IWGIA.

van Duijl, Marjolein, Wim Kleijn, and Joop de Jong. 2012. "Are Symptoms of Spirit Possessed Patients Covered by the DSM-IV Or DSM-5 Criteria for Possession Trance Disorder? A Mixed-Method Explorative Study in Uganda." *Social Psychiatry and Psychiatric Epidemiology* 48 (9): 1417–30. https://doi.org/10.1007/s00127-012-0635-1.

Viveiros de Castro, Eduardo. 2009. *Métaphysiques cannibals: Lignes d'anthropologies post-structurales.* Paris: Presses Universitaires de France.

Wilber, Ken. 2007. *Espiritualidad integral: El nuevo papel de la religión en el mundo actual.* Translated by David Gonzáles Raga. Barcelona: Kairós.

CHAPTER FIVE

Between Matter and Spirit: An Unfinished Journey

STEFANO VARESE

At age sixteen, I became a hesitant traveler in search of my father. That initial trip to the distant shores of Peru soon became an existential journey marked by the constant dialectics of materiality and spirituality, or what I perceived to be the unsolvable tension between tangible matter, which occupies concrete space and time, and the intangible but undeniable presence of other entities that are spaceless and timeless. In the following pages, I will use the Campa-Asháninka and other Indigenous peoples' cosmologies as metaphors—or rather, emotional devices—for my attempts to unravel and understand these paradoxes. I am choosing this unusual method and narrative, a kind of intimate memoir, more as a tribute to the Indigenous peoples of the Americas and their uncanny way of seeking knowledge than as an intentional critique of my modern Western, rational background. Therefore, I am not going to support my text with numerous sources and careful reviews of academic analyses. Rather, I will pursue what theoretical quantum physicist F. David Peat (2002, 2014) has identified as Native/Indigenous people's approach to knowledge, which he calls "coming-to-knowing."[1] In this approach, "Knowledge . . . is not a dead collection of facts. It is alive, has spirit, and dwells in specific places . . . . Coming-to-knowing means entering into relationship with the spirits of knowledge, with plants and animals, with beings that animate dreams and visions, and with the spirit of the people" (Peat 2002, 65). I am the storyteller of my journey and my occasional brush with the sacred, the mysteries of incorporeal spirits, manifestations of the numinous, and occurrences of "acausal" coincidences or synchronicities that occurred in my life.

The Early Stages

My life began as a Catholic child in Torino, Italy, raised by an absentee mother who had to work too much for our survival and by an anticlerical grandmother and a socialist grandfather who had turned into petty-capitalist entrepreneurs running a bookstore. My Catholicism ended abruptly when I turned twelve or thirteen and a school friend invited me to a *Festa del Lavoro* (Festival of Work) organized by the postwar Italian Communist Party. I gradually started to see my life in Torino through the lens of the working-class children of the public school I attended. By the time I decided to look for my father in Peru, a vague sense of class awareness—not really a fully developed class consciousness—had become my way of perceiving and representing the world. I started to nourish this vision with leftist and socialist readings of all types, actively seeking to position myself to the left of any discussion with my family or friends. It helped that I lived ten years of my young life with my grandparents in the realm of their bookstore, where I could always find some book that would confirm my intuitions. My newly found materialism was rent, however, by a fissure, possibly left by my early Catholic indoctrination. The word "religion" is derived from *re-ligare*, to reconnect, and indeed, some years later, I reconnected with spirituality when I decided to live a few months with the Asháninka people of the Gran Pajonal in the Selva Central of Peru, learning from them how to be an anthropologist. After having spent more than a decade committed to materialism and political economy as instruments for understanding social and cultural issues, especially those related to social and environmental justice, I ended up writing a dissertation in which I ascribed the success of the Asháninka people in resisting colonial oppression to their deep spirituality and their hylozoistic cosmology.[2]

After the Asháninka spiritual interlude, I was dragged back into the Anthropocene by way of the material and commoditized world of Peru's underdeveloped capitalism, where the ideals of social and ethnic justice for everyone in the country could only be enforced by a radical socialist and short-lived revolution. I took part in this uprising equipped with the prestigious and celebrated ideas of political ecology, anti-imperialism, socialism, decolonial politics, and Indigenous peoples' autonomy and self-government. I paid little attention, if at all, to the inscrutable force of Indigenous peoples' spirituality, even though I had gazed upon it while staying with the Asháninka. Soon I relegated it to the archives of impractical and suspicious forms of idealism.

However, it was Eros—also glimpsed among the Asháninka—that came to me through the beauty of my partner and wife Linda and restored my trust in the ethical strength of the beauty, truth, and goodness of *kamétsa*, the Asháninka's central principle of life. I later realized that both *kamétsa* and Linda's love were expressions of what Carl G. Jung described as the Platonic notion that Eros—the desire for beauty—might well be the first condition of all knowledge, including the essence of the sacred (Jung, cited in LaFontaine 2018). My relation with Eros, however, was not devoid of ambiguities; plenty of questions about sexuality had to remain unanswered in the darker corner of my sexually abused past. In my repeated encounters with Eros among the Asháninka or Andean peoples, I could have confirmed Jung's dictum that "Men are rarely split off from sexuality, because it is too evident for them, but what they lack is Eros, the relational function" (Jung, cited in LaFontaine 2018).

Asháninka Eros and Sexuality[3]

When I think of Pashuka, I remember her absent, almost nostalgic gaze and her capable hands preparing cassava[4] and fish, combing her daughter Irerene's long hair, attentively searching for intrusive lice, while the songs she sang recalled the mysteries of ancient poems and long histories. It was February of 1967, near the Unini River. She told us,

> Long ago, the little Tzia bird was human. He spent his time tormenting Mamántziki, his mother, wanting to go to bed with her and asking her to give him the coca that she held hidden in her vagina. They say that Oriátziri, the sun, had asked his wife Mamántziki to protect the coca and not let anyone have it. But Tzia finally took his mother Mamántziki by force, raping her and running off with the coca. When Oriatziri returned to their hut, he asked Mamántziki for water to drink. She was desperate and began vomiting coca leaves. Oriátziri knew what had happened because he saw coca leaves stuck to Tzia's penis. Furious, Oriátziri burned his son Tzia, threw his wife out of the hut, and rose to heaven, never again to return to earth.

In all the months of the years I spent in the Great Pajonal and on the banks of the rivers that surround and cross it, I was continually trapped in an erotic tension that pervaded all the relationships I established with Asháninka women and men. I found myself attracted through many senses: to their visual voluptuousness and the sensuality of their scent through the mystery of its origin; to the sounds of a flute in the emptiness of the forest, heard on the dim trail, knowing that it might be played by a single woman eager to shelter a man; or to the warm noise of the clattering engraved bones hanging from the strap crossing from shoulder to breast of a young nursing mother, forbidden and unreachable despite her constant flirtation. Or the penetrating scent of red annatto anointing someone's cheeks, mixed with the sight of the black *huito* lines that crossed the face, attracting the seduced's gaze and sense of smell toward the neck and shoulders adorned with fragrant seeds. And then there were the voices and songs that referred to passions, nostalgia, past and future dreams, and concerns expressed through a symbolic density of references I did not understand, allusions to animals, birds, herbs, and landscapes that were deeply mysterious to me[5]:

With him I'm playing in Hantani
I followed him to his hut
Pipóki, pipóki ("come, come")
Now I've thrown my snail
And we are always going to see each other
You are going to wait for me in my place, that is beautiful
And we are always going to see each other
I'm going to take my fire and you're going to wait for me alone
Look at me. Can't you see I'm drunk?
Let's go both swimming.

As an indiscreet observer, I became both the observed and the involuntary participant in the thread of local stories braided into the audacity of the fermented *masato* they drank. The ethnographer became a subject anthropologized by the Asháninka, such as Pashuka, who in her songs invited me to reconsider my return to Lima:

You look at the Asháninka as they are
But you're going to be crying,
Tears are coming out
The *Wiracocha* is crying,
At home he will cry
Come, come
Look at me, look at me
Let's go to your place, to your house as soon as possible.
I accompany you, do not wait for other people
With us will go my *katári* and the *chowántzi*
There you will look, you will look for me
And you will not see me
Because you will already be with your little snail.

One afternoon, lying on the dais of the *kaapa* (the house for men and guests) with Juancito, Irerene, and Pashuka, I had to dismantle my petty-bourgeois decencies in an instant and face the challenge of understanding and expressing my sexuality in front of a young teenager and his mother and older sister. Without fully realizing the symbolic charge of the term *totzíro* (snail), I began to comment that I would soon return to Lima to meet with *nototzíroti* (my snail). Pashuka and Irerene, in accordance with the Asháninka kinship and marriage system, asked me if I had given *totzíro* to my brother.

"No," I answered with little anthropological sensitivity, "Because it is not our custom."

"Will your wife have given her *totzíro* to others?"

"That would really bother me a lot," I answered, insisting on my Euro-Christian morality.

"*Pikóye itótziro ashéninka*?" They all laughed cheerfully at Pashuka's question about whether I would like to sleep with an Asháninka.

Just then I realized with amazement that a joke that young Juancito made to me a few days earlier—*Pitzi teroche*, "Let's make love"—might have been more than simple mockery. Was Juancito a *berdache*, someone who is "two-spirited" and combines gender attributes, such as those found among the Native people of the North American prairies and who so scandalized puritanical Anglo-Americans? Was he bisexual? Or was he simply a young man who had gone through the puberty ceremony, received his penis cord and new Asháninka name, and was now waiting for his first companion while exploring and practicing his new sexuality? Perhaps it was no more than a cultural case similar to that of the adolescent Nambikwara boys whom Claude Lévi-Strauss photographed in Mato Grosso and who freely made love in front of everyone, including members of the community and foreign visitors, on the white sands of the Brazilian Altiplano. Lévi-Strauss's well-known book, *Tristes Tropiques* ([1955] 2012), which I read during those years, came to my aid: my petty-bourgeois and socialist morality was in good repair under the learned gaze of the great French anthropologist.

How the Asháninka World Came to Be and Was Almost Destroyed by the *Wiracochas*

Toward at the end of 1969 in the district of Chorrillos, to the south of Lima, I met a young Asháninka-Nomatziguenga man, a former soldier now transformed into a construction worker. We had a short, intense, and finally devastated friendship due to our opposed political positions. Our relationship, however, allowed us to dwell on our respective life stories and cultural itineraries. Moisés Gamarra was originally from the Selva Central, born and raised in San Ramón de Pangoa, in the province of Satipo. He had been forcibly recruited and trained by the Peruvian army, turned into a semiliterate second-class citizen, later banished from his land in the rainforest, and deported to the miserable urban existence of coastal Peru. I knew only some fragments of his childhood and his life in the military, but his knowledge of the Asháninka-Nomatziguenga language, culture, and cosmology was remarkable, especially since he constantly compared them to the deeply racist urban criollo society of Lima. Moisés did not agree with my opinion about the good intentions of the revolutionary government of the armed forces and President Velasco Alvarado. For him, the Peruvian military revolution was just another version of the same system of oppression and exploitation that had dominated the country for centuries. For me, by contrast, the Velasco revolution had opened an initial, if weak, fault line in the Peruvian neocolonial social structure, which the government attempted to use to advance a progressive political agenda.

Moisés and I would gather in the evenings in Miraflores in a small, poorly lit office in the Instituto Raúl Porras Barrenechea of the National University of San Marcos. There I would review my field notes brought back from the Gran Pajonal, and with his help, I would transcribe the recordings of songs, poems, myths, and interviews. When we started working on one of the myths of creation that Pashuka had narrated to me in the Gran Pajonal, Moisés discussed it by adding details that revealed a deep knowledge of the Asháninka cosmology. With his assistance, I transcribed a version of the Asháninka cosmogonic myth and elaborated an analysis of it. A few years later, it was published as an article coauthored with Moisés (Varese and Gamarra 1976). Although his name was listed as a coauthor, I was not able to let him know or give him copies. By then, we had broken our friendship and taken different paths. In some flyers that appeared in 1974 along the Ene River Valley, Moisés accused me of being an agent of the Velasco "dictatorship" and of being linked to the CIA. I never knew if, in those early days, Shining Path was forming cells in the jungles of Ayacucho, which later, in the 1980s, became one of the main theaters of operation for the armed movement. Even now, after many years have passed, I refuse to believe that Moisés could have been indoctrinated into such an ideology that manifested a vision of time, space, and the future of humanity so far removed from the entire Asháninka and Andean cosmology and their utopian hopes. Asháninka communities and villages along the Tambo, Ene, Perené, Pichis-Palcazú rivers and even further into the interior of the Gran Pajonal suffered losses of thousands of human lives at the hands of Shining Path, the Tupac Amaru movement, and the Peruvian army and police forces.

I have wondered for years how to understand how Indigenous people can forget their roots and being seduced by a biblical speech of vapid parables and formulas of prophetic vulgar Marxism and banal allegories. Their Native world is imbued with visions that integrate materiality with sacredness and ethics with aesthetics and that assign to the history of the world and the universe a privileged place in which no biological or physical, earthly or astral, tangible or mysterious entity has superior hierarchy. How could "the Fourth" or "the Fifth Sword" proclaimed by the mestizo leader of Shining Path as symbols of Maoist liberation make any sense to people in the mountains and jungles of Peru? In what remote and arcane symbols buried in what collective subconscious did Shining Path find the meaning of hanging dogs on poles or trees as a political message to traitors? Why behead and dismember human bodies and then bury them in an unworthy jumble of destroyed hopes? Or were those macabre rituals, practiced equally by the army and the insurgents, an inchoate attempt to mark the conquest of Peru, its terror, and the Indian exegesis of Inkari, the sacred principle dismembered and buried by the conquistadores and awaiting a renewed Indigenous Peru for a triumphant divine return?

Rainy and cold evenings in the Gran Pajonal lend themselves to storytelling. On one of those evenings sitting on the *chonta* dais of the *kaapa* longhouse, I heard Pashuka talking of Kásiri-Moon, of when, a long time ago, at the beginning of time, he was a *tzirámpari*, a human being:

> Can't you see how he wears his *amatérintsi*, his crown, on his head like the Asháninka? The crown he has when he appears in the night sky, after he goes fishing on the Ucayali River and is no longer seen. There he makes his fishing *nasa* (basket) and gathers fish, not only fish but also dead Asháninka: those who commit suicide, those who die of influenza and measles, those killed by arrows (*los flechados*), and babies who die within days of birth. There are other Asháninka he does not collect, he lets them walk the world . . . but everyone (those he has collected and those he leaves behind) is given to his wives Mamántziki and Inankáuti. Inankáuti only accepts the dead. Mamántziki does not want to know about anything that is dead, she does not want to see them. Mamántziki is alive and not dead, she is like us. Inankáuti instead is dead and receives the dead, who are like her children, suckles them, and raises them all together; when they are grown up, they come together and have their families.

Pashuka kept telling me how in the old days, in the beginning, Kásiri-Moon was a *sháawo* (agouti) and ate cassava[6]:

> One day, Kásiri-Sháawo tells Mamántziki's father, "Follow me, come to my house to collect fish. When the fish is finished, I, Kásiri, will come to your house to visit and pick cassava, saying, *Naare tátzimi kániri Naani*, 'I have come to get some cassava.'" The next day, Mamántziki's whole family goes to Kásiri's house, up into the sky, following a large, clean, and straight trail. They stay for a while, and when they finish the visit, they go to Kásiri's house, but Mamántziki can no longer find the way back. She is lost and cannot go back, and so she stays there. But Mamántziki is alive, and that's why she does not receive the dead.

Two years later, in Lima, Moisés told me that the moon is actually called Manchákori and that his relationship with the young teenager, whose name is perhaps Mamántziki, is much more complex. Moisés narrated that the whole family of the teenage girl who had her first menstrual period had gone to the forest to seek "earth-soil" for their food. The girl was left in the little menstrual hut built for young Asháninka girls who have their first period. There the young girl kept busy spinning cotton and waiting patiently for the end of the bleeding.

> The young girl heard a noise near the hut, and without thinking much, she spat out some plants she was chewing to regulate her cycle.
>
> "You made a mistake, why did you spit on me?" said Manchákori-Moon. "From now on, when I'll be up in the sky, you will see spots on my face." Other Asháninka say that the spots that we see on the moon are because when Manchákori-Moon slept with the young

girl, she lovingly stroked his face with hands blackened from fire soot and marked him for eternity.

"Do your parents have nothing to eat, only dirt? Now open your eyes, look, from now on, there will be food for the people." And Manchákori-Moon made all the fruits of the forest appear. "When your parents return, you are going to say that there, in the forest, are the fruits. When they want to eat them, they must gather them in order, without taking them all, without mixing them, without cluttering what I just made."

When her parents came back, they heard the story of the young girl and rushed to the forest to reap the fruits. They did it without care, jostling, eating to satiety, and disordering the planting that Manchákori-Moon had so carefully arranged in the jungle. When Manchákori-Moon returned, the Asháninka were frightened and started running in all directions. Those who fell to the ground were transformed into insects like *katzíkori*, the kind of ant we now eat. Those who were perched in trees were transformed into monkeys, sloths, and other beings that are now all forest animals. That is what happened at that time and that is why all the fruits of the jungle are now scattered and mixed everywhere.

Only the mother and young daughter kept their human life. After a while, the young woman realized that she was pregnant. When the time came to give birth, Manchákori-Moon told the young woman, "Hold tight onto that tree so you will not feel pain or burn yourself."

The tree that Manchácori-Moon pointed to was the *tziritosi*, or tucunay tree, but the young girl mistakenly grabbed the oropel tree. The Asháninka say that they know when the dry season begins because the *tziritosi* starts to flower, and when the flowers dry out, it is time for the rains to begin.

When her baby boy was born, it was not human like us, but instead a piece of burning ember that completely burned his mother. Katziminkáitiri-Sun was born with a very long tail, like a star. It was his umbilical cord. Manchákori-Moon cut it into pieces saying, "This is white, this is black, this is good, this is bad, and this last little bit is Asháninka." And eventually the Asháninka became few. Then he blew and dispersed the whites, blacks, and Asháninka so that everyone would have their own land and their place in the world.

Later, Manchákori-Moon summoned the father of the young woman, and Maonte appeared in the world. "Come, hurry up, carry this burning tinder far from here and bury it so that later it will rise up into the sky. If it burns, do not put it out with water but only with dirt where it burns." While Maonte was carrying him, Katziminkáitiri-Sun kept growing and burning too much. Maonte put water on him in order to continue carrying him and finally bury him. He then returned to Manchákori-Moon, who asked, "Did you pour any water?"

"Not at all."

"We'll see you later."

> Soon there was light, and darkness disappeared as the sun was born in the east. But as the sun rose, mists also appeared. And then Manchákori-Moon told Maonte, "You must have poured water, and from now on there will be fog. You have lied and there will be rain. There would have never been rain if the fog had not collected water. But your mistake was not to have taken the water everywhere, and now there will be a lot of rain in one place and too little in another."
>
> At dawn, as the sun rose, Maonte said, "I have made a mistake." And sitting in his hut, he looked at the sun from dawn to dusk every day. That is why his back is black. Manchákori-Moon asked Maonte, "Are you sad?" and then he changed Maonte into a night-bird for lying about his mistake. That is why Maonte sings only during full-moon nights. "Well," said Manchákori-Moon, "now you have someone who gives light in the day. Now you need someone who will illuminate you at night." The nights were too dark because there were no stars in the sky. Before jumping into the sky, Manchákori-Moon left Maonte's youngest son, Tasorintzi, to be his replacement. Moon began to rise skyward and got larger and larger until we could see the spots on his face.

In the Gran Pajonal, Pashuka and others, such as Lidia, Sebastian, and Coronado, told me that many beings, which at the beginning of time were people, were turned into animals by Oriátziri, who got angry about the theft of his coca leaves and ascended to the heavens. He wanted to take all his people up with him. Onkiro the mouse said that he had a toothache and could not go to heaven with him. He actually lied. His mouth was full of corn kernels, and his wife had filled her mouth with pumpkin seeds. They slashed and burned their garden preparing it to sow their seeds, but the smoke rose up to the sky and annoyed Pawa (Father) Oriátziri, who transformed Onkiro into a little mouse with two teeth in front of his mouth. Ever since that time, Onkiro eats corn, pumpkin, and cassava from the Asháninka gardens because he thinks that they are his.

It took me some time to understand that for the Asháninka, the loss of the divine and sacred status of humanity/animality/earthiness, which was all integrated into a unitary cosmos, is always the consequence of errors. It was the mistake of Tzia, the incestuous defiler, that caused Oriátziri-Sun to ascend to Inoki, the heavens. The errors of the Asháninka are the cause of the loss of their human condition and their transformation into earthy humans/"other-than-humans"/animals/plants that now and forever populate the universe. Our Kipátzi-Earth, which used to connect Inoki-Heaven with the Mekori-Cloud World and the Kivinti-Underworld, is now Kamaventi-The Land of Death, because we are not immortal anymore, as we were at the beginning of time in the "before-Poerani," the phrase used by the Asháninka at the beginning of all stories. It took me even longer to develop a comprehensible representation of Asháninka cosmology. Many years later, in 1975, I could read and compare my interpretation of the Asháninka world with that proposed by the anthropology of that time.

Between 1963 and 1964, I traveled from the Gran Pajonal to the confluence of Perene and Ene rivers. From there, I embarked with two Asháninka men in a motorized canoe down the Tambo River. On the first evening of navigation, we reached a small Asháninka hamlet located on the left bank. When we turned off the outboard motor, the first thing we heard was one of the Brandenburg Concertos of J. S. Bach. A thin young man greeted us with the circumspection of someone who has lived for many months in isolation from his own culture and in the liminal space of the barely unveiled culture of another people. He was Gerald Weiss, a doctoral student in anthropology at the University of Michigan. I understood his suspicions because I myself had felt it repeatedly when some outsider would get too close to my place of residence in the Gran Pajonal. Moreover, Weiss had gone through the devastating experience of having lost all his field notes, recordings, photos, plant collection, and part of his equipment on a canoe wreck going downriver to Atalaya and Pucallpa. Our meeting, if I recall it correctly, was brief: a couple of days of circumspect talks about our respective studies of the Asháninka. I understood clearly that Weiss had accumulated an extraordinary empirical knowledge of the riverine Asháninka. I secretly admired his skills, his free handling of the conversation in the Campa language—as we then called the Arawak language of the Asháninka—all the while keeping control over the envy I felt by the abundance of technology and resources that Weiss had at his disposal in the middle of the Amazon rainforest. In my precarious, underdeveloped, and leftist Peruvian student status, I rationalized the insufficiency of my doctoral research project by invoking the academic "dependency theory" and sociological hypothesis of the "unequal and combined development" suffered by us Latin Americans in the Third World. And yet the encounter with this gentle and bright "agent of Yankee cultural imperialism" motivated me to try to understand more about the Asháninka of the Gran Pajonal, of whom Weiss confessed to know very little.

When Weiss finished his doctoral thesis (1969), I had already completed mine two years earlier and transformed it into the book *Salt of the Mountain* (Varese [1968] 2004). In reading his dissertation, I realized the many parallels in our studies but also the deep ethnographic and theoretical differences between the two of us. Something similar would happen with the dissertation of another American anthropologist, John Bodley, who published his texts on the socioeconomic adaptations of the Campa-Asháninka in the 1970s. My friendship with Bodley became a bit closer than that with Gerald Weiss, maybe because I recognized in Bodley a strong social commitment with Indigenous peoples both in the Amazon and around the world, while Gerald Weiss had undertaken a more strictly academic approach, which somehow I interpreted as more isolated from the specific needs of the Asháninka people. And yet I had to change my opinion about Weiss's alleged isolation when I read the lines he later wrote (Weiss 2005) about his visits to the Asháninka communities in 1977 and 1980. Here he pays tribute to the resilience of the Campa-Asháninka and their resistance to the oppression exerted by Peruvians of all political stripes and economic creeds, and he recognizes that the fundamental goodness of the Asháninka people is responsible for their survival and stubborn attachment to their own cultural values.

Asháninka Ethics and the Kirinka-Wiracocha Dystopia

The Campa-Asháninka cosmology—their vision or perception of the world—recognizes only partially and reluctantly the existence of evil. Back in the 1960s, I grasped the essence of this vision and approach to life thanks mainly to the conversational narratives that I could barely understand when I recorded them, which later I transcribed with the help of Juancito, Pashuka's talented young son. Among the Asháninka, conversational narratives occur between two individuals in the presence of others. They can be claims of debts arising from reciprocal bartering[7] or they may be cosmological and ethical restatements expressed metaphorically in anecdotes of hunting, fishing, or bartering travels, or they may even obviate the need of metaphor by retelling fragments of cosmological narratives that reaffirm the inherent order of things.

I confess, however, that I did not arrive at these understandings purely on my own. It was necessary for some time to pass and for me to read other interpreters of Campa-Asháninka culture, and especially for me to carefully review my field notes and engage in subsequent dialogues with Asháninka and other Native peoples elsewhere in order to arrive at these moral exegeses. When I wrote *Salt of the Mountain* ([1968] 2004), I understood the central cosmological and philosophical principle of the Asháninka—which I dared to call "gnostic"—a principle that ignorance and error cause loss, leading to the absence of goodness and beauty. *Kamétsa asaike*, which means the "beauty in life," cannot be translated simply as a secular, materialistic "good life" but, rather, as an "existence in beauty and harmony." Evil (*kamári*) exists and manifests itself in negative and malevolent beings—considered demonic in our Judeo-Christian-Mediterranean Islamic ideologies—whose appearance causes terrible damage to people. However, this damage can be repaired through the intervention of "our people, our brothers" (Nosháninka-Noshéninka), the *amatsénka*-human spirits that surround us, and the *manínkari*-occult entities, intangible and spiritual beings that populate the universe in all its cosmic levels. Especially crucial is the saving intervention of the *shiripiári*, the shaman, the healer of the world, that exceptional human capable of "marriage" for life to *tsiri-sheri*-tobacco, the powerful teacher spirit and sacred plant and substance of the Asháninka. With *tsiri-sheri*-tobacco and other master teachers and sacred plants—the *kamarampi-* ayahuasca, *horova*-leaves of the *nato* (floripondio) and *saáro* (genus *Datura*)—the *shiripiári* is responsible for repairing human error and restoring the balance that was temporarily lost. The spirits of sacred plants, called Master Teacher Plants, heal the world. The *shiripiári* is the intermediary between the goodness of plants, the goodness of intangible Asháninka who populate the universe, the goodness of the other sacred beings transmuted at the beginning of time—*in illo tempore*—and humanity, which often makes mistakes, deviates, confuses, and sickens itself and the world through its stumbling.

Kamári-evil is the negation of life. There is no explanation for its existence. The Asháninka know that *kamári* exists and manifests itself with particular violence through diseases of the body and spirit and that Wiracochas (creole Peruvians), some Chori (Andean mestizos), and Kirinko (foreign gringos) are particularly likely to be bearers and disseminators of *kamári*. Stories they tell have confirmed this truth every time: from the times of Spanish conquistadores and their companions the missionary priests to the colonial militias that went to fight the Indigenous rebellion led by Juan Santos Atahualpa in 1742 and to the Peruvian armed forces sent to exterminate the guerrillas of the Revolutionary Left Movement (MIR) in the 1960s and 1970s with indiscriminate bombing, summary executions, and practice of state terrorism, all the outsiders have ever brought is *kamári*: death, destruction, and evil.

The Kirinko-gringos began to appear long before the American military advisers who settled in Mazamari in the 1960s to defeat the guerrillas. There were already various Kirinko, tall, blond men and women who lived in Indigenous villages, learning their languages to translate the same stories or similar ones that Jesuits, Franciscans, and Dominicans had been touting for some time in the past. Now the missionaries of the Summer Institute of Linguistics (SIL), those strange men and women (often mysteriously single women who were not at home raising children) have added some more rules that go beyond the requirements of Catholic priests. They say that the Asháninka should no longer drink the fermented *masato*, the cassava beer that obscures reason and enhance the emotions, that they must also stop eating fish without scales and forget the use of sacred plants: the *shiri*, the *kamarampi*, the *koka*, the *sááro*, and all the rest of the spiritual healing herbs (*ivénki-piripiri*) governing pregnancies, healing broken-hearted amorous misadventures, defending the Asháninka from the murderous bullets of the military, police, and oppressive bosses, and serving as the best helper for enchanting elusive lovers. These are very great sacrifices to ask to communities of good people who should not have to pay for the guilt of unknown and alien ancestors of a remote and unheard-of biblical Middle East.

In the time I spent in the Gran Pajonal, especially in the Franciscan mission of Oventeni, which stood next to the residence of two SIL evangelical missionaries, I lived the intensity of witnessing the deaf strife between two versions of vulgar

Christianity inflicted on Asháninka children and adolescents in exchange for candies and promises of heavenly beatitudes in the afterlife. At the time, I was unable to fully grasp the gravity of the Franciscan practice of forcing these girls and boys into boarding schools at the Oventeni and Puerto Ocopa missions, a practice that tore them away from their families and their culture. The alternative strategies used by the SIL missionaries to convert the Asháninka were no less ethnocidal. The two sectarian and dogmatic expressions of a patriarchal, authoritarian, and fundamentally racist ideology benefited from each other's work of cultural destruction of the Asháninka by producing human remains thrown afterward into the *kamári* world of Wiracochas and Kirínko as domestic servants, abject peonage, cheap labor, or "informants" for evangelical and pastoral work. The cycle of such sordid coloniality has thus been completed through the humiliation and grief inflicted on the Asháninka and the replacement of their emotional and intellectual world by a dull parody—one in which the spirituality of *kamétsa* (beauty, goodness) has been displaced by the misery of *kamári* (evil), unleashed and thus outside the domain and control of the *shiripíari*.

The Teachings of Shiri

When I met Poshano, the *shiripiari* who lived in the wooded hills of Chenkari, beyond the savannas of Kishimasháwo, I did not imagine that this tiny, thin man with dark, penetrating gaze and a passion for teaching was going to bring about a radical change in my vision of the Asháninka people. For months, I had asked Coronado and Omaga, members of the Asháninka family with whom I was living on the edge of the Chitani River, to take me to meet this famous *shiripiari* who had threatened missionaries, soldiers, and settlers with all kinds of mystical hostile actions. His reputation, which had reached the settlements of Andean farmers and all the way beyond the territories under Franciscan administration and those controlled by the Summer Institute of Linguistics evangelists, had captured my imagination to the point that my life in Chitani had become somewhat agitated. My intuition told me I had to meet with Poshano to shed some light on parts of my understanding of the Ashninka that were still vague and obscure. It turned out that I was right: Poshano became my teacher and initiator, the gentle whisperer of some ideas that started to make full sense only years later when I thought I could learn nothing more about the Asháninka people.

In my field diary entries of February 25, 26, and 27, 1967, I wrote the following:

> It's almost impossible to describe the impression that Poshano caused me. Perhaps what I had imagined beforehand influenced my positive opinion. I had imagined a pompous individual, introverted, somewhat arrogant . . .. Who said shamans must be neurotic, mentally unstable, or at least epileptic? Poshano inspires a sense of security and peace. It seems

to be like facing a psychologist, a therapist. The old man is dirty, with long, tangled hair that disguises the beginnings of a balding head. He carries a thin headband made of *chuchuhuasca*. He walks straight, with firm steps that seem light, almost as if he does not touch the ground. And his smile is always there, as if he does not take himself or me too seriously. And yet he is neither vain nor superficial, and he shows it when we start to talk and ask questions. What keeps me amazed is his extreme simplicity . . . his deep, ironic, warm eyes . . . his affability . . . there are no mysteries to protect; nothing of what I ask seems to bother him. What amazes me in this wise man is his extreme openness and clarity. He seems to know much of what exists beyond the rational, sensible, and physical world. During the recording of sacred songs, Poshano ask me, "When I die, are my songs going to stay?" I suddenly fear I will not have time to learn from him . . . . There is something in all this so deep and mysterious that I cannot imagine how my studies can help me. How can I understand any of this? . . . As we walk to his gardens behind the *intómoe* to gather the sacred plants to treat and travel the universe, he whispers, "I jump to the sky. And I always return."

Coronado, Omaga, and I left the house of Poshano early in the morning. "Poshano will accompany us on our journey back," Omaga told me. After a first stretch of trail down the hill of Chenkari, we stopped to rest in a *purma*, an old cultivated plot where the Asháninka come back to gather fruits and to hunt small game. A woodpecker came flying by, confidently going to pose on the stump of a felled tree on which I was resting. Omaga told me that the bird was Poshano. On two other occasions during our journey back to Chitani, we saw a woodpecker. After these three instances, I never saw a woodpecker for the rest of my stay in the Gran Pajonal, and neither did I see Poshano again.

Forty Years of Silence

Whereof one cannot speak, thereof one must be silent. (Wittgenstein, *Tractatus*, quoted in Peat 1987, 140)

I had to wait forty years before I could dare to write about this encounter with the spirit of Poshano on my way back to the Chitani River. My embeddedness in the materialistic and empiricist professional discipline of anthropology had instilled in me the fear of being branded esoteric and irrational by the anthropological community and especially by the leftist social activists to whom I had committed my loyalty. We academics are "people of reason" and not "people of customs," to use the taxonomic dichotomy employed by the Indigenous people of southeastern Mexico. Newtonian and Cartesian reason cannot accept a return to the occult or the admission that mystery exists and requires an epistemic shift from "calculative" thinking to a "contemplative" one (Tedlock and Tedlock 1975, citing Heidegger 1966, passim).

I do not dare say that my experience on the way back from the hill of Chenkári in the company of a transmuted Poshano into a *torónkoti* or *chamanto* woodpecker was really an episode of mysticism or a numinous experience, that is, an inexplicable brush with the occult, a brief unconscious step into the depths of the Asháninka world inhabited by *kamétsa* (beauty, goodness) and the *amatsénka*, our spiritual brothers, as well as the hidden and mysterious *manínkari*. And yet I find no explanation in my "modern reason" for this experience. Nor can I satisfy the logic of materialistic empiricism or psychoanalysis to understand a dream that visited me on the afternoon of July 1991, a few hours after my friend Guillermo Bonfil Batalla died in a car accident, unbeknownst to me. In my dream, he came to meet me at the broad Avenue of the Pyramids of the Sun and the Moon in the Indigenous metropolis of Teotihuacan, the two of us dressed in modern clothing walking along leisurely, surrounded by pre-Cortesian Mesoamericans clothed in the white robes of the Aztecs. We were chatting as we had been doing for the past decades, slightly indifferent to the surrounding Teotihuacans, celebrating a long and intense friendship. My dream took place while I was in the bedroom of my house in Oaxaca, six hundred kilometers from Mexico City, in the torpor of a warm nap that made it difficult to recover from this phantom oneiric visit, even less so when, a few minutes after I awoke, our mutual friend Salomón Nahmad called me by phone to tell me that Guillermo Bonfil had died at dawn in a road accident on the way to the airport.

Contemporary "sensory anthropology" is revisiting questions that had already been explored in the first half of the twentieth century by C. G. Jung, Wolfgang Pauli, Paul Radin, Ernesto de Martino, and other scholars of religions, namely, that in many societies around the world, certain human senses receive individualized attention and selective privileges.[8] I am not interested nor do I care now to go into the labyrinth of empirical scientism that seeks to resolve these paradoxes in laboratories and statistical surveys that tend to impute archaism to any form of perception and representation of reality that does not involve the sensory hierarchy used in the West. In this respect, my interest is hermeneutical rather than heuristic, since I am seeking meanings rather than "rational" explanations. On which of my senses did Poshano and Guillermo Bonfil rely to say goodbye without being physically present? Which of my senses served as host to the visitors?

My parasensorial experiences of Poshano and Guillermo Bonfil were separated in time by a quarter-century and in space by thousands of kilometers and drastically different geographical yet analogous cultural landscapes. In much the same way, during the rest of my secular life in Peru, Mexico, Central America, and California, other uncanny phenomena dotted my life.

During one of the earliest nights I spent with my wife Linda in a small house in the Andean piedmont town of Chaclacayo, Perú, I was lying by her side in profound sleep when I was suddenly awakened by the silent vision of an ethereal, ashen, long shape ascending and disappearing through the ceiling. No explanation seemed plausible to the two of us, newly found friends and lovers. However, when Linda moved in with me in the rental house, she insisted in sealing the whole perimeter of the house with salt, claiming, to my astonishment, that she did not feel safe in that single-room house. Linda, whose maternal ancestry may have been Native American from Montana or Western Canada, had been traveling in the Amazon region looking for Indigenous medicinal plants and had met an eminent *ayahuasquero* healer or "shaman,"[9] who gave her a bottle of ayahuasca (*Banisteriopsis caapi*) to be shared with me. That *ayahuasquero*—or maybe a different reincarnation of him—reappeared later in my life through the writing of Amazonian poet and novelist César Calvo, who describes him in "Don Hildebrando Reads from the Air a Book by Stéfano Varese" (Calvo [1981] 1995). Calvo's literary chronicle of the conversation with the healer(s) Don Hildebrando and/or Don Javier reveals that Amazonian healer(s) do not need to personally know an individual or an author in order to know and understand what they think and say.

> I have never seen that book, but I know it—I know it very well .... The thoughts of well-meaning people live in the air. They inhabit the air as we do our houses. Before they are placed in books, and only by thinking them, and even if they are never written, they already live in the air ... an eternal beginning .... Because the air belongs to everyone. Perhaps it is the only thing that nowadays belongs to everyone. The sound of life. And without our knowing it, without our understanding it is in our heads, the ideas and the souls of ideas that inhabit the air nourish us and encourage us. (Calvo 1995, 86–89)

César Calvo's book and poetic account of the Amazonian shaman's way of understanding humanity's position in space/time can be interpreted as an alternative and complementary explanation of the complex and debated phenomenon of synchronicity proposed by C. G. Jung and others as an acausal connecting principle in which events in the external world might align to the experience of the individual, reflecting personal emotions, thoughts or entire sets of events (Peat 2014). The specific lived spiritual experiences that Linda and I went through in Amazonia, although separated in precise times and locations, came together in our fortuitous encounter in Peru. It was a coincidental precursor of love and friendship sealed by Don Hildebrando and/or the gift from the healer Don Javier, a bottle of ayahuasca, the Master Teacher Plant, which we drank at last in the secrecy of our Andean house and which brought us the numinous and life-changing experience of ontological Total Oneness.

In the mid-1960s, during a two-night-long Asháninka *masateada* (communal festival) in the Chitani longhouse, drunk with fermented cassava beer, I experienced the worst tobacco intoxication of my life, not exactly the numinous epiphany I had expected to be brought to me by the *shiripiari*'s "wife" and Master Plant. During two days, the whole world was spinning around out of control, no matter what position I assumed: standing, sitting, and finally lying down in total misery besieged by my own secretions. Toward the end of the second day, I dragged myself along the trail to the evangelical mission and asked for some kind of allopathic medicine. The response of the SIL missionary woman was a stern refusal to help me, the atheist sinner who obviously preferred the company of "savages" to that of Christians. My disappointment, nevertheless, was not with the evangelical truculence but, rather, with the harshness of *shiri*, the shaman's "wife," which in my naiveté I had imagined to be a compassionate and enlightening entheogen plant open to the spiritual inquiry of an honest young Wiracocha. I knew little at that time about the harshness and severity of the long initiation journey required to know and respect the spiritual power of this master entheogen plant, which we call *Nicotiana rustica*. After those two days of chaotic and unrestrained sensations and disturbing visions caused by the ingestion of the *shiripiari* tobacco juice, I did not return to Asháninka tobacco until the beginning of the third millennium, when tobacco was blown over my head and whole body by an Amazonian shaman while he was singing three sacred songs, or *icaros*, aimed at showing me the way back to my place in the cosmos.[10]

Should I see a sign of synchronicity, an acausal coincidence, in my ambiguous and traumatic first encounter with the *shiripiari*'s tobacco (*Nicotiana rustica*) and my decision to stop smoking the commercial and profane species (*Nicotiana tabacum*) after years of juvenile addiction, following the gentle invitation of my beloved Linda? What am I to make of how, at the same time, she convinced me to reconsider the spiritual power of salt that surrounds the home of our ethereal and physical body with an impenetrable shield of protection? Why did I center my understanding of the Gran Pajonal Asháninka life and history around *tzivi*-salt—that of *The Salt of the Mountain* (Varese [1968] 2004)—one decade before I ever met Linda? In August 2015, I had another significant encounter, separation, and re-encounter with *tzivi*-salt. During a nine-day healing diet (*ieta*, as they say in Amazonian Spanish), I was totally deprived of salt and nourished myself twice a day with watery unsalted rice and boiled unsalted plantains, a strict diet accompanied by three daily ingestions of infusions of Master Plants. The lack of salt, the Amazonian shaman explained to me, opened my spiritual and physical body to the healing power of the three Master Plant infusions each day. My open spirit/body, however, also became open and vulnerable to negative energies, thus it was necessary to protect me and my nourishment with tobacco smoke. My separation from *tzivi* and my renewed friendship with *shiri* accompanied my spiritual regeneration and the healing emotional power gained through the ayahuasca session. On the ninth day of my retreat, when I savored the first bite of salt, I felt that I was circling myself and my rediscovered memories of the protective energy that *tzivi* had always given me and my wife and children ever since my times in Tsiviniki and Tsiviriani in Asháninka territory.[11]

Small Deaths as Reminders

Death has visited me in a few occasions in dreams and visions, especially in one apparition aided by the Master Plant *uspawashasanango* in the Takihuasi, the House of Silence and Songs, surrounded by friendly trees, a few monkeys and birds and plenty of noisy and meticulously punctual cicadas. In the middle of the night, the dark, short figure of a familiar woman who had died two years earlier came to visit me, standing silently in a diagonal line of sight, close to the door of my shed. She did not address me but simply stood there, faceless but recognizable, reminding me of her disturbing presence in the depths of my memory. How did the *uspawashasanango*—the "memory of the heart"—bring her to me? How was the Master Plant able to convoke, animate, and give corporeal substance to a person who had died physiologically two years before and who, for me, had expired emotionally many years earlier? How could this spirit attain substantiality once again in such a way that my visual sense became the vehicle of perception? At the time of this event, I was following the strict Amazonian shamanic diet of Master Plants in isolation and silence. This healing observance does not allow the practitioner to have access to a watch or any electronic device, only reading and writing materials. My devotion to Western rationality was left intact by the shamanistic cure, so I seized the opportunity to complete some academic works that required tranquility and isolation.

Since my early childhood, I have associated death with animals and other-than-human entities. I must have been four or five years old when World War II forced my father to move the family out of the port of Genova, which was under constant bombardment by the British navy, to a small farming village in the Ligurian Alps. This village was safe from the Allied forces' bombs but lay in the crossfire of the partisans' resistance and the Nazi-Fascist army. One late afternoon, walking through the village, I stepped into a courtyard from which a terrible human-like scream, accompanied by children's laugher, was assaulting my ears. I saw that a pink-skinned pig was being slaughtered and bled on top of a wooden table. I stood paralyzed in terror before the scene, not understanding why the children were happily dancing and shouting in the courtyard as if the death of a pig were a celebration of life. I believe that the archetypical ritual and its violent crudity became engraved in my subconscious as evidence of human arrogance and its separation from the rest of living beings. I spent the rest of my adult life in the company of animals or seeking to be in the company of the pluriverse, perhaps trying to expiate the experience of witnessing the original killing. Ever since, the slaying of innocent animals has haunted me, from the northern Amazon, where the hunting dogs of the Awajún young hunter became the executioners of the capybara,[12] all the way to a Mexico City neighborhood, where I had to participate unwillingly in the cruel death of a rooster buried alive as a sacrificial victim in an alleged ritual to secure the fidelity of a slippery_lover. My Brazilian friend, who requested the ritual, never recovered her lover anyway, perhaps confirming that the alleged Nahuatl shaman was, rather, a malevolent and vulgar sorceress.

Then again, in the Zapotec region in the Northern Sierra of Oaxaca, I got trapped into having to sponsor a village festival that began with the festive and not very ceremonial slaughter of a cow. My already disturbed omnivore conscience made one final leap toward an increasing commitment to vegetarianism, although never fully achieved. It is paradoxical that my anthropological calling took me to Amazonian peoples who are mostly horticulturalists, hunters, and foragers. They rely heavily on fishing, hunting small animals, and gathering insects to obtain animal fat and proteins. Some of the killings of game are as violent and cruel as our Western industrial treatment of "domesticated" animals. The difference lies in the scale, the intense labor output required to obtain game flesh, and the extreme ritual demands on the hunters who often have to fast, abstain from sex, salt, and hot peppery foods and to enter into a sacred, reciprocal agreement with the owners of the animals. Hunting—and for that matter, any other agricultural and foraging activity of Indigenous people—is a celebration and reiteration of the membership of each individual in the larger kinship of humans, other-than-humans, and the visible and invisible entities that people the cosmos.[13]

During the last fifty years, I have lived with dogs and cats, most of them rescued from desperate situations of abuse and cruelty or at least indifference. All of our family animals survived and enjoyed years of good life and peaceful transitions to death. Two of our earlier dogs, one an Andean male that accompanied Linda and me as expatriates to Mexico, and a small Oaxacan male that demanded to be accepted in our house, are now buried in the garden of our San Felipe house. The rest of our dogs and cats were cremated, and their ashes are stored in the depths of our bookshelves. Thus, death and the final transition have been constant occurrences in our family, with Linda as well as our children, both when young and as adults, participating in the final farewell rituals in some adapted form of the way that Indigenous and mestizo commoners of Oaxaca perform it for their dead. The Day of the Dead (*El Dia de los Muertos*) ceremony became part of our family calendar, complete with an "altar," offerings of food, drink, and tobacco, and marigold flowers adorning the trail to the altar so that our departed relatives, friends, and animals could find their way to their preferred meals and beverages.

The belief that the dead (*los muertos*) come back to visit us on the first and second day of the month of November of every year became more than a credence one November night thirty years ago, when Linda and I decided to visit the San Felipe del Agua cemetery with our two children and take part in the festive occasion with members and friends of the community of San Felipe. It was a cold night, lit by plenty of fires burning in between the tombs covered with marigolds, colored tablecloths, all sorts of foods, fruits, and drinks of mezcal and beer, with children playing "table games" on top of the tombs. There were two sources of music: the traditional brass band of San Felipe and a quasi-rock ensemble of young villagers. That night, Linda and I took an entire roll of color pictures with our Kodak camera: we wanted to record our children's experience in this highly traditional Mexican festivity. A few days later, the colored prints of the film came back from the photographic laboratory, and, to our bewilderment, every shot taken at the cemetery showed grayish auras floating around and in the background of the image.

In April, two years ago, Linda and traveled back to Oaxaca and spent a few days at the house in San Felipe. Chaparro, the oldest dog of the household and the loyal protector of Señora Josefina, the housekeeper, was in such a terrible shape that he had to be euthanized. That same night, we helped Josefina to mourn Chaparro in the same way that dead relatives are taken care of; the body of Chaparro was surrounded with marigold flowers and by candles and some of his preferred food. We turned off the lights and sat around drinking hot beverages, eating some light food, and telling stories of the happy life of Chaparro, the ferocious small male dog that could always recognize good people and identify those with bad intentions, "like that time that Chaparro did not accept at all the presence of a worker who later turned out to be a robber." As Josefina and her two children were sharing these stories with Linda and me, sitting in the covered patio of the house, we heard a noise coming from the roof. Suddenly, an unknown animal of the size of Chaparro started to walk quickly but calmly along the top of the wall toward the garden, heading in the direction of the small patio where Chaparro had spent most of his last years of life. The animal stopped for a few seconds and looked at us, maybe at Chaparro's body, with curiosity. All five of us saw the animal, which we later identified as some sort of fox. Actually, it turned out that it was an exceptionally rare and endangered *cacomistle*, (*Bassariscus sumichrasti*), a nocturnal, arboreal, and omnivorous member of the carnivoran family *Procyonidae*. This particular one was a *cacomistle* out of place, since they are not common at all in semiurban areas or in the region of San Felipe. My rational explanation was that the *cacomistle* had learned to visit Chaparro at night to take advantage of some leftover food. Josefina, however, with her Indigenous ancestry of Zapotec people from the nearby Sierra of Ixtepeji and as the daughter of a herbalist and shamanic healer, had an alternative explanation. Hers was a synchronic view aligned with the Mesoamerican concepts of *nahual* and *tonal*, alter-ego duplicates that accompany all living beings throughout life and death.

Cosmic Dialogues

My journey between the secular and the sacred has been marked by small insinuations, allusions, miniscule epiphanies, revelations, and a few fleeting encounters with the unknown and the inexplicable, which have combined with holistic insight to constitute the symbolic weaving of the hammock of my life. I find myself living and resting, reflecting and imagining time and space in a loose web of unrelated and acausal events that allude to mysteries rather than to algorithms and issue a call to emotional intelligence rather than to the Cartesian *logos.* I could not have grasped the uncertainties of life in the company of animals, plants, other-than-humans, and the quantum ontology of the cosmos if I had not shared periods of my life with Asháninka men and women of the Gran Pajonal, with the Andean and Amazonian Quechua, with Zapotecs and Chinantecs and Mixes from southeastern Mexico, with Guatemalan Mayan war refugees in Mexico's camps, and with undocumented Mixtecs and Triques farm workers in the agricultural fields of California. The juncture of activism and social science has opened multiple windows of perception and ethical awareness that otherwise would have been obscured by the obsessive modernist epistemology of material empiricism.

So in my own experience, knowledge increasingly involved a process analogous to the Indigenous way-of-knowing: the acknowledgment that we humans, animals, plants, mountains, rocks, water, air, sun, moon, stars, all other celestial entities, and all invisible beings are locked together in a cosmic contract of existence that took place at the beginning of time, in what the Asháninka call *poerani*, like the *in illo tempore* of my own tradition, when all manifestations of life began. This cosmic pact implies a cosmic ethics binding each and every entity—tangible and intangible, visible and invisible—into a complex polyphonic dance that seeks harmony beyond and even through conflicts, a dance that disregards human-created hierarchies or static taxonomic formulas or immutable laws devised by scientists. A few years ago, after a personal epistemic crisis, I began to use my own neologisms, "cosmocentrism" and "cosmocentric," to indicate the radical difference between the modern Western cosmology that I identified as anthropocentric and the Amerindian cosmologies that I branded polycentric or cosmocentric (Varese 2001, 2003, 2006; Varese and Grofe 2007).

As I kept exploring the Amerindian way-of-knowing or Indigenous epistemologies, I had to contend with the demon of my own Western formation, deeply rooted in Judeo-Christian monotheistic cosmology, Aristotelian philosophy, and modern science. Anthropocentrism means the separation of humanity from nature, the hierarchical positioning of the former at the pinnacle of an unquestioned "world order," the allegedly "divine" or secular mandate it has to rule over everything else in existence, and lastly, the increasing commoditization of the universe now devoid of its sacredness and ethical constraints.[14] Furthermore, it became clear to me that the supposedly self-regulating market of the capitalist economy is not bound by any ethical mandate, being by definition amoral and hence lacking any obligation toward "other relatives," human or otherwise. After a few weeks of life among any Indigenous community, one realizes that any social practice based exclusively on materialistic principles is essentially destructive of nature, now degraded to the status of "resource": a series of commoditized and unrelated entities that have no intentionality or intelligence and no cosmic reason to be in existence. I can argue that since modernity and its expression in the capitalist market economy took hold initially in Euro-America and increasingly in the rest of the world, the cosmological representation of life and its social practice abandoned all remaining vestiges of spirituality and even of humanism. Modernity has reduced our contemporary value system—our axiology—to a parody of the post-Marxian binary option between use-value and exchange-value, implanting in our minds a restricted code of conduct driven by self-centered, individualistic monetary interests.

No matter how long one's individual pilgrimage has been, in a world dominated by exchange value and the overpowering commoditization of everything, it is extremely difficult for modern individuals to perceive and acknowledge the wholeness and interrelatedness of the apparently disparate entities that inhabit the universe. Unless, of course, one has had the privilege of living among Indigenous people who, despite their material scarcity, share the wealth of their wisdom, which in its ubiquitous presence can be perceived only by engaging the heart and the mind, the soul and the spirit, through the mystical vision induced by meditation and Master Plants or through the dreams dreamt in the silence of the longhouse. In this Indigenous world—this cosmocentric universe—dwell animate and inanimate beings engaged in a cosmic conversation that acknowledges each entity as an interlocutor with intelligence, will, sentiments, and finality, that is, its own participatory teleology.[15]

The metaphor that best represent the cosmocentric conception and practices of Amerindian people can be drawn from their own historically rooted ecological economies. For the last thousand years, Amerindian communities of the entire hemisphere have been practicing a social economy that combines foraging techniques, hunting, fishing, agroforestry, and agriculture. Contrary to any reductive social-Darwinist hypothesis, these techniques and technologies of production, consumption, distribution, and trade have coexisted in a bounded system even in the face of the encroaching capitalist market economy, which may penetrate some parts of the social body without fundamentally altering the ethical principles that support the entire Indigenous system. The domestication of plants that occurred at least ten to twelve thousand years ago was undertaken by realizing that biological diversity—and its intentional maintenance and encouragement—was a sacred gift of fertility and eternal renewal that could not be ignored and desecrated. The *milpa*, the gardens, and the Three Sisters, are polycultures of agricultural production and foraging systems found all over the Amerindian continent. The principle of biological diversity—which extends to the acceptance of cultural and linguistic diversity—supports the second essential principle, that of reciprocity: a symmetrical or asymmetrical system of exchange exercised between all the entities of the cosmocentric world. For the contemporary Popoluca of southwestern Mexico, the seeds of corn and beans, which are planted together with squash seeds in the soil of the *milpa* at the beginning of the rainy season, are the mythical twins who, at the beginning of time, were sacrificed for the good of humanity. In fact, the name of corn in the Popoluca language is *homshuk* (*hom*, "new"; *shuk*, "bean"): beans and corn are planted together in the *milpa* because they benefit each other, with the beans returning nitrogen to feed the corn and both tender plants growing under the generous protection of the big leaves of the squash (Varese 1985). This specific architecture of diversity and reciprocity stated symbolically in the *milpa* and the Three Sisters is accompanied by the third principle of complementarity: whatever is not available in one community-system-unit can be reciprocated for a similar or equivalent entity from another community.

The relational character of the "sociocosmic community" (Descola 1998) requires the acknowledgment that the realm of social relations and the ethical code of conduct include the total domain of the universe: hunters communicate with the animals they hunt, and the animals respond to them by adjusting their conduct, similarly, agriculturalists and foragers talk to plants and trees in a mutual learning process based on a true communication of information (Pollan 2013; Pollan and Jacobs 2014). In the Andes, the mountains are former humans and thus have to be treated as such by living people through specific rituals and conversations conducted by the local spiritual priests/shamans.[16]

The languages and the ethics of the various manifestations of cosmocentrism—those of the many local Indigenous communities having great historical depth in the tangible and intangible landscapes—may be considered in a "relativistic" or "perspectivistic" manner by us anthropologists, external observers bound by our own binary ethical code (nature/culture or nature vs. culture). However, my journey in search of meanings rather than practical results, of spirituality rather than material goods, has confirmed that, as part of humanity, I am not standing apart from the cosmos in some insular, empowered position, nor am I struggling against plants, animals, rocks, waters, amoebas, subatomic particles, and invisible enemies. I stand attentive to the whispers of the universe, to the sound of the wind in the foliage, to the barking of my dogs, the howls of coyotes and the answers of crows, and mostly to the sound of my soul when, awake at night, I touch Linda's shoulder, and I am grateful of a well-lived life in cosmic company.

Notes

1. In this chapter, I use F. David Peat's most recent (2014) of three books bearing the main title of *Synchronicity*, each with different subtitles. I used to have his first one (1987), but in a true

synchronic event, it was stolen from my Indian shoulder bag in the Kechwa Otavalo market plaza in July 2016—I hope by some aspiring shaman.

2. The ethnonym of this large Indigenous nationality of the Peruvian Amazon is Asháninka for the entire group of more than seventy thousand people, while the ethnonym Ashéninka is used for the subgroups that live in and around the plateau of the Gran Pajonal.
3. Most of the information regarding the Asháninka is based on my 1968 doctoral dissertation materials, field notes, interviews, and further comparative analyses with other ethnological sources, incorporated into Varese (2004).
4. Cassava, yucca, mandioca or manioc (*Manihot esculenta)* are some of the vernacular names of the main staple of people who live in the tropical rainforest, lowlands, and savannas of South and Central America.
5. The botanical name of red annatto or achiote is *Bixa orellana*, and that of black huito is *Genipa americana*.
6. Agoutis (*Dasyprocta punctata*) live as couples in ten to twelve acres of forest, feeding mostly on fruits and seeds they bury for use in times of scarcity. This practice increases the reproduction and growth of bushes and trees.
7. In my book, I called this the *ayumpari* system of reciprocity (Varese [1968] 2004).
8. See de Martino (1948); Radin (1956). My references to Carl Jung and Wolfgang Pauli are based largely on my readings of Peat (2002, 2014).
9. I use the term "shaman" fully aware of the process of erosion and distortion that the word has undergone in the secular New Age environment of frivolous exoticism.
10. For a good description of Asháninka and Amazonian tobacco chemistry and its cultural sacred use, see Tindall and Plotkin (2008). For a thorough study of the ethnobotany of sacred plants, see Schultes and Hofmann (1992).
11. *Tsivíniki* is the Asháninka name of the Cerro de la Sal (the Mountain of Salt), and *Tsiviriani* is the Asháninka name of the Río de la Sal (River of Salt). On the fundamental cultural role of the salt among the Perené Ashéninka, see Mihas, Peréz, and Rodríguez (2014).
12. Capybara (*Hydrochoerus hydrochaeris*) is a large gregarious rodent that lives in tropical areas close to rivers and wetlands.
13. For superb work on Amazonian people, see Rival (2002). My friend and colleague Frédérique Apffel-Marglin has enlightened me with her profound analyses of the role of rituals in the Indigenous communities of Quechua farmers of the northeastern Amazon of Peru (Apffel-Marglin 2011).
14. The Industrial Revolution is the defining moment of materialism as the dominant mode of interaction between society and nature-world. Materialism and the effects of machines over production, distribution, and consumption on a commercial society, with the resultant commoditization of everything existing on earth, have been analyzed brilliantly by Karl Polanyi ([1944] 2010).
15. Between the 1970s and 1990s, a series of studies of Amazonian peoples opened a window in the understanding of the "ecological" relations that Indigenous humanity has established with what in the West we call "nature." These new approaches to the study of nature and culture were based on the initial contributions of Euro-American and Native American/Indigenous intellectuals who, at the turn of the twentieth century, began a continental dialogue that in 1945 peaked in Pátzcuaro, Mexico, at the First Inter-American Indigenous Conference, where hundreds of delegates from Indigenous communities and "tribal" peoples presented their cases regarding their cultural sovereignty and ethnic autonomy in relation to the nation states and the

market economy. Between 1976 and 1987, Gerardo Reichel-Dolmatoff published a series of studies on the shamanistic knowledge of the Tukano of Colombia, studies that were taken with a grain of salt by mainstream academia (see, e.g., Reichel-Dolmatoff 1987). In 1986, Philippe Descola's doctoral dissertation was published as *La nature domestique*, later translated as *In the Society of Nature* (Descola [1986] 1996). After this watershed contribution followed a panoply of his texts, which soon became the paradigm of the study of cultural ecology (see inter alia, Descola [1993] 1998; Descola and Gísli Pálsson 2004; see also Ingold 2000; Narby 2006; Rival and Whitehead 2001).

16. See chap. 7 in this volume, by Guillermo Delgado-P., "Andean Entification: *Pachamamaq Ajayun*, the Spirit of Mother Earth."

References

Apffel-Marglin, Frédérique. 2011. *Subversive Spiritualities: How Rituals Enact the World.* Oxford: Oxford University Press.

Calvo, César. (1981) 1995. *The Three Halves of Ino Moxo: Teaching of the Wizard of the Upper Amazon.* Translated by Kenneth Symington. Rochester, VT: Inner Traditions International.

de Martino, Ernesto. 1948. *Il mondo magico: Prolegomeni a una storia del magismo.* Torino: Einaudi.

Descola, Philippe. (1986) 1996. *In the Society of Nature: A Native Ecology in Amazonia.* Translated by Nora Scott. Cambridge: Cambridge University Press.

Descola, Philippe. (1993) 1998. *The Spears of Twilight: Life and Death in the Amazon Jungle.* Translated by Janet Lloyd. New York: The New Press.

Descola, Philippe, and Gísli Pálsson, eds. 1996. *Nature and Society: Anthropological Perspectives.* London: Routledge.

Hidegger, Martin. 1966. *Discourse on Thinking*, trans. John M. Anderson and E. Hans Freund. New York: Harper & Row.

Ingold, Tim. 2000. *The Perception of the Environment: Essays in Livelihood, Dwelling and Skill.* London: Routledge.

LaFontaine, Lewis. 2018. "Carl Jung and 'Eros'—Anthology." *Carl Jung Depth Psychology: Life, Work and Legacy of Carl Jung* (blog). August 24. https://carljungdepthpsychologysite.blog/2018/08/24/carl-jung-and-eros-anthology/.

Lévi-Strauss, Claude. (1955) 2012. *Tristes Tropiques.* Reprinted. Translated by John Weightman and Doreen Weightman. London: Penguin Classics.

Mihas, Elena, Gregorio Santos Pérez, and Delia Rosa Rodríguez. 2014. *Upper Perené Arawak Narratives of History, Landscape, and Ritual.* Lincoln, NE: University of Nebraska Press.

Narby, Jeremy. 2006. *Intelligence in Nature: An Inquiry into Knowledge.* New York: TarcherPerigree.

Peat, F. David. 1987. *Synchronicity: The Bridge between Matter and Mind.* New York: Bantam Books.

Peat, F. David. 2002. *Blackfoot Physics: A Journey into Native American Universe.* Boston, MA: Weiser Books.

Peat, F. David. 2014. *Synchronicity: The Marriage of Matter and Psyche.* Pari, Italy: Pari Publishing.

Polanyi, Karl. (1944) 2001. *The Great Transformation: The Political and Economic Origin of Our Time.* Boston: Beacon Press.

Pollan, Michael. 2013. "The Intelligent Plant." *The New Yorker*, December 23. http://www.newyorker.com/magazine/2013/12/23/the-intelligent-plant.

Pollan, Michael, and Jesse Jacobs. 2014. "Tesis, antithesis y fotosíntesis." *El estado mental* 5 (November). https://elestadomental.com/revistas/num5/tesis-antitesis-y-fotosintesis.

Radin, Paul. 1956. *The Trickster: A Study in Native American Mythology*. Commentaries by K. Kerenyi and C. G. Jung. New York: Schocken Books.

Reichel-Dolmatoff, Gerardo. 1987. *Shamanism and Art of the Eastern Tukanoan Indians*. New York: E. J. Brill.

Rival, Laura M. 1996. "Blowpipes and Spears: The Social Significance of Huaorani Technological Choices." In *Nature and Society: Anthropological Perspectives*, edited by Philippe Descola and Gísli Pálsson, 145–64. London: Routledge.

Rival, Laura M. 2002. *Trekking Through History: The Huaorani of Amazonian Ecuador*. New York: Columbia University Press.

Rival, Laura M., and Neil L. Whitehead, eds. 2001. *Beyond the Visible and the Material*. Oxford: Oxford University Press.

Schultes, Richard Evans, and Albert Hofmann. 1992. *Plants of the Gods: Their Sacred, Healing and Hallucinogenic Powers*. Rochester, VT: Healing Arts Press.

Tedlock, Dennis, and Barbara Tedlock, eds. 1975. *Teachings from the American Earth: Indian Religion and Philosophy*. New York: Liveright.

Tindall, Robert, and Mark J. Plotkin. 2008. *The Jaguar that Roams the Mind: An Amazonian Plant Spirit Odyssey*. Rochester, VT: Park Street Press.

Varese, Stefano, ed. 1985. *Agua, Mundo, Montaña Narrativa Nahua, Mixe y Popoluca del Sur de Veracruz*. Técnicos Bilingües Unidad Acayucan. Tlahuapan, MEX: Premia Editores.

Varese, Stefano. 2001. "The Territorial Roots of Latin American Indigenous Peoples' Movement for Sovereignty." *Hagar International Social Science Review* 2 (2): 201–17.

Varese, Stefano. 2003. "Commentary: Language of Space: The Territorial Roots of the Indigenous Community in Relation to Transcommunality." In *Transcommunality: From the Politics of Conversion to the Ethics of Respect*, edited by John Brown-Childs, 129–45. Philadelphia: Temple University Press.

Varese, Stefano. (1968) 2004. *Salt of the Mountain: Campa Asháninka History and Resistance in the Peruvian Jungle*. Translated by Susan Giersbach Rascon. Norman, OK: University of Oklahoma Press.

Varese, Stefano. 2006. *Witness to Sovereignty: Essays on the Indian Movement in Latin America*. Copenhagen: International Work Group for Indigenous Affairs.

Varese, Stefano, and Michael Grofe. 2007. "Notas sobre territorialidad, sacralidad y economía política *bennizá/binnigula'/beneshon*." *Mexican Studies/Estudios mexicanos* 23 (2): 219–51.

Varese, Stefano, and Moisés Gamarra. 1976. "Deux versions cosmogonique campa: Esquisse analytique." *Annales: Économie, société, civilisations* 31 (3): 469–80.

Weiss, Gerald. 1969. "The Cosmology of the Campa Indians of Eastern Peru." PhD dissertation, University of Michigan.

Weiss, Gerald. 2005. "Los Campa Ribereños." In *Guia etnográfica de la alta amazonía*, Vol. 5, *Campa Ribereños/Ashéninca*, edited by Fernando Santos and Federica Barklay, 1–74. Lima: Smithsonian Tropical Research Institute/Instituto Francés de Estudios Andinos.

CHAPTER SIX

To Dwell in the Thick Darkness

The Sacred Dark in Jewish Thought

RABBI FERN FELDMAN

> It came to pass, as the trumpeters and singers were as one, to make one sound heard, praising and thanking Adonai, and when they lifted their voice with the trumpets and cymbals and instruments of song, praising Adonai, saying, For God is good, for God's loving kindness endures forever, then the house, the house of Adonai, was filled with a cloud; And the priests could not stand and serve because of the cloud; for the manifesting presence of Adonai had filled the house of God. Then Solomon said, *Adonai has chosen to dwell in the thick darkness.* (II Chronicles 5:13–6:1)

When I explore my own nature or experience the sacred, most often I find darkness. Although dominating theologies assert binaries in which light is holy and darkness is evil, a recognition of the multivalent nature of all that is can evoke awareness of wave upon wave of dark and light.

Some say they want to "embrace the dark" when they mean embrace the grief, anger, and suffering in the world and be present with it rather than denying, ignoring, or hating it. But that is not the aspect of sacred dark that interests me most. What interests me is how in darkness all separation dissolves into oneness. Darkness is depths, cave, womb, soil that sprouts seeds, soothing shade, nighttime during which we dream, grow, and make long-term memory. Darkness can be a source, essence, innermost being, transcendence, embodiment, nothingness, emptiness, mystery.

Darkness is often associated with the earth for a number of reasons. The sky can be dark or light, but the earth has no light of its own except the molten core that shows its fiery light when it erupts. Under the surface of the earth, of course, it is always dark. So, too, we find that darkness is associated with depths while light is associated with heights.

The dark of the womb and the common association between women and the earth are aspects of a metaphorical connection between women and the dark. European colonialism and white supremacy have been invested in associating dark skin and dark hair with negative metaphors of darkness. When we discount the power of darkness, we devalue all one might associate with it—dark skin, women, and the earth. Audre Lorde explains it this way:

> These places of possibility within ourselves are dark because they are ancient and hidden; they have survived and grown strong through darkness. Within these deep places, each one of us holds an incredible reserve of creativity and power, of unexamined and unrecorded emotion and feeling. The woman's place of power within each of us is neither white nor surface; it is dark, it is ancient, and it is deep. (Lorde 1984, 36–37)

It is essential to be aware of the places of possibility Lorde describes. *We need the holiness and the liberating power of deepening into the dark. And this particular historical moment requires of us the dismantling of the negative dualist metaphors of blackness and darkness, both for the sake of justice and for the sake of the life of the planet.*

In his book, *The Dark Center: A Process Theology of Blackness*, Christian theologian Eulalio P. Baltazar writes about what he calls "Western color symbolism," based on "white Christianity and white theology" (Baltazar 1973, 7–8). These developed out of Apollonian thought, combined with Aristotelian dualism, and led to the problem mentioned above, in which white is associated with good and black with evil. However, Baltazar claims, up through the Middle Ages, that set of beliefs about metaphysical colors did not transfer into prejudice toward black-skinned people. It was not until the philosophy of empiricism was developed that a belief arose that

> man is his appearance, or man is as he appears. It was this new anthropology that facilitated the transference of the color symbolism from the soul to the body . . . . But even this transference would have remained purely in theory and not in fact if the economic colonization of Africans and the need to justify slavery were not present as reinforcing factors. In other words, the economic superiority and dominance of the Europeans confirmed their belief in the positive theological values attached to white skin, and conversely, the negative theological values attached to people with dark skins. (Baltazar 1973, 29)[1]

Baltazar also argues that European Christian culture "integrated sexuality with blackness" (1973, 35) and points out that,

> From a psychoanalytic point of view, darkness or blackness is the symbol of the unconscious; whiteness or light of the conscious. Western color symbolism in psychoanalytic terms is precisely the expression at the conscious level of the flight from the unconscious . . . . Thus the Western psyche is . . . split, for the ego is separated from the id from which

> it flees. This results in an abstraction which is then projected into everything that is considered the nonself: nature, making of it a "harlot" to be used, wasted, deadened; and nonwhites made invisible. (Baltazar 1973, 57, 63).

At issue, then, in addition to the broader political implications, is the opportunity for integration of the psyche—which, of course, also leads to further societal healing.[2]

Ellen Davina Haskell, in writing about the Zohar, a sacred Jewish text, discusses an image that is "offering a startling juxtaposition specifically designed to promote contemplation and mental reordering for its reader" (Haskell 2012, 11). In speaking of sacred darkness, I hope in this chapter to promote contemplation and a mental reordering. How we imagine, what images we make, reorders our awareness, and reorders the world. So, dear reader, I invite you to contemplate, and let the reordering begin.

Baltazar develops what he calls "a processive theology of blackness," in which a "theology of blackness symbolizes the Supreme Reality as Divine Darkness and Faith as a saving darkness" (Baltazar 1973, 2). Catholic theologian M. Shawn Copeland tells us "To come to terms with blackness means to come to terms with the failure of Western metaphysics and ontology" (Copeland 2013, 634). She suggests,

> Perhaps a route [that] theology might take side by side with the symbol of blackness to a future with authentic and luminous possibility emerges from the ancient mystical tradition of apophasis, the *via negativa* or negative theology. In this posture, rather than attempting to overcome the opacity of the symbol, theology draws us near to it and into its meanings, its agonies, and its ecstasies. Theology worked out on apophasis eschews easy harmonizations, questions every similarity and dissimilarity even as it holds these in creative tension, and resists simple closure. Negative theology acknowledges the inability and poverty of language to express any experience of awe, of the holy, of divine Mystery. Always, there is more—a dense and fruitful residue that can never be grasped or uttered or rendered absolute. Experience of divine Mystery eludes the very structures of language; such experience is beyond words, beyond saying. (Copeland 2013, 635)

Certainly, the notion of sacred darkness evokes experience of that mystery that is beyond saying. Perhaps a pause here is in order, to do just that—a moment for awe, for acknowledgment of the ungraspable—and then to make an attempt, not to grasp, but to taste the mysteries of the sacred dark.

Often, we think the paths to holiness are to be found by ascending into the light. Certainly, light is an evocative metaphor for that which gives life, and for the flow of divine presence. But there are numerous levels to the metaphors of dark and light. On one level, dark can be seen as that which blocks the flow of light. This can be limiting when we imagine that light is the source of the good or the sacred. But the blocking of light can be protective, as when light is destructive or too much to handle, or if time is needed to hibernate or grow. A further level to the metaphor is the darkness *beyond* the light. That darkness is the place where all separation dissolves into oneness, and there is a taste of what it means to be an integrated part of all that is.

The portions of the brain that register physical sensations are greatly reduced when experiencing mystical states (Begley and Underwood 2001, 52). In the dark, one may be more open to these states. Astrophysics researchers are also encountering the power of the darkness. The theory of dark matter proposes that there is a dark, unseen substance that provides most of the gravitational pull that is literally holding the universe together (CERN 2012). What an evocative metaphor for the divine this is—that which holds it all together. Rabbi Marcia Prager discusses insights by the Chernobyler Rebbe, who noted that one of the most common Jewish terms for God, *Adonay,* shares a root with the word *adanim*,

> usually translated as "ball joints" or "sockets." A ball-joint, the Rebbe mused, is a mechanism for flexible connection. Just as the flexible *adanim* held the upper and lower sections of the *Mishkan* [traveling sanctuary] together, so too *Adonay* holds the lower and higher worlds together. (Prager 1998, 105–6)

One might, then, imagine God somewhat like dark matter, holding the worlds together.

Jewish tradition evokes many forms of sacred darkness, and this will be the focus of the rest of this chapter. Nighttime study brings a thread of loving-kindness into the world. Divine presence can be a sheltering shade. Revelations occur at caves. Torah was received in darkness, formed of black fire on white fire, and still the ink is black. The source of everflow (*shefa*) is imagined as a burning black coal or a deep spring. Before God said "Let there be light," there was already darkness, the darkness of wisdom and beyond. These images, along with the texts that hold them, are openings that take us deeper into the sacred.

Night

The Talmudic sage, Resh Lakish (BT Hagigah 12b) comments that "Whoever engages in Torah by night, the Holy One draws down upon that person a cord of loving-kindness by day, as it is said 'By day, Adonai[3] commands His loving-kindness' (Ps 42:9). What is the reason? Because 'And by night His/Her song is with me' (Ps 42:9)."[4] The Talmud goes on to comment that *some* say that Resh Lakish elaborated and compared night to this world, and day to the world to come. In acknowledging that only *some* make this association, the passage recognizes two levels of darkness—on one level, night is to be compared to this imperfect world, in contrast to the world to come; on another level, night is the sacred time in which our Torah study brings forth loving-kindness.

The Zohar explains that this cord of loving-kindness to which Resh Lakish referred comes from the original light of creation that was hidden away for the righteous to receive at the end of days, that is, at the time of the final redemption.[5] And yet, it says,

> Had it been hidden away altogether, the world would not have been able to exist for one moment. But it was only hidden like a seed which generates others, seeds and fruits, and the world is sustained by it . . . whenever the Torah is studied by night, a little thread of this hidden light steals down and plays upon them as they are absorbed in their study. (Zohar II 148b–149a)

Earlier, the Zohar explains that the original light of creation

> issued from the darkness which was carved out by the strokes of the Hidden One; and similarly from that light which was stored away there was carved out through some hidden process the lower world darkness [that is, night], in which light resides. (Zohar I 31b–32a)

Light, here, issues from the darkness that *precedes* creation, the place of the "Hidden One," the one that cannot be seen or known directly. And night, that "lower world darkness," can carry us back to a deeper level of consciousness, to that "darkness which was carved out by the strokes of the Hidden One." This text teaches us about the holiness of the night and its potential to connect us to the deepest level of the sacred imaginable. It also evokes the awareness that there is not just one type of darkness and one type of light, but layer upon layer, wave upon wave, of dark and light, making diffraction patterns that ripple in all directions.[6] Slightly further on in the passage, the Zohar explains,

> The difference by means of which light is distinguished from darkness is one of degree only; both are one in kind, as there is no light without darkness and no darkness without light; but though one, they are different in color. Their differentiation is not of kind, but gradations of levels of color. (Zohar I 31b–32a)

Thinking of dark and light as gradations of color, one might understand that this variation is part of what makes things beautiful, like the threads of a tapestry. There is also not a complete separation; "their differentiation is not one of kind." All is one, without all being the same or homogenous. In exploring the reverberations of dark and light, we hold the awareness that we are not talking about any sort of essentialism or any inherent binary, in which all associations can be piled one on top of the other to create sets of false associations, as is so often done by thinking male=light=good=mind, and female=dark=bad=body.

Another Zohar passage describes what happens at midnight as well as mentioning the thread of loving-kindness that comes down for those who, in this case, play with Torah. This passage explains that the Holy Blessed One arises to play with the righteous ones in the Garden of Eden. And more:

> At midnight, when the Holy One, blessed be, enters the Garden of Eden, all the plants of the Garden are watered more plenteously by the stream which is called "the ancient stream" and "the stream of delight," the waters of which never cease to flow. When a person rises and studies the Torah at this hour, the water of that stream is, as it were, poured on his head and he is watered by it along with the other plants of the Garden of Eden. (Zohar I 92a, Matt 2012 translation)

Quite strikingly, the passage continues further on:

> It is written "Midnight I will rise to give thanks to You because of Your righteous judgments" (Psalms 119:62). Since the word "at" is omitted, we may take "Midnight" as an appellation of the Holy Blessed One, who is addressed thus by David. (Zohar I 92a, Matt 2012 translation)

That is to say, midnight is a name of God.

Redemption

The spiritual opportunities of midnight abound in Jewish tradition. One aspect of its power is implied by the events that take place in the middle of the night in Torah. Rabbi Shalom Noach Berezovsky (sometimes called the Slonimer Rebbe, or simply the Slonimer), in his multivolume work *Netivot Shalom* (2012) points out that both of the Israelites' moments of redemption during the Exodus from Egypt occurred at night. Pharaoh "arose in the night" (Exodus 12:30) and told the people to leave; thus, the Exodus itself (the first night of Passover) was at night. Almost a week later (the seventh night of Passover), the Reed Sea was split during the night. Not only was it at night, but the Israelites were assisted in the process by a pillar of cloud: "Thus, there was a pillar of cloud with the darkness and it cast a spell upon the night" (Exodus 14:20).[7] The Slonimer explains the reason redemption came at night, through a commentary on a verse from Psalms: "To tell in the morning of your loving-kindness, and your faithfulness in the nights" (Psalms 92:3). The reason, the Slonimer says, that the verse from Psalms says "nights" in the plural is to remind us of these two nights of redemption, the first and seventh nights of Passover.

He explains that we merit loving-kindness in the day, which he parallels with redemption, due to faithfulness during the night, which he equates with exile. While in exile in Egypt, the Israelites were unable to purify our own ethical qualities. The faith we expressed purified our souls and bodies for us, to prepare us to cleave to the Holy Presence when it was revealed in the Exodus and at the Reed

Sea. Once the Presence is explicitly revealed, it becomes impossible to experience deep faith, because faith is unnecessary unless there is cause for doubt. The Slonimer is thinking of a level of darkness that is a lack of light, saying that even though it is painful, it is also an opportunity to manifest virtue; the night makes faith possible (Berezovsky 2012).

The Slonimer's perspective on darkness somewhat parallels the second of the two opinions in the Talmud passage discussed above, that is, that night can be correlated with this world and day with the world to come. The Talmud balances that opinion, however, with the view that the loving-kindness we experience in the day has its roots in our Torah study, God's song, in the night. On this level, the darkness is connected with the concealment of divinity. The Zohar adds another dimension: *it correlates the concealed world with a higher level of divinity and with the world to come*. In the Zoharic perspective, the revealed world, also called "this world," is the "lower Mother," also associated with the Shekhinah, the manifestation of divine presence, while the concealed world is associated with the "Upper Mother," the realm of *Bina*, or Understanding, which is also "the world to come" or "the world that is coming" (temporality is non-linear in Jewish thought, as well as in Hebrew grammar). This results in an opposite set of associations from that of the Talmud: whether we associate revealment with this world and concealment with the world to come, or vice versa, is situational. Associations shift and move; there are no binary fixed positions but, rather, multidimensional shifting valences, where identities are not inherent but arise out of the intra-actions of which they are a part.[8]

In his essay on Chanukah, Emmanuel Lévinas writes:

> Before the miracle of generous light, and as a condition of this miracle, another miracle took place: a dark miracle that one forgets. One forgets it in the blaze of lights triumphantly burning brighter. But if, in the Temple . . . one had not found in a little flask of pure oil bearing the seal of the High Priest, which, ignored by everyone but unchanging, had remained there throughout the years while the candelabra remained empty, there would have been no Hanukkah miracle. There had to be preserved somewhere a transparent oil kept intact. Oh! nocturnal existence turned in on itself within the narrow confines of a forgotten phial. Oh! existence sheltered from all uncertain contact with the outside . . . a clandestine existence, isolated, in its subterranean refuge, from time and events, an eternal existence, a coded message addressed by one scholar to another . . . . Oh! miracle of tradition, conditions and promise of a thought without restraint that does not want to remain an echo, or brief stir of the day. Oh! generous light flooding the universe, you drink our subterranean life, our life that is eternal and equal to itself. You celebrate those admirable hours, which are dark and secret. (Lévinas 1990, 230)

The Zohar teaches that one "who desires to penetrate to the mystery of the holy unity should contemplate the flame which rises from a burning coal or candle" (Zohar I 50b, Matt 1995 translation). When we look at the center of a candle flame, we can see the traces of the dark miracle of the oil of which Levinas speaks; that oil held inside and protected bleeds through into the dark center of the flame. We sense the *pnimiyut*—the innerness—the divine shelter holding the most precious. And when we look closely, we also see the dark that surrounds the flame. And in this seeing is awareness that there are layers upon layers of dark and light; not a binary but a beautiful, diverse, multivalent, living wholeness that can in no way have any of the rigid mind's reified binaries laid upon it. Dark within the flame; dark around the flame; dark between each flame. And finally, dark beyond the flames. As Emmanuel Levinas says of the menorah at the outset of his essay, "One light the first evening, two the following day, three the day after, and so on up until the triumphant blaze of light on the final evening—up until the strange and mysterious night that will surround the candelabra after this final illumination." Beyond the story, beyond the eight lights, holding them in its endlessness, is the mysterious night. Night, *Laila*, the word that the Zohar says is a name for God, holds all that we know—before and after the lights, and deepest within, the sacred dark, sustaining it all, holding it all.

The Hidden Face and the Hiding Place

Other Jewish images of hiddenness also reflect a multivalent view of the dark. There is a concept called *hester panim*, the hiding of the face, based on Deuteronomy 31:15–18 and developed in the Prophets and Psalms. Deuteronomy 31:17 reads, in part: "Then my anger shall be kindled against them in that day and I will forsake them, and I will hide my face from them, and they shall be devoured, and many evils and trouble shall befall them." This concept is used to explain why negative events can happen—they are a result of the hiding of the divine face, of the presence. In the language of light and dark, *hester panim* is the blocking of the divine light that leaves us abandoned and alone in the destructive dark. But there is a flip side to this concept, which is discussed by Herbert Levine (1995): the *seter*, the shelter or hiding place. The root *samech-tav-resh* (or *s-t-r*) has to do with hiddenness: it is used both as the root of the term *hester*—the hiding of God's presence—and as the root of the word *seter*, the hidden divine shelter. One of the sources for this image is in Psalm 18, which contains numerous images of the sacred dark, including these:

> He bent the sky and came down,
> Thick cloud beneath His feet.
> He mounted a cherub and flew,
> Gliding on the wings of the wind.
> He made darkness His shelter (*seter*);
> Around Him, His *sukkah*,[9]

Dark thunderheads, dense clouds of the sky. (Psalms 18:10–12)

Levine, referring to a later portion of Psalm 18, writes, "For moral support, the psalmist petitions that all God's devoted ones be hidden, from the root *s-t-r*, within the divine presence, treasured in God's *sukkah* (v. 21)" (Levine 1995, 162). He explains: "God's *sukkah* is a liminal doorway that transforms vulnerability into protection. God may at times seem hidden, but one can be concealed even with the hidden God" (1995, 162–63). That is, protection comes from being hidden with/in God. Levine examines numerous other psalms that contain related imagery, including Psalm 27, which Jews recite twice daily for about six weeks in preparation for and during the Days of Awe (the season of the Jewish New Year). He writes:

> having entered into the concealment of God's tent, *seter 'oholo*, the psalmist remembers the opposite feeling of God's hiding his face, *hester panim*: "do not conceal Your face/presence from me" (*'al taster paneka*, v. 9) The expression of both security and vulnerability through the root *s-t-r* indicates the necessary interrelatedness of these emotions and the pivotal, unifying role played by *s-t-r* in the psalm. Each emotion must be played out in full. (Levine 1995, 168)

The issue in regard to hiddenness, then, is whether we are within or without the hiding place. In Hebrew, the root for "face" and the root for "within" are the same, *peh-nun-heh*. The hiding of God's face or presence and being within God's presence are thus linguistically related in two ways. If God is within and we are without, then we feel loss. If we can also be within the hiding place, then we experience protection and connection with the divine. Perhaps when we are deeply in the experience of being hidden from sacred presence, we are uniquely situated to cry out to be held within the divine shelter. That can be a great and compassionate gift in moments of despair.

Shade and Shadow

Psalm 91:1 expands the image of a shelter further and includes the reader within the shelter: "Oh you who dwell in supernal shelter (*seter*) will pass the night in the shade (*tzel*) of *Shaddai*." Shaddai is itself an interesting God name: it is often translated into English as "Almighty," but the root is related to a word for mountain and a word for breast. In the Talmud, it is understood homiletically to mean "enough":

> *El Shaddai* [means] I am the One who said to the world: "Enough!" Resh Lakish said: When the Holy One, blessed be He, created the sea, it went on expanding, until the

> Holy One, blessed be He, rebuked it and caused it to dry up, for it is said: He rebukes the sea and makes it dry, and dries up all the rivers. (BT Hagigah 12a)

This hearkens back to the idea of the God name Adonai having something to do with holding the worlds together. In numerous places in Psalms, the concepts of *seter*, *sukkah*, and shade or shadow are connected (the latter two are the same word in Hebrew). Psalm 91 starts with an image of being sheltered in the shade of Shaddai, the holy mountain, or divine breast, as the power that holds the world together. What does it mean to be in the shade?

There are multiple levels to the metaphor. If we need the sun to grow or see, being completely blocked from it could be a problem. But shade felt so vital to people in the ancient Near East that the psalmist actually equated it with God: Psalm 121 says, "Adonai is your keeper; Adonai is your shade, at your right hand" (Psalms 121:5). The psalms also repeatedly refer to the shadow of God's wings. In verse 4, Psalm 91 implies this sense of shade: "He shall cover you with His feathers, and under His wings you will find refuge." This image of the divine as a protecting bird, likely a mother bird, is central in Judaism. Each morning, right after putting on a *tallit* (prayer shawl), which is draped over the head and enfolds the person engaging in prayer, Psalms 36:8 is recited: "How precious is your loving kindness, O God! The children of humanity take refuge under the shadow of your wings." One can imagine the *tallit* as the sheltering, shade-giving wings of the divine.

The traveling sanctuary, the *Mishkan*, which was the predecessor of the Temple, is also imagined as a place of sacred shade. The craftsman who designed it was named B'tzalel, a name literarily meaning "in the shadow of God"; that is, the sanctuary itself is the place of divine shade as well as divine presence, for whenever there is a shadow, the source of the shadow is nearby. And there are many other levels to the image. The ancient Jewish philosopher Philo of Alexandria saw the divine shadow as the presence of God that could be experienced—the "Divine Mind, the Idea of Ideas . . . the pattern of all creation and the archetype of human reason" (Winston 1981, 26). Furthermore, the Zohar, along with later mystical teachings and practices, calls the *sukkah* "the shade of faith (*tzila dim'heimnuta*)."[10] The *sukkah* is understood as an evocation of the clouds of divine presence that protected and guided the Israelites as they wandered through the desert on the way from slavery to freedom, so calling the *sukkah* the shade/shadow of faith evokes this experience of divine presence. As with a physical shadow, where the shadow necessitates a presence casting the shadow, when we feel ourselves to be in the shadow of God, we can feel the presence of the source itself. That is, there is some link in our awareness between the experience of the sacred darkness and the sacred itself. Rabbi Levi Yitzchak of Berdichev takes this idea a step further:

> "God is your shadow (*tzilcha*)" (Psalm 121). That is, just as a person's shadow does whatever a person does, so does the blessed Creator, as it were, do whatever a person does.

> Consequently, a person needs to do *mitzvot* (commandments and good deeds) and give *tzedakah* (righteous acts and donations) and have compassion on the poor, so that the blessed Creator will do likewise with them. (Yitzchak, in Meir n.d., 70b)

The shadow is a manifestation of one's own soul as well as a link between that soul and the Holy One. And it is more: that dark, ephemeral companion is a reminder that what we do here creates a pattern that is repeated on other levels, beyond what we can imagine.

Revelation as Endarkenment

The experience of the Israelites at Mount Sinai took place in the shadow of the mountain. Shortly before the Ten Speakings[11] are given, Exodus 19:17 states: "And they stood in the underside of the mountain." The Talmud tells the following Midrash[12]: "Rabbi Avdimi bar Hama said: The verse implies that the Holy One overturned the mountain upon them, like an inverted casket, and said to them: If you accept the Torah, it is well, if not, your grave will be right here" (BT Shabbat 88a). Alluding to this, the *Pesikta de Rav Kahana* says, "Israel accepted the Torah that was given out of darkness." (Piska 7, in Mandelbaum 1962) The Torah was given out of darkness, that is, the darkness of the shadow where they were standing, because the mountain was being held over their heads." At this point, the Torah relates that the people experienced thunder, lightning, dense clouds, *shofar* blasts, smoke, and the shaking of the mountain; then they heard the divine revelation. The Israelites then become too fearful to continue the intensity of this encounter, and they asked Moses to go on alone. The Torah tells us, "So the people stood at a distance, and Moses came near the thick darkness (*'arafel*) where God was" (Exodus 20:18). Philo wrote:

> Moses entered into the darkness where God was, that is into the unseen, invisible, incorporeal and archetypical essence of existing things. Thus he beheld what is hidden from the sight of the mortal nature, and in himself and his life displayed for all to see, he has set before us, like some well-wrought picture, a piece of work beautiful and godlike, a model for those who are willing to copy it. Happy are they who imprint that image in their souls. (Philo et al. 1961, 158)

The Torah was given in the darkness because it gave Moses the experience he needed to transmit the Torah, which is itself a pathway back to the essential darkness. Rabbi Menachem Mendel, the nineteenth century Kotzker rebbe is quoted in the P'ninei haTorah as saying that the *'arafel* is the essence (*ikkar*) and the innermost part (*pnimiyut*), and that is why the divine presence was there. And the medieval rabbinic authority, Jacob ben Asher, comments on Exodus 20:18 by saying, "the thick darkness (*'arafel*): in numerologic equivalence (*gematria*), the Divine

Presence (*haShekhinah*)" (ben Asher n.d.). That is, he equates the thick darkness that Moses entered to receive Torah with the Divine Presence itself.

So, at the time of the giving of the Torah, Moses and the Israelites were both having experiences of the sacred dark, although of different sorts. But later, the people had the experience of thick darkness (*'arafel*) in the Temple, the same experience that Moses had at Sinai. When King Solomon was building the Temple, he stated, "God has chosen to dwell in the thick darkness (*ba 'arafel)*" (I Kings 8:12 and II Chronicles 6:1). So the place of our most powerful communal religious experiences is described as holding the same type of sacred dark as that which Moses entered to receive the Torah. The core experiences of manifestation of Divine Presence—in the transmission of Torah and in bringing holiness into the world through the Temple—were both experiences of thick darkness.

Caves, the Temple, the Holy of Holies, and Divine Manifestation

After traveling up Mount Sinai during the initial revelation of Torah, Moses ascended again to receive the written version of the Ten Speakings on stone tablets. This time, the mountain was covered with cloud, which the Torah associates with the divine presence (Exodus 24:15–16). When Moses came down, the Israelites were worshipping the golden calf, and Moses broke the tablets he was carrying. He ascended again to receive a second set of tablets. Moses was full of doubt: "And he said, 'Please show me Your presence'" (Exodus 33:18). In response, God said

> As My Glory [or Presence] passes by; I will put you in a cleft of the rock and shield you with My hand until I have passed by. Then I will take My hand away and you will see My back; but My face must not be seen. (Exodus 33:22–23)

This revelation thus takes place in the dark: Moses is in a cleft of rock, covered by the divine hand. When he is allowed to see, what he sees is the back of the divine head. The Talmud comments thus: "R. Hama b. Bizana said in the name of R. Simon the Pious: This teaches us that the Holy One, blessed be, showed Moses the knot of [God's] *tefillin*" (BT B'rachot 7a).[13] So the image of the divine that Moses sees in this central epiphany is blackness—the black leather knot of *tefillin*.

When Moses descended from the mountain with the second set of tablets, the skin of his face was shining with rays of light (Exodus 34:29, 30, 35). The Midrash asks:

> From where did Moses derive these rays of splendor? The sages said: From the cave [i.e., the cleft of rock] as it says, As My Glory passes by, I will put you in a cleft of the rock and shield you with My hand until I have passed by. R. Berakhiah the priest said in the

> name of R. Samuel: The tablets were six handbreadths in length and six in breadth; Moses grasped two handbreadths and the Shekhinah [the Divine Presence] another two, two handbreadths being left in the center, and it was from them that Moses derived those rays of splendor. R. Judah ben Nachman said in the name of R. Simon ben Lakish [Resh Lakish].[14] A little ink was left on the pen with which Moses wrote the Ten Speakings; when he passed this pen through the hair of his head, the rays of splendor appeared. (Exodus Rabbah 47:6)

On one level, this Midrash is suggesting that there are three different sources of the rays of light shining from Moses's face. But on another level, the three possibilities are being paralleled, equated, or diffractively read through each other.[15] The cave, the divine presence, and the black ink of Torah have something in common—perhaps their sacred darkness—that makes them all possible sources for the way Moses's face looked when he came down the mountain. This reading may remind us of the (somewhat later) Zoharic teaching that the light of creation came forth out of the more primordial sacred darkness, referred to earlier in this chapter (Zohar I 31b–32a). I will come back to this concept later.

For now, let us return to the image of the cleft in the rock. The Zohar (I 84b) comments on Song of Songs 2:14 with this observation: "'O my dove in the clefts of the rock, in the covert of the hill,' saying: 'In the covert of the cliff'—this is the place that is called 'the Holy of Holies,' the heart of the entire world." The text then goes on to say that the Shekhinah secluded herself there. So the covert, or the cave, is once again where the Shekhinah dwells. No wonder revelation happens in caves. And the cave is the Holy of Holies, the most sacred spot in the Temple, which is described as a place of thick darkness (*'arafel*).[16] Herbert Levine describes the Holy of Holies thus: "The innermost sanctum was a place of deep darkness, where God's awesome presence could be experienced without any earthly distractions" (Levine 1995, 41). He describes the Temple as "a bridge between two worlds" and a "liminal zone" (1995, 43). In the Ba'al haTurim, Rabbi Jacob ben Asher[17] noticed the repetition of a word in two different verses: the first is in Leviticus 16:12, which describes what Aaron, the High Priest, should do in the inner sanctum of the *Mishkan*, the precursor to the Holy of Holies in the Temple: "And he shall take a panful of glowing coals scooped from the altar before Adonai, and two handfuls of finely (*dakah*) ground aromatic incense, and bring this behind the curtain" the second is in an epiphany of Elijah at a cave: "And after the earthquake a fire; but Adonai was not in the fire; and after the fire a still small (*dakah*) voice" (I Kings 19:12). The Ba'al haTurim explains that the common word *dakah* (fine, thin, or small) is telling us that in both cases, "the glory of Adonai appeared," that is, the divine presence manifested. The cave, the Holy of Holies, the liminal space of the deep darkness of divine presence, is an opening for healing and transformation. In the place where boundaries are no longer determinate, reality can shift. Moses not only had a revelation in a cave; he is also described by the Midrash as *being* a cave:

> You find sometimes, "And the Adonai spoke to Moses," and, "And Adonai said to Moses"; so also you find, "And Moses said to Adonai" and also, "And Moses spoke to Adonai." It can be compared to a cave situated by the seashore into which the sea once penetrated, and having filled it, never departed, but was always flowing in and out of it. So it was that Adonai spoke to Moses, and Moses said to Adonai. (Exodus Rabbah 45:3)

In addition to the Torah telling of Moses's epiphany at a cave, and the prophetic section of the Hebrew Bible telling of Elijah's epiphany, the Talmud (BT Shabbat 33b) tells the story of the great mystic, Rabbi Shimon bar Yohai, who spent twelve years with his son in a cave, hiding from the Roman authorities after he was reported for criticizing them. They used the time to develop mystical skills. In the foundational texts of Kabbalah, the Zohar, the character of Rabbi Shimon bar Yohai is the greatest master and teacher of mysticism. In the Zohar narratives, caves are places of revelation of the maternal aspect of God's presence and of light coming from the darkness.[18]

Perhaps the most famous cave in Jewish tradition is the cave of Machpelah. Abraham bought it to bury his wife, Sarah, and eventually it became the burial place of Isaac, Ishmael, Rebecca, Jacob, and Leah. Genesis Rabbah (58:8) tells us that the primordial Adam is buried there, too. The root of the name Machpelah in Hebrew is *khaf-peh-lamed*, which means "double," so the Midrash hypothesizes about what is doubled.[19] The Talmud (BT Eruvin 53a) wonders whether there are two chambers, an upper and a lower, or one within another. The Midrash (Genesis Rabbah 58:8) tells us that "the name signifies that the Holy One, blessed be, bent [the primordial, and very tall] Adam double, and buried him within it." The midrashic text Pirkei de Rabbi Eliezer (chap. 36) recounts that Abraham chased a stray calf who had run into the cave of Machpelah, and there he found Adam and Eve buried. The Zohar (I 127a–128b) adds to this story that when Abraham followed the calf into the cave, he saw a river of light emanating from it—yet another image of light emerging from the sacred dark. This same Zohar passage also addresses the issue of doubling. The cave and the field it is in are both referred to in the Torah as Machpelah. The Zohar says:

> The term Machpelah belongs properly neither to the cave nor to the field, but to something else with which both were connected. The cave belongs to the field, and the field to something else. For the whole of the Land of Israel and of Jerusalem is folded up beneath it, since it exists both above and below, both of the same pattern. The Jerusalem above has a twofold attachment, above and below; similarly the Jerusalem below is linked to two sides, higher and lower . . . . Further, the esoteric implication of the term Machpelah [which is referred to in the Torah as *HaMachpelah*—the Machpelah—with a letter *heh* at the beginning and at the end] relates it to the Divine Name, in which the letter *heh* is doubled, though both are as one. (Zohar I 127a–128b, in Abelson, Simon, and Sperling 1984)

The Zohar is presenting an image of a cave that is a pathway between the worlds, a liminal zone. One enters the dark cave of the first *heh* of *HaMachpelah* and winds up coming out the second *heh* into another level of reality.[20] It is by passing through the dark, where boundaries are blurred, that one can make this transition. Machpelah is the cave in which the ancestors are buried, so it is a place for souls to make a transition out of this life and into something else, but it is also a place for transition from the more mundane level of reality to somewhere both literally and metaphorically deeper. This liminal power of the cave of Machpelah, connected with its double nature, is brought forward in a variety of tales, including a modern one by Howard Schwartz, based on traditional sources, in which Talmudic sages entered the cave where it was geographically located, in Hebron, and saw, through the far end, where the cave opened out into the Garden of Eden (Schwartz 1993, 276–79). Another version of this idea is brought by Itzchak Buxbaum, who recounts a story that took place during the Ottoman empire, in which a midwife named Fruma Riveleh entered the cave of Machpelah and was locked in it. She wandered through the cave, meeting the spirit of King David, and walked out the far end, which opened out onto her home street in Jerusalem (Buxbaum 2002, 207–11). These stories are clues to us that the image of the cave of Machpelah is a meditative tool, where one can visualize entering into the dark depths and emerging transformed.

Another aspect of the double *heh* of the cave of Machpelah, has to do with the way that the letter *heh* ה is shaped a bit like a pair of legs and a pelvis, and is associated, particularly in the Tetragrammaton, with the female aspects of the divine.[21] So the image of entering a cave, particularly one that is associated with the letter *heh*, and being taken through into another world, evokes the image of passing through the womb. The imagery of the *heh*, the cave, the earth, the womb, and the Shekhinah are all overlaid in Jewish thought. The *Sefer Yetzirah* (1:13), a second century CE mystical text, takes the first three letters of the Tetragrammaton, the *yod heh vav*, and associates each of the six possible permutations of these three letters with the six directions (above, below, east, west, south, and north). In imagining these letters as the cardinal directions, one is surrounded with the letters of the Divine Name (Kaplan 1991). What does not get mentioned explicitly is what is missing: the final *heh* of the Tetragrammaton and the seventh direction—the center. That is, the final *heh* is the center—the earth, the place of the self, surrounded by the rest of the directions. The most manifest level of reality for us, then, the earth, is also the innermost part—that which is surrounded on all sides. This is like the womb, which is in the center of the body but brings forth into existence. The womb is the place of the darkness that pre-exists the light in the creation of every mammal. In Jewish thought, this is related to the darkness that preceded the creation of the world, which I will discuss shortly.

First, however, let us consider another perspective on the double *heh*s and the cave of Machpelah, which comes from Rabbi Isaac Luria, the sixteenth century eponymous creator of Lurianic Kabbalah, whose work was based on deep contemplation of the Zohar. His teaching is brought forward by contemporary philosopher and rabbi, Marc-Alain Ouaknin, who explains that Luria called the doubling of the *heh* in the Tetragrammaton (*yod heh vav heh*) "The Cave of Machpelah" (Ouaknin 2000, 388–89). He goes on to explain that the shape of the *heh* ה can be written in two ways, either comprised of a letter that looks like three connected points, describing a plane—the letter *dalet* ד—and a letter that looks like two connected points, describing a line—the letter *vav* ו. Alternatively, the *heh* can be comprised of a *dalet* ד and a letter that looks like a dot—*yod* י. If the first *heh* is formed with the *dalet* and *vav*, and the second *heh* is formed with a *dalet* and *yod*, then the Tetragrammaton *heh* ה

> turns into an explosion of the point (*yod*) into two points that are arranged in space as a mirror image: *yod-dalet-vav/vav-dalet-yod*. The Tetragrammaton is the movement of a point that returns to the point.
>
> The name withdraws at the same time it is given .... An analysis of the graphics of the name ought thus to make it easier to understand its construction as a deconstruction based on the point. In fact, as we have said: "The point returns to the point."
>
> The name is a meditation on nothingness that becomes a being, and which returns to nothingness. It is entry into movement and an infinity of time. (Ouaknin 2000, 389)

Nothingness is not determinately empty. Rather, quantum field theory teaches that there are continual fluctuations of the vacuum. As Ouaknin puts it in the above quotation, "nothingness that becomes a being, and which returns to nothingness," and as Barad expresses it when speaking of the quantum vacuum, "Nothingness is not absence, but the infinite plenitude of openness .... Infinity is the ongoing material reconfiguring of nothingness" (Barad 2012: 16). In introducing the idea of the Tetragrammaton as the cave of Machpelah, Ouaknin has this to say: "Seeing the four-letter name is to be engulfed in the nothingness of the senses, to penetrate into an annihilation of consciousness, to experience a vacuum, a void, the infinite" (Ouaknin 2000, 386). Rabbi Alan Lew wrote,

> the Great Temple of Jerusalem was an elaborate construction surrounding nothing. There at the sacred center, at the Holy of Holies ... is precisely nothing—a vacated space, a charged emptiness, mirroring the charged emptiness that surrounds this world, that comes before this life and after it as well. (Lew 2018, 221)

So we might see the journey of the cave of Machpelah, the contemplation of the Divine Name, and the entry into the Temple's Holy of Holies as an evocation of awareness of the vibrating life force of all that is, the continual entering into and returning from the void in endless waves of diffraction patterns of dark and light, dark and light.[22]

The Darkness before Creation

Genesis, the creation story, ends like this: "the spirit/wind of God hovering over the face of the water" (Gen. 1:2). The Jewish mystical tradition likens the waves of emanation of creation to the patterns made by a stone thrown in a pond.[23] Creation is a wave-making process: sound waves, water waves, diffraction patterns of all kinds. We can still feel these waves vibrating deep inside the body. We might imagine them as wave upon wave of dark and light—first, a primordial darkness, out of which comes the light of creation. Out of that light, the deep darkness we call night, and out of that, the light of day. Out of the darkness we call night, which includes the depths of divine thought, says the Zohar, comes voice and speech, which reveal the depth.

In Genesis, before God said "let there be light," there was "darkness over the face of the deep, the spirit/wind [*ruach*] of God brooding/hovering over the face of the water" (Gen. 1:2). Biblical poetry is often structured with two parallel stitches in a verse. In this case, "over the face of the deep" parallels "over the face of the water," and "darkness" parallels "the spirit/wind of God." There is something profoundly holy about this darkness, which Genesis tells us pre-existed what we think of as creation.

The creation story starts by saying "*B'reishit bara Elokim et hashamayim v'et ha'aretz*," a strange grammatical structure saying something like, "With a beginning of, God created the heavens and the earth." The midrashic collection from the fourth and fifth centuries known as Genesis Rabbah (and subsequent Jewish tradition) interprets this "beginning" to be wisdom, *Hokhmah*,[24] that is, the wisdom with which God created the heavens and the earth. In the proof text for this interpretation, *Ḥokhmah* is envisioned as a crone, standing at the crossroads[25]:

> It is Wisdom calling, Understanding raising her voice. She takes her stand at the topmost heights, by the wayside, at the crossroads .... Adonai created me at the beginning of His path (*reishit darko*). (Genesis Rabbah, Proverbs 8)

It is a vision of ancient dark female wisdom, assisting in the birthing of the world. In Jewish thought, wisdom, *Hokhmah*, differs from understanding, *Bina*, in that it precedes analysis. It comes from the level of realization, the "aha" moment before it has been articulated in distinct words. The darkness of wisdom where all boundaries dissolve is a pathway that can take us beyond our individual selves into something bigger.

The Black Ink of Torah

Hokhmah is also associated with the black ink of Torah (Kaplan 1991, 75). In his brilliant commentary on the ancient mystical text, *Sefer Yetzirah*, Rabbi Aryeh Kaplan cites Exodus Rabbah:

> Three things preceded the creation of the world—water, wind/breath,[26] and fire. The waters conceived and gave birth to thick darkness; the fire conceived and gave birth to light; the wind/breath conceived and gave birth to wisdom, and with these six things the world is maintained: with wind, wisdom, fire, light, darkness, and water. (Exodus Rabbah 15:22)

Here (as in the Genesis creation story itself), there is an association between water and darkness. The *Sefer Yetzirah* (1:11) calls *Hokhmah* "water from breath." The breath of life that God breathed into the world condensed on its way. *Hokhmah*, the wisdom that comes from going beyond the individual's boundaries to a place beyond reason, beyond analysis to synthesis, is like the rain, which does not differentiate but falls on everything without distinction. Kaplan (1991, 74) explains that this idea is rooted in the prophecy of Isaiah:

> But as the rain and snow descend from heaven
> and return not there
> without watering the earth
> making it bloom and bud
> giving seed to the sower and bread to he who eats.
> So the word that emanates from My mouth
> shall not return to me empty-handed
> without accomplishing that which I please
> and succeeding in its mission. (Isaiah 55:10–11)

The *Sefer Yetzirah* goes on to say of *Hokhmah* that "with it He engraved and carved [22 letters from] chaos and void, mire and clay." Here watery wisdom, *Hokhmah*, is related both to the chaos that preceded creation and to the mire that provides the ink for the Hebrew letters. The letters, then, are another way of taking us beyond ourselves. And it is no coincidence that they are black. Exodus Rabbah (47:6), we recall, suggests that the black ink that got into Moses's hair from the reed with which he wrote the Torah may have been the source of the beams of light that emanated from him.

The Burning Bush, Black Fire, and Burning Coal[27]

Moses had a formative encounter with the sacred dark earlier in his life as well. The Torah reads: "An angel of Adonai appeared to him in a blazing fire out of a bush. He gazed, and there was a bush all aflame, yet the bush was not consumed" (Exodus 3:2). The burning bush that is not consumed sparked the movement for liberation from slavery, which birthed the Israelite people and continues to inspire Jewish and other liberation movements. It can also inspire *us* and give us guidance in how to take further steps in our own movements toward liberation in whatever forms we are called upon to do so.

So, what is the burning bush here to teach? The Midrash in Exodus explains:

> From this they derived that the heavenly fire shoots out branches upwards, burns but does not consume, and is black in color; whereas fire used here below does not branch upwards and is red, and consumes but does not burn. (Exodus 3:2)

It may be hard to imagine a black fire; perhaps it is counterintuitive or paradoxical. So let us examine it further. The fire metaphor itself is multifold. It implies something awesome, powerful, something with the potential to give life or death. And then there is the concept of blackness or darkness. In seeing light, one sees rays bouncing off surfaces—one might get the impression that the world is made up of separate, inherently bounded entities. But when we are in the dark, it is easier to sense that all boundaries are situational—to feel beyond the limits of the self with a small S and to become aware of being part of a larger whole that includes all that is, and even all that is not. This may be why it is said, a few verses after seeing the burning bush, that Moses hid his face: "Moses hid (*vayaster*) his face in awe from looking[28] at God" (Exodus 3:6). This is usually interpreted in a negative way, that it is unfortunate that Moses was not willing to see what would have been revealed to him had he not hidden his face. But perhaps in hiding his face, he avoided focusing on the sense of sight, the sense that may induce a belief that we are separate individuals, with boundaries that are revealed as light bounces off them. With face hidden, eyes closed, one enters the realm where boundaries disappear, and it is easier to sense being part of a whole that is all that is, easier to sense how we are not separate from God or from anything else.

In seeing the black fire of the bush, Moses was brought into a realm beyond the usual boundaries. He was empowered to move beyond what he thought possible, and he was given the means to do it through the ability to connect with a power beyond his individual self. This is in keeping with another image of black fire, which comes from a teaching brought in the Talmud and Midrash. The Talmudic form of the teaching says the following:

> Rabbi Pinchas [said] in the name of Rabbi Shimon ben Lakish: the Torah that the Holy Blessed One gave, its hide is white fire, its ink is black fire; it is fire mixed with fire, carved from fire, and given from fire: "at His right hand a ritual of fire for them" (Deuteronomy 33). (Talmud Yerushalmi, Sotah 37a)

The shape and flourishes of the letters of the Torah scroll, calligraphed in a particular style called Ashurite script, are very reminiscent of flames, and not coincidentally. An additional layer of meaning comes from the midrashic telling of the same teaching, which has another phrase at the beginning. It says, "Rabbi Yochanan said: One who engages in Torah should see oneself as if he were standing in fire" (Yalkut Shimoni, Brachah, 951). That is, we are not to experience the sacred fire of Torah just from the outside: we are meant to experience it from the inside. We are meant to be immersed in it. There is another teaching saying that each Israelite is a letter of the Torah.[29] These teachings together invite each individual to experience themselves as a letter of Torah, a letter made of black fire.

Medieval Kabbalist Rabbi Isaac the Blind called the black fire of Torah "the world to come" and associated it with the Oral Torah—that is, Talmud and Midrash. He explains:

> It is the hue of a black fire on white fire, which is the Written Torah. Now the forms of the letters are not vowelized nor are they shaped except through the power of black, which is like ink. So too the Written Torah is unformed in a physical image, except through the power of the Oral Torah. (cited in Dan 1986, 75–76)

He also calls the black fire the "crown of the kingdom," *keter malchut*. This is an unusual concept—*keter*, or crown, is the highest or most transcendent *sefirah*, or sphere of reality, in the Kabbalistic Tree of Life, while *malchut*, or kingdom, is the lowest, the most manifest. Yet in this royal imagery, the crown sits right on the head of the king. So the whole system has a different topology from what we usually think of: it is not a linear hierarchical system but, rather, circles that turn back on themselves.[30] Rabbi Isaac explains this concept with another image of black fire, that of the burning coal, in his commentary on the ancient text, *Sefer Yetzirah*, which says, "Ten *sefirot* of Nothingness. Their end is embedded in their beginning and their beginning in their end, like a flame in a burning coal. For the Master is singular; He has no second, and before One, what do you count?" (*Sefer Yetzirah* 1:7, cited in Kaplan 1991, 57). Rabbi Isaac the Blind explains this verse by saying,

> Their end is (found) in their beginning: just as many threads come out of the burning coal, which is one, since the flame cannot stand by itself, but only by means of one thing; for all (the) things (that is, *sefirot*) and all (the) attributes, which seem as if they are separate, are not separate (at all) since all (of them) are one, as their beginning is, which unites everything in one word. (Isaac the Blind, cited in Gottlieb 1989: 410–11)

The burning coal, then, reminds us of that from which all else emanates. "Their end is embedded in their beginning" explains what Rabbi Isaac said in his previous text, that the black fire is the crown of the kingdom, that is, the ultimate source, which one might think is farthest from the manifest world but which is actually closest to it. And all of it is rooted in that black coal, without which the flames that emanate could not exist. *Our source is in the darkness. Without it we do not exist. And although that darkness is the most transcendent we can imagine, it is also closer to us than anything else ever could be. It is our innermost being.* That dark source is where we go to experience the sacred, where we go both when we want to feel safe, and when we are challenged to go beyond the beyond—the innermost, the deepest, the furthest, the closest. Kaplan also explores another aspect of the burning coal image, the fact that it troubles the binarism of cause and effect:

> A flame cannot exist without the coal, and the burning coal cannot exist without the flame. Although the coal is the cause of the flame, the flame is also the cause of the burning coal. Without the flame, it would not be a *burning* coal. Since Cause cannot exist without Effect, Effect is also the cause of Cause. In this sense, Effect is the cause and Cause is the effect. Since beginning and end are inseparable, "their end is embedded in their beginning, and their beginning in their end." (Kaplan 1991, 57)

Contemplating the burning coal can lead to awareness of the nonbinary: cause and effect, transcendence and immanence, up and down, light and dark are interwoven in multivalent, multidirectional ways and waves. They cannot be essentialized or dualized.

The Dark Depths

The burning coal in *Sefer Yetzirah* is associated, as mentioned above, with *keter*, what is usually considered the "highest" or most transcendent of the *sefirot*, spheres of being, in the Kabbalistic Tree of Life (Kaplan 1991). Often in Western thought, up and high are privileged over down and low or deep, leading to some of the problems we have mentioned, such as devaluing the female, the earth, and the subconscious. But in Jewish thought, *keter*, in addition to being imagined as the highest, is also imagined as the dark depths. The Zohar pictures *keter* as the source of a spring, which flows into an ocean, from which rivers flow that are the *sefirot*, the channels for the flow of divine into the world:

> He has no attribute, no image, and no form. It is like the sea. The waters that come from the sea cannot be grasped, nor do they have form. But when the waters of the sea spread themselves over a vessel, which is the earth, an image is formed, and we can then make a calculation as follows: the source of the sea is one; a spring comes from it as it spreads in the vessel, in a circle, which is a *yod*; and so we have the source—one; and [together with] the spring that comes from it—two. After this, He makes a huge vessel, like someone digging a great pit that fills with the water that comes from the spring. This vessel is called "sea," and is the third vessel. And this huge vessel is split into seven vessels, like long vessels. Thus the waters from the sea are spread out into seven streams. And so, we have a source, and a spring, and a sea, and seven streams, making ten. (Zohar II 42b–43a, Raya Mehemna)

In this image, the source is in the depths, and the flow goes up. Here, the closest that can be imagined to the source of the emanation of all creation is the dark point at the depths of the deepest spring. It is possible in awareness to follow the steps described here back in the opposite direction, from the rivers into the sea, then back to the spring, and finally to the source. Perhaps it is divine compassion that took the difficult path by emanating upstream in order to allow us simply to float on the rivers' currents to return to the source in the sea.

The Zohar refers to *keter* as the "depth of the well" (*'amika d'beyra*) (Zohar III 289b, Idra Zuta). Medieval rabbi Bahya ibn Pakuda uses a similar metaphor to describe an inner journey:

> "Counsel in a man's heart is deep water; but a man of understanding can draw it out" (Proverbs 20:5). The meaning is that wisdom is implanted in man's nature, in his character and in his powers of perception like the waters that are hidden in the depths of the earth. The intelligent and discerning man will try to tap into his own inner potential for wisdom, uncover it, and bring it to expression, drawing it out of his own heart much like the search for water in the depths of the earth. (Bahya ibn Pakuda 1996, 25)

Rabbi Bahya compares wisdom to water in the depths of the earth, in the depths of the heart. One needs to reach into these depths to draw on the source of wisdom. The deeper we can go into the dark, into our own inner depths, and into the infinite depths, the closer we get to the source, to wisdom. The Zohar also teaches:

> Every person who presents his request before the King should focus mind and will on the root of all roots, to draw blessings from the depth of the well, so that it will gush blessings from the spring of all. And what is that? The place from which the river issues and derives, as is written: A river issues from Eden . . . (Genesis 2:10) . . . . This is called "out of the depths"—depth of all, depth of the well, springs issuing and flowing, blessing all. This is the beginning of drawing blessings from above and below. (Zohar II 63b, Matt 2012 translation)

The Zohar invites us to focus our awareness in the deepest depths, to the place from which creation emanated, the source of all blessings. And the more we can draw the flow of blessing from the deepest of the deep, the more the world will be filled with the presence of the sacred.

Acknowledgments

My deep gratitude goes to Rose May Dance, Starhawk, and the rest of the Reclaiming Collective in San Francisco, California, from whom I learned to explore the sacred dark; to my teachers, Rabbis Marcia Prager, Miles Krassen, and Elliot Ginsberg for helping me find the sacred dark in Jewish tradition; to my teacher and rabbi Reb Zalman Shachter-Shalomi z"l, who supported this work, and invited me to write my final paper for rabbinic ordination, "Toward a Theology of Darkness," which is an earlier version of this article; and to Dr. Ibrahim Farajaje, may his secret be sanctified, who loved the sacred dark as much as I do.

Notes

1. Although Baltazar's book was published in 1973, it did not come to my attention until 2016, and I read it after completing most of this paper. Had I read it earlier, I would have found it very influential, as several of the images of the sacred dark that I have developed and have never or rarely seen mentioned elsewhere were touched on in his book. I feel much gratitude for the work he did, and I direct readers to this book for many reasons, especially for an excellent historical analysis of how Western color symbolism led to racism and a theology of the sacred dark in Christian thought.
2. Another way of framing the racial implications of color symbolism comes from Amoja Three Rivers, an American-born African, Choctaw, Tsalagi, Ojibwa Jew, who, in her booklet *Cultural Etiquette: A Guide for the Well-Intentioned* (2018), asks, "How about instead of 'the pot calling the kettle black,' you say, 'the pus calling the maggot white'?"
3. Adonai is the most common name used to express the inexpressible, most sacred name of God in Judaism, spelled *yod heh vav heh*, sometimes indicated YHVH. It is spelled without vowels and so is not pronounced as written. For more about the meaning of the name, see Prager (1998).
4. Depending on how one reads the *k'rei/ketiv* (the words as they are to be read vs. the written version) in this verse, one could read "Her song" or "His song."
5. Redemption is one of the most central concepts in Judaism and can be understood to mean communal liberation from suffering.
6. The image of diffraction as a way of understanding the making of difference—in contrast to reflection, which is a making of sameness—comes from the work of Karen Barad (2007).
7. I am indebted to the Jewish Publication Society Tanakh for this understanding of the root *aleph-resh-resh*, "to cast a spell," rather than the more common translation, "to light up."
8. The term "intra-action" is a neologism by Karen Barad. For a brief explanation, see Kleinman (2012). For a deeper discussion, see Barad (2007).
9. A *sukkah* is the temporary outdoor booth where Jews dwell during the harvest festival of Sukkot.
10. See Zohar III 103a and elsewhere.
11. The Hebrew term *Aseret haDibrot*, often translated as "the Ten Commandments," literally means "the Ten Speakings."
12. The term Midrash refers to the interpretive rabbinic tradition of stories and legal interpretations based on oral traditions related to the Hebrew Bible. Some use the term only to apply to ancient and medieval rabbinic texts, while others include what might be called "modern Midrash," sacred stories that are still being crafted.
13. Tefillin are sometimes translated into English as "phylacteries," but often the Hebrew word is used in English.
14. It is not a coincidence, I believe, that the Talmudic sage Resh Lakish is cited as the source for teachings about both the power of the dark of night and the dark of the ink of Torah. As we will continue to see, he had a developed understanding of the sacred dark.
15. See Barad (2007), chap. 2, on diffractive reading.

16. See I Kings 8:12 and II Chronicles 6:1. Interestingly, Thorkild Jacobsen (1976) reports that in Ancient Mesopotamia, the Temple's "holy of holy, the god's private apartment shrouded in darkness was the 'dark room' . . . which 'knows not daylight,' its ritual vessels 'no eye is to see'" (Jacobsen 1976, 16).
17. Commenting on Leviticus 16:12.
18. See, for example, Zohar III 149b–150a.
19. Genesis Rabbah claims that the reward of anyone who is buried there is doubled.
20. I owe my understanding of the doubled cave being the two *hehs* in *haMachpelah* to the teachings of R. Elliot Ginsburg, with whom I studied this Zohar passage in a course he taught at Elat Chayyim in the summer of 1998.
21. See my article (Feldman 2018) for more on the two *hehs* of the Divine Name.
22. See Barad (2012).
23. Baumann (1979), 23, teaching about the thought of Isaac Luria.
24. The word *reishit* is found in a grammatical form called a "construct," which means it should be the first of two nouns in a row, where it would be translated as the <first noun> **of** the <second noun>. Thus, *reishit* should mean "beginning of"; however, there is no second noun. To address this problem, the tradition looks for other places where the same word is used to clarify its meaning. Proverbs 8:22 refers to wisdom/*Ḥokhmah* as "*reishit darko,*" "the beginning of [God's] path," thus the association of "with a beginning" with wisdom.
25. Similar images of wisdom as a dark crone standing at the crossroads occur elsewhere in Mediterranean traditions as well, for example, in the Greek figure Hekate. See the work of the greatest recent scholar of Proverbs, Avigdor Hurowitz, who writes, "*Ḥokhmah* is depicted as a woman. There are those who say that there isn't anything here except a metaphor to concretize the availability of *Ḥokhmah* to anyone who wants to become wise . . . . Others believe that there is here an image of mythological divinity (*elohut mitologit*)" (Hurowitz 2012, 247) (my translation).
26. The Hebrew word *ruach* means variously breath, wind, and spirit.
27. This section is a slightly edited version of a talk given at Kehillah Community Synagogue, December 21, 2013, which was also published as Feldman (2015).
28. Often translated as "because he was afraid to look." "Fear" and "awe" use the same word in Hebrew, and the infinitive can be translated in multiple ways, since there are many fewer tenses in Hebrew than in English.
29. See Zohar Hadash Shir HaShirim 74d.
30. For more on this topology, see Feldman (2018).

References

Abelson, J., Maurice Simon, and Harry Sperling, eds. and trans. 1984. *The Zohar*. Brooklyn, NY: Soncino Press.

Agnon, Shmuel Yosef, and Michael Swirsky. 1994. *Present at Sinai: The Giving of the Law*. Philadelphia: Jewish Publication Society.

Baltazar, Eulalio R. 1973. *The Dark Center: A Process Theology of Blackness*. New York: Paulist Press.

Barad, Karen. 2007. *Meeting the Universe Halfway: Quantum Physics and the Entanglement of Matter and Meaning*. Durham, NC: Duke University Press.

Barad, Karen. 2012. *What Is the Measure of Nothingness? Infinity, Virtuality, Justice*, bilingual edition. 100 Notes, 100 Thoughts, Documenta 13, Book 99. Ostfilder, DEU: Hatje Cantz.

Baumann, Saul. 1979. *Sefer Miftehe Hokhmat Emet*. Brooklyn: Mekhirah Ha-reshit, Kelilat Yofi.

Begley, Sharon, and Anne Underwood. 2001. "Religion and the Brain." *Newsweek*, May 7. http://2000-2013.gustavorol.org/newsweek.htm. (accessed on May 22, 2019)

ben Asher, Jacob. n.d. "Kitzur Baal Haturim on Exodus." *Sefaria*. Website. https://www.sefaria.org/Kitzur_Baal_Haturim_on_Exodus.20.21.1?lang=bi&with=all&lang2=en (Accessed May 22, 2019).

Berezovsky, Sholom Noach. 2012. *Netivot Shalom*. Jerusalem: Yeshivat Bet Avraham.

Buxbaum, Yitzhak. 2002. *Jewish Tales of Holy Women*. San Francisco: Jossey-Bass.

CERN. 2012. "Dark Matter." *CERN Accelerating Science* (blog). December 1. https://home.cern/about/physics/dark-matter.

Copeland, M. S. 2013. "Blackness Past, Blackness Future–and Theology." *South Atlantic Quarterly* 112 (4): 625–40.

Dan, Joseph, and Ronald C. Kiener. 1986. *The Early Kabbalah*. New York: Paulist Press.

Feldman, Fern. 2015. "The Burning Bush and Black Fire." *Tikkun*, December 29. https://www.tikkun.org/newsite/the-burning-bush-and-black-fire.

Feldman, Fern. 2018. "Gender and Indeterminacy in Jewish Mystical Imagery." In *Unsettling Science and Religion: Contributions and Questions from Queer Studies*, edited by Lisa Stenmark and Whitney Bauman, 199–219. Lanham, MD: Lexington Press.

Freedman, H., and Maurice Simon. 1983. *Midrash Rabbah*. London: Soncino.

Gottlieb, Freema. 1989. *The Lamp of God: A Jewish Book of Light*. Northvale, NJ: J. Aronson.

Haskell, Ellen Davina. 2012. *Suckling at My Mother's Breasts: The Image of a Nursing God in Jewish Mysticism*. SUNY Series in Western Esoteric Traditions. Albany: State University of New York Press.

Hurowitz, Avigdor. 2012. *Proverbs: Introduction and Commentary*. Tel Aviv, ISR: Am Oved.

Idel, Moshe. 1988. *Kabbalah: New Perspectives*. New Haven, CT: Yale University Press.

Jacobsen, Thorkild. 1976. *The Treasures of Darkness: A History of Mesopotamian Religion*. New Haven, CT: Yale University Press.

Jewish Publication Society. 1999. *JPS Hebrew–English Tanakh: The Traditional Hebrew Text and the New JPS Translation*, 2nd ed. Philadelphia: Jewish Publication Society.

Kahana, Rav. 1962. *Pesikta De Rav Kahana*. Translated by Bernand Mandelbaum. New York: Jewish Theological Seminary of America.

Kaplan, Aryeh, and Neḥunya Ben Ha-Kanah. 1989. *The Bahir*. York Beach, ME: Samuel Weiser, 1989.

Kaplan, Aryeh. 1991. *Sefer Yetzirah (The Book of Creation): In Theory and Practice*. York Beach, ME: S. Weiser.

Kleinman, Adam. 2012. "Intra-actions: An Interview with Karen Barad." *Mousse Magazine* 34: 76–81.

Krauss, Lawrence Maxwell. 1989. *The Fifth Essence: The Search for Dark Matter in the Universe*. New York: Basic Books.

Ḳovalsḳi, Shalom Dov, Yitshak Figenboim, and Shalom Hayim Parush. 1997. *Sefer Penine Ha-Torah: U-vo Mivḥar Perushim 'al Ha-Torah 'arukhim U-mesugnanim Be-śafah Ḳetsarah, Ḳola'at Ṿe-'inyanit*. Jerusalem: Y. Ha-Kohen Figenboim.

Levine, Herbert J. 1995. *Sing Unto God a New Song: A Contemporary Reading of the Psalms*. Bloomington: Indiana University Press.

Lévinas, Emmanuel. 1990. "The Light and the Dark." In *Difficult Freedom: Essays on Judaism*. Baltimore, MD: Johns Hopkins University Press.

Lew, Alan. 2018. *This Is Real and You Are Completely Unprepared: The Days of Awe as a Journey of Transformation*. New York: Back Bay Books/Little, Brown.

Lorde, Audre. 1984. "Poetry Is Not a Luxury." In *Sister Outsider: Essays and Speeches*, 36–9. Trumansburg, NY: Crossing.
Matt, Daniel Chanan. 1995. *The Essential Kabbalah: The Heart of Jewish Mysticism*. San Francisco, CA: HarperSanFrancisco.
Matt, Daniel Chanan, trans. 2012. *The Zohar*. Stanford, CA: Stanford University Press.
Meir, Levi Isaac Ben. N.d. *Sefer Kedushat Levi: 'al Ha-Torah*. Bnei Brak: Heichal Hasefer.
Ouaknin, Marc-Alain. 2000. *Mysteries of the Kabbalah*. New York: Abbeville.
Pakuda, Bahya Ben Joseph Ibn. 1996. *Duties of the Heart*. Translated by Daniel Haberman. Jerusalem: Feldheim.
Pakkuda, Bahya Ben Joseph Ibn, Moses Hyamson, Torat Hovot Ha-levavot, Sefer Torat Hovot Ha-levavot, and Yehudah Ibn Tibon. 1970. *Duties of the Heart*. Jerusalem: Feldheim.
Philo of Alexandria. 1980. *The Contemplative Life, the Giants, and Selections*. Translated by David Winston. London: Paulist Press.
Philo of Alexandria, Sa'adia ben Yoseph, Yehudah Ha-Leví, Alexander Altmann, and Isaak Heinemann. 1961. *Three Jewish Philosophers*. Cleveland, OH: Meridian.
Prager, Marcia. 1998. *The Path of Blessing: Experiencing the Energy and Abundance of the Divine*. New York: Bell Tower.
Schorr, Gedalia. 2000. *Sefer Or Gedalyahu*. Jerusalem: Keren Zikhron Gedalyahu.
Schwartz, Howard. 1993. *Gates to the New City: A Treasury of Modern Jewish Tales*. New York: Avon.
Simon, Maurice, and Isidore Epstein. 1990. *Hebrew–English Edition of the Babylonian Talmud*. London: Soncino.
Stenmark, Lisa, and Whitney Bauman, eds. 2018. *Unsettling Science and Religion: Contributions and Questions from Queer Studies*. Lanham, MD: Lexington Books.
Three Rivers, Amoja. 2018. *Cultural Etiquette: A Guide for the Well-Intentioned*, Rev. ed. Minneapolis, MN: Athena Persephoni.
Winston, David. 1981. *Philo of Alexandria: The Contemplative Life, the Giants, and Selections*. Mahwah, NJ: Paulist Press

CHAPTER SEVEN

Andean Entifications

Pachamamaq Ajayun, The Spirit of Mother Earth

GUILLERMO DELGADO-P.

In memoriam

Elías Delgado M., Eugenio Bermejillo, Rodolfo L. Meyer

Qallarinapaq: To Begin

In this article, my purpose is to foreground recent discussions on reontologizing the notion of "socionature" as a way of critiquing modern thought and practice, which subverted the original hylozoism of Andean peoples. The term "socio-nature" acknowledges a critique of anthropocentrism. As we know, the introduction of modes of ecocidal extractivist capitalism as an economic strategy of "accumulation by dispossession" legitimized the deontologization of nature by making it dispirited and deprived of agency. Such a process instigated disruptive colonialist "acts against nature"; the perpetrators were and are humans who, as inheritors, provoked and sustained ecocidal interventions on Indigenous peoples' nature(s) considered alive. Nowadays, the Anthropocene scores recognition of such intrusion(s).

A conscious retrieving of pluricultural and multinatural worlds thus involves a rejection of the full commoditization of nature(s), an ethics that inspires a vibrant environmentalist consciousness that re-establishes dialogues with variants of radical Western deep ecology, Gaia stories, GMO critics, and Indigenous thought regarding global warming today. This attitude, based on the assumption of *many natures* and one culture (rather than one nature and *several cultures*), is ethical (i.e., derived from ethics) in the sense that, as a grammar of the dignity of socionature, it proposes a larger spiritual and radical (that is, from the "roots") reappraisal that hopefully re-evaluates humanity's stance on "eARTh."

We find substrata of this decentering thinking and of issues of permanence on "eARTh" if we especially highlight parts of the world where Indigenous cosmicentric and chorographic convictions were, and are, retained despite colonialism's fury layered as anthropogenic heritage. Here I am referring to Karen Barad's discussion on "agential realism":

> All bodies, not merely human bodies, come to matter through the world's performativity—its iterative intra-activity. Matter is not figured as a mere effect or product of discursive practices but rather as an agentive factor in its iterative materialization, and identity and difference are radically reworked. (Barad 2011, 125)

The Indigenous peoples of the Andean area of South America offer clear examples of this "agentive factor," and the term "entification" which I added to the title, illustrates nature's ontological status, considered to be alive rather than inert. Beware that ontology, in Western philosophy, refers to the specificity of *being* human, *ente* (from the Latin *ens* and the Greek *onto*; in Spanish, *ente*) but without nature; in other words, it is anthropocentric. Such centrality can be seen as an accident, as in "the accidental culture" or as in "accidental rationality." On the contrary, the notions of entification, hylozoism, and perspectivism constitute the corresponding cosmicentric concepts that define the sensory character of non-Western, Indigenous philosophies around the world.[1] These terms expose a cosmicentric world in its spatial-chorographic depth of intra- and interconnections or relationalities. Here, I will analyze four instances of Andean hylozoistic concepts as contemporary examples of an ontologized nature. Given that Andean cultures, and for that matter, all Indigenous peoples, resisted or negotiated the imposition of European modernity and the full deontologization of nature, such quincentennial resistance materializes in the actual recurrence to and sustenance of hylozoistic convictions.

Despite unrelenting presence of Christianity in its varieties, the Andean area has preserved clear hylozoistic spiritual principles that are perceived to coincide with notions of the sacred shared by other non-Western religions. Transmitted throughout these last centuries, the reaffirmation of notions of pre-Columbian origins is detectable,[2] contesting the one-directional nomos and centrality of the modern, with its exhausted teleology of progressionism and development (as high risk, uncertainty) we identify as the Anthropocene.

Sumaq Orqun Illakuna: The Energy of Mountains

Taking a bird's-eye view of the Andean territory, what else, hidden or visible, do the materiality of mountains feature? Mountains are considered to be the residence of gods, enlivened by *K"ama*, spaces of energy that animate the *Illi* or *Illa*[3] of the telluric gods *Achachila, Apu, Wamani,* and *Jillaranataka*. In ancient times, mountains were like people: they enjoyed themselves, they fought with each other; they visited and talked to each other; and today, in memory of the past, several are still called by their proper names. Ancient Andean space designers adopted innumerous *Wak'a* and *Apacheta* throughout the mountain range, and spiritual leaders *Yatiri, Chamaq"ani, Laiqa, Paqu, Jamp'iri*, which have been mistakenly homogenized as shamans, were (and are) in charge of keeping the *Wak'a* alive. The *Wak'a* can have different aspects and needs, but they are generally natural sculptures, prominent boulders, stones, and rock outcrops (*Wank'a*) that have unusual shapes, often revealing to the naked eye images of majestic birds (such as the *Kuntur*), heads of llamas or alpacas, human faces, human or animal body parts, gigantic ants, a mythical petrified snake, or a large toad. The *Wak'a* marked the ancient Inkan *ceque* system that perceptually radiated from Cuzco, the navel (*Q'usqu*) or center of the world. *Ceque*s are imaginary lines that cohesively membered four *Suyus* or four constitutive parts of the Andean territory of the Inkas. The *Wak'a* signaled the interconnected relationality of the *ceques*, a whole ecology of telluric selves. Several scholars who have discussed the *Wak'a* have treated them merely as remnants from the past, devoid of life, and references to something that existed centuries ago—in short, as archeological "ruins" (Bovisio 2011, 53–84; Bray 2015).

Archaeologists continue to climb and reach Andean mountaintops, often to "discover," or rather, uncover the *Wak'a* and ritual offerings that involved human beings in ancient times. Today, it is scientists collecting DNA who are interested in retrieving the remains of human sacrifices. However, not being direct beneficiaries of such taphonomic contents or remains—interrupting affairs that disturb the *Ch'ullpa*s or ancient burials—the *Yatiri*, Andean specialists of the sacred, think that removing such ancient buried remnants undermines the telluric energy of the mountains, their *Illa* and *K"ama*. Note a coincidence: the tragic receding of Andean glaciers due to global warming has been observed by many *Yatiri* who are conscious of foreign hands rummaging and removing ancient offerings. Such actions are interpreted as causing emaciation of the *Achachila* or the *Apu* found in mountains. The *Yatiri* are not so wrong (and they never were), because their reasoning departs from the perspective of relationality. If the *Yatiri* recognize the animism of a mountain, they know that offerings enliven the *Illa* making them be; the *Illa* is a symbol of their and our lives, its *K"ama* or energy.

Seasonally, Andean peoples adopt the task of caring for the *Wak'a* or the *Apacheta* (Figure 7.1). They become attached to one of them (and vice versa), presumably reproducing a historical correspondence of place and space (what we might call "sp[l]ace"), and human, other than human, and nonhuman community (Castro and Aldunate 2003, 73–79).

To keep a *Wak'a* alive,[4] Andean devotees travel long distances to salute and feed them following the commendations of spiritual leaders. For example, Mount Illimani, in today's city of La Paz, Bolivia, is considered to be the *Jach'a Wak'a, Jach'a Apu, Inka Achachila* (major *Wak'a*, major Inka *Achachila*) amid the ones found in the area who have proper names: *Mururata*, *Illampu*, *Wayna Potosí*, *Janq'uma*, *Sajama*, *Chörölque*, *Quimsa Cruz*, *Chakalt"aya*, *Ch'uruq"ella*, *Siqa-siqa*, and the like. Healers, or specialists of the sacred,[5] called *Chamaq"anis* are in charge of rituals offered to such entities, since they have the power to talk and to listen to them in the dead of night—hence the term *Chamaq"ani*, is translated as "those who sense animated energy *K"ama* in darkness." All these ritual activities resonate with the desire to heal the sustained damage inflicted by enduring anthropogeny. In this sense, the healers' activities, in their many specialized forms, end up affirming the earth's politics of mutual bioregeneration.[6] The healing of the eARTh reflects the spiritual awakening of the human but no longer as the center of a "dominated" planet.

Figure 7.1. A *Wak'a* of the Pachapurichej on the Oruro–La Paz route. Sojourners, herders, and modern travelers stop to offer libations (Photo: Guillermo Delgado-P.).

This awakening can be detected amid Andean peoples, who had seemed to be destined to disappear in the *mare magnum* of globalization. Areas of uninterrupted habitation that anthropologist Stefano Varese calls bioregions or ethnoregions, such as the Andean and Mesoamerican cradles of civilization, are showing a remarkable resilience. The archaeologist Brian M. Fagan observed that in the Andes,

> The Altiplano was [and is] a cosmic landscape, the home of Pachamama, Mother Earth. Local farmers have always believed the earth gives them crops. In return, they give elaborate offerings to her. For all this harshness, two ancient cities flourished here in antiquity, Pukara inland from the northern edge of the lake, and later Tiwanaku to the south. Lake Titicaca [Titiqaqa] deep waters trap and release warmth, which reduces frost damage on agricultural land close to the lake. (Fagan 1995, 213)

Today, the Aymara people of the *Ayllu* Qaata in the Bolivian Andes studied by anthropologist Joseph W. Bastien suggest that in their perspective, their *Ayllu* assumes a shape of a human body, and they think of themselves as constitutive part of such body-territory. Recently, Bastien's research documents the cultural resilience of the Uru-Chipayans, described as "a small group of people who continue to practice rituals and tell stories that are concerned less with Protestantism and Catholic beliefs and more with Lake Coipasa, flamingos, irrigation canals, towers, earth shrines, graves, and ancestors" (Bastien 2012, 195). They offer us clear signs of the cosmicentric nomos, cultural resilience, and adaptation of this ancient Andean population, which during the 1980 and 1990s, faced alienating forces and displayed disturbing anomic trends (analyzed in Wachtel 1992). However, they have revitalized their culture and are alive.

Kawsaq, Wañuq: Life and Death, an Ontologized Nature

As I stated earlier, my purpose here is to engage in recent discussions about the reontologization of socionature, specifically among Andean people. Socionature can be considered a tenacious perspective and critique of Western modernity that contests and destabilizes the centrality of anthropocentric assumptions, which constitute such a detrimental episteme that, in thought and practice, has challenged Andean hylozoism.

The *Wak'a*[7] are tutelary stone shrines spread throughout the Andes, located at prominent rocks or boulders on specific mountain sites. People establish reciprocal exchanges of onomastic recognition with such rocks or boulders. As we saw, each of the *Wak'a* bears a proper name and serves as a reminder of a nature that is alive rather than inert. Periodically, the Quechua and Aymara pay offerings as a way of keeping the *K"ama* or energy of such shrines alive. This illustrates that the spirit world is "both beyond the phenomenal world as well as integral with materiality," as the introduction to this volume affirms.

When I revisited a *Wak'a* in the Bolivian Andes in August 2015 and 2018, two issues caught my attention. One was a landscape infested with discarded plastic bottles and bags tossed out by careless bus passengers on the La Paz–Oruro–Cochabamba–Santa Cruz route. The other one was the announcement that the high plateau saline Lake Poopó[8] (90 km by 32 km, with an airway surface liquid of 12,500 feet) had dried up as of December 2015 (Casey 2016). It is now dying. High altitude environmental impacts (Baer and Singer 2014, chap. 5) are dangerously unfolding here and elsewhere to such an extreme we are now talking about thermal expansion and "climate refugees" (Orlove 2016). As recently as August of 2018 as I traveled through the area, I could feel how much the temperatures had decreased, which local inhabitants have been experiencing so excessively that their rates of morbidity and mortality are rising due to extreme weather conditions. Nevertheless, underground mining continues relentlessly, evidence of the last neoliberal arrangements to exploit slowly exhausted ore veins. Whereas rivers in several parts of the intermountain valley of Cochabamba are chock-full of plastic refuse or contaminated by mining sludge.

Landscape ethnography, part of the anthropology of mining, is about knowing all the inhabitants of such landscape, which entails humans, nonhumans, and other-than-humans, such as *Wak'a*, *Apus*, *Achachilas*, and so on. Andean landscape ethnography is a recent posthumanistic proposal that represents a critique of the earlier Andean ethnology that was so formidably anthropocentric. In fact, Andean ethnographers have always paid attention to the environment, biomes, and ground-level forms of agriculture, so much so that the late John V. Murra once told me that "Andeans have learned to domesticate the cold (*aprendieron a domesticar el frío*)." Likewise, Joseph Bastien thinks that our understanding of Andean occupancy must not underestimate the majestic mountain system of the Andes and the exuberant Amazon as influential forms of ontologized nature turned ethical (from the term "ethos"); the originary inhabitants who continue to sustain persistent cosmicentric views claim them as practice and episteme, that is, as an experience of full relationality: human, nonhuman, other-than-human, as reconfirmed by Denise Y. Arnold recently (Arnold 2004, 145–79). It is not in the scope of this article to revisit the encyclopedic *pensum* of Andean ethnology in its chorographic density; instead, I want to focus on the hylozoistic element emphasized in the ethnography of landscape as "an ecology of memory," which is also a matter of contemporary archeological research. In a sense, ontology is foregrounded but not from a human angle. Previous modern perspectives that have considered nature as inert or unresponsive, that is, deontologized and essential, are being challenged. Instead, reontologizing pertains to the recognition of entification and relationality.[9] In 2009, the Bolivian and Ecuadorean constitutions acknowledged the rights of Mother Earth (*Pachamama*), possibly the first legal documents to enshrine such recognition as a subject of rights. This nominal acknowledgment still needs to be translated into proposals and policies centered on sustainable practices; as yet, they remain declarative, in need of active upholding and implementation.

Mining is an old footprint of the Anthropocene: minescapes are usually carved onto Andean mountains and *Wak'a* often protrude on such mountainscapes. Quechua and Aymara underground miners consciously think about their intrusive and extractivist tasks. Released from the sweating walls of dark and deep galleries, a latent lugubrious concern dwells in their minds, wrapped up in the dense underground odoriferous mist they inhale. Almost instinctively, they become worried and apprehensive, revealing guilt and feeling mournful because they have disturbed nature; they can often foretell an end or outcome of tragic consequences. And so, when geological faults slide or ceilings collapse, miners yell out, "*Aysa!*" ("Landslide!"), and immediately warning calls follow, alerting everyone that nature as *Pachamama* is now shaking, angry.[10] The chaotic aftermath convokes the whole mining community to restore equilibrium, restoring health by offering special *T'inka* (reciprocal exchanges) to the mine, to *Pachamama*, and to the *Achachila* and *Apacheta* mountain gods. This is because the miners know that they have hurt *Pachamama*, Mother Earth. And so when a tragic *Aysa* occurs, the lethal presence of death invades the mining camps, prompting a call to enact rituals of restoration and regeneration,[11] reaffirming an interwoven togetherness, an entangled life of mine, animal, and human, since miners do not perceive an ontological separation among them. This type of tragedy happens, however, when long-held moral economies and ritual codes have been broken or disrupted, having to do with a sort of fault that destabilizes and "sickens" the relations between the spirit world, humans (*Runa*), nature (*Akapacha, Jallpa*), and other-than-human spirits (*Ajayu*). These relationships are reconstituted through the offering rituals.

Tragic death, in particular, can be devastating for underground miners because they think that the dead are trapped inside *Pachamama*'s womb, wandering around, seeking a place to rest in peace. When tragic death is a matter of a fatal accident, it is even more complicated because dynamite, which is normally used to blast rocks after daily drilling, can, if it unexpectedly goes off, tear a human body to pieces. This aspect of accidental death is hard to bear, as miners believe that a miner's *Ajayu* (soul) wanders around in the dark galleries, looking for human limbs to recompose a dismembered body.

As a vital concept, the Aymara concept of the *Ajayu* has acquired geomythical meanings in Andean socionature and language for millennia. In Andean beliefs, the dead are never dead; rather, they are said to "continue living on in an eternal place" (*markaparuw sarxi*). Their loved ones call upon the dead, who come back to visit. As the dead drift around, the forlorn survivors await them on their ritual days to feed them again. In this celebration of life-death and death-life, family and community members indulge and treat the deceased with offerings, serving their favorite foods and drinks. The dead are free from the impacts of time, inhabiting a sort of no-time time. They are truly free and they come back, *re-membered*, to be with the living, both human and nature.

The geological warnings that mountains radiate before they shake up constitute a sign that healers[12] must feed the *Achachilas*, *Apus*, and *Apachetas* of the Andes, offering them "mesas," a sort of ritual banquet served to appease the mountain spirits. These include *Wilanchas*, sacrificing llamas and smearing their *Wila* (blood) on mountains and humans; *Ch'allas*, drinks sprinkled around; *Kharak'us*, feeding the earth and the community with *Sullu* offerings of unborn llama offspring that are preserved and offered for regeneration and life-giving powers—a clear symbol of "holotropic" fertility. Only then do the *Aysas* settle down and relax; the old *Chullp'as* and *Awicha*, spirits of ancient mountains,[13] are comforted and fed; and *Supay,* the chthonic god, becomes happy (Fernández Juárez 1995). *Supay* (or *Tío*) expects a *Kharak'u*, a banquet, before he is able to reveal the riches again.

Here, an element of Andean cosmicentric ideology is retrieved: *Wak'a* are reminders of how nature is alive rather than inanimate. Industrial mining activities menace, disrupt, and interfere with nature; its extractive and destructive work provide us with more than tangible meaning of the Anthropocene. Humans must recognize that we are not the main actors in the world and acknowledge the agential realism of matter. It is crucial to restore a sense of reciprocity with this live entity, *Pacha*, since, as humans, we belong to a set of relations that includes nature.[14] We are not superior, and nature is not there to serve us.

Uk"u Pachapi: The World of the Underground

Mining work requires squads of workers to enter earth's entrails on daily schedules of three shifts each, also called *Mit'a* or the *Veinticuatreo* (the twenty-four-hour cycle of a day). They form paired units (*-uq-*) that usually work in cadres or *cuadrillas* of ten miners. Before entering to the depths of *Uk"u Pacha*—the dark World of Below—each and every miner sprinkles libations and offers cheologic rites. As they descend into the womb of *Pachamama*, all miners stop and share the reciprocal act of *Akhulli*, chewing coca leaves (*Erythroxylum coca*) and smoking tobacco (*Nicotiana sp.*) together at an underground niche where a shrine holds a clay figure called *Supay* or *Tío*, a priapic and telluric force, expecting daily offerings that join the numen and the human into a single unit (*-uq-*). Miners identify the *Supay* or *Tío* as the real owner of the ores they extract from the underground, so before they go to toil in their daily *Mit'a*, they come together in paired units (*-uq-*) and ask *Supay* or *Tío* for protection. The same character is known in Andean Peru by other names (*Tata Muqui, Muyki*, or *Chinchiliqu*) and is often represented by a large block of stone that may be given human proper names (*Wank'a*). They are carved with capricious shapes, forming a mix of petrous human and animal sculptures (Figure 7.2).[15]

Each of these sculptures contains the stories of what is rightly called an "ecology of memory," in this case rocks, boulders, tunnels, passages, and chutes that miners know how to read, retelling stories of interconnection between humans, nature, and detritus (Delgado-P. 1996). Fernando Santos-Granero reminds us about this kind of Indigenous practice: "*Topographic writing* constitutes a kind of protowriting system common to other Amerindian societies . . . an important means of preserving historic memory and consciousness in nonliterate societies" (Santos-Granero 2005, 175; emphasis mine).

If we scratch through the layers of mythical time, *Supay* is related to an ancient Andean god *Huari*, whose other name is *Qöqena*, a fertility god sculpted in the shape of Andean camelids who, confronted by the arrival of a new god in the sixteenth century, chose to hide the riches in the *Uk"u Pacha*, the World of Below, to survive and rule. *Supay or Tío* is therefore a guardian of the riches, a clear symbol of fertility and reproduction, a phallic deity. The underground god mediates a reciprocal exchange of damage for restoration, by granting riches and being fed.

Figure 7.2. *Tío* or *Supay*, placed 380 meters down in the Moroqoqala tin mine in Bolivia (Photo: Guillermo Delgado-P.).

Miners who enter the underworld before performing their daily extractive work stop at the priapic shrines to feed these figures with coca leaves, make them smoke tobacco, pepper them with sweets and coca, and sprinkle alcohol libations, bestowing this numen with *Q'aytu* or *Wira*—that is, vigor or energy, *K'allpa*. Often, alcohol is considered to be *Tío* or *Supay's* sperm that fertilizes the mine. As Haraway explains, "In the Anthropocene, the chthonic ones are active, too; all the action is not human, to say the least. And written into the rocks and the chemistry of the seas, the surging powers are dreadful. Double death is in love with haunted voids" (Haraway 2016, 295).

In the many versions of his multiple characterizations, *Supay* or *Tío* is depicted with an erect penis because he is ready to engage the *Awicha*, the Old Mountain, who is seen as a female symbol that gives birth to the riches of the mine, which are delivered to the miners. An open, craving mouth represents *Tío* or *Supay*'s hunger and thirst, who smokes a cigar and expects to be fed as if were a child, an ancient deity in need of care. Humans take care of this numen, and the numen reciprocates. Although ignored by most Western technicians and geologists, *Supay* or *Tío* is placed on specific geological axes that sustain the mountain's tectonic layers or stratigraphy—a classic example of traditional Andean geoecological knowledge that allows miners to drill the rock and block cave the mines in search of ore deposits that *Supay* and *Awicha* will reveal. Ore deposits are called *sirqa*, meaning "vein," and miners follow such veins to extract the rich ores. The mountain has its veins, and the riches circulate through its body, just as blood circulates in the human body (Figure 7.3). But *Supay or Awicha* can be tricky and often friskily hide the *sirqa*, reminding the miners to reciprocate by feeding the chthonic deity before *Supay* will reveal the *sirqa* to them.

Supay or *Tío* is sculpted of clay and mineralized mud by Indigenous artists. As soon as it is completed, it is secretly placed at a spot notable for its strong geological axes. Usually the sculptor remains unknown to the rest of the mining squad. The artistic expression is numinous and mysterious, an anonymous sample of creativity rather than of alienation. Soon a *Yatiri*, a healer who understands the meaning of these underground deities, is invited to perform appropriate rituals, sprinkling libations and feeding the telluric shrines, establishing an emotional fusion of mountain, llama, and human being, which has been described as relationality, socionature, or the pluriverse.

When an underground miner suffers a tragic and violent death, it is said that *Supay* or *Tío* has taken away the deceased's *Ajayu*, or soul, to ensure that others do not face the same deadly risk, at least not on a daily basis. Often, failing to feed[16] the shrine results in the demise of *Pachamama* herself, since her own *Ajayu* faces the menaces of ecocidal activities that are further endangered by *Supay* or *Tío*'s anger. The miners feed the shrines underground by offering not only food, coca leaves, confetti, tobacco and alcohol but also ore samples, called *T'urus*, which are

Figure 7.3. A tin miner shows a *sirqa* or "vein" of ore. Moroqoqala mine, Bolivia (Photo: Guillermo Delgado-P.).

considered to be "seeds" for rich ore sources. This act regenerates both the earth (*Pachamama*) and the benevolence of *Supay* or *Tío*. Once the *T'uru*s are selected, they are deposited and exhibited at Tio's feet as if such ore samples were gifts to the god. It is said that *Tío* or *Supay* reproduces them in the veins of the mine "like seeds" to be spread around the *Kay Pacha*, the world of the present. This illustrates the dynamics of environmental destruction and regeneration settled in the consciousness of underground miners. The *T'uru*s or *Tinka*s symbolize the mutual exchange or reciprocity between humans, animals, nature, and nonhumans to assure the reproduction of ore deposits and humans. One cannot take out without putting in. Daily libations ensure the regeneration of the *Pachamama*, and miners see themselves as being part of this numen rather than separate from it.

Pachaq Ajayun: The Spirit of the Nature-World

In what follows, I wish to offer one explanation for why the notion of full proletarianization of Andean miners, that is, the synchronized industrial timing and culture of Western characteristics, is not a fait accompli. Proletarianization, in the words of the Bolivian sociologist René Zavaleta Mercado (1986, 177), suggests the full substitution of one worldview by another. I contend, however, that an element of hylozoistic conventions has not disappeared completely from the cultural beliefs of these Andean underground miners. A large percentage of this rank and file originates in the intermontane Andean valleys, where the notion of land and territory as a living entity constitutes ancient heritage that capitalism has not been able to destroy by fully deontologizing it. Likewise, nature-time is slow and subsists in defiance of the synchronic, homogenizing, and artificial time of clocks we call "industrial time." Rituals of regeneration subvert disciplinarian industrial time and speed; it is evident that these two concepts of time collide. Here, we must retrieve an important observation that pertains to nature-time in the relation of the words *Ayllu,* and "seed." In the Aymara language, *Jilaña* means "to grow," and its derivatives are: *Jilakata*, (authority) and *Jilata* (brother). *Hatha*, on the other hand, is the Aymara word for "seed" (*Muju* in Quechua), the origin of life. The *Ayllu* (the community at large, including nature, *Akapacha*, *Sallqa*) expresses this mutual relationality; the *Ayllu* grows from a seed (called *Ispalla or Malki*), which, when treated with care, becomes a plant[17] with an *Illa* (spirit, energy). The word *Malki* also means corpse and fetus. When a process of Western deontologization entered the Andes, what we have is a history of systematic disturbance rather than the previous sporadic form of disturbing the land. Since the era of colonial mining, Andean mountains have been worked and reworked by uprooted human communities. Disturbance shows layered stratigraphy, enough to lead to the archaeology of mining as part of that disturbance, intensified with colonial extractive mining and worsened by contemporary transnational industrial fracking. In the process, nature, humans, nonhumans and other-than-humans have embraced each other through death and dying and have been buried together ever since, rendering tangible and unfinished, the meaning of the terms "primitive accumulation" and "accumulation by dispossession" (Harvey 2003). Rituals are enacted to heal human, animal, nature, and nonhuman (souls such as the *Ajayu*) together, since they do not exist separately from each other.

By adding their own contributions to the process of disturbance, other elements participate in intensifying this anthropogeny. Think of palynologists, who study pollen, both living and fossilized, telling us about mercury remnants in today's dust of the *Sumaq Urqü*, the Potosi mines so well known to the world market of three centuries ago, as documented by Nicholas Robins (2011). A recent report written by essayist William Finnegan (2015, 56) about a new gold mining fever at La Rinconada, Peru, states that "Mercury poisoning can affect the central nervous system, causing tremors, excitability, insomnia, and a grim range of psychotic reactions. Crime and violence in La Rinconada [Mine] are often attributed, on no medical basis, to mercury poisoning."

Found in the vocabulary of underground mining taking place today, the Quechua term *Qüpaqira* ("liquid trash," from the Quechua root term *Qüpa*, "trash," Hispanicized as *copagira)* designates the contaminated water runoff that is leached through drilling that transforms it into chemical substances, releasing acidic water through long-term, continuous filtering. If this caustic liquid gets on the clothing and rubber boots worn by miners, they are eventually "eaten up," ripped apart, harming workers and the environment alike. This acidic water circulates in small brooks and ends up dyeing the underground with colorful edges: deep yellow, greenish, reddish, black, an obvious association with ores such as copper, antimony, zinc, tin, silver, iron, and so on. As anthropologist Anna L. Tsing (2015, 133) puts it, "capitalist commodities are removed from their life-worlds to serve as counters in the making of further investments." In the Andean case, disruption and disturbance offer us a toxic colonial heritage to breathe, since mercury is still found in the dust of Potosi today. Tsing points out that

> Disturbance brings us into heterogeneity, a key lens for landscapes. Disturbance creates patches, each shaped by diverse conjunctures. Conjunctures may be initiated by nonliving disturbance (e.g., floods and fires) or by living creatures' disturbances. As organisms make intergenerational living spaces, they redesign the environment. (Tsing 2015, 161)

Being that mining as extractive economy has redesigned such environment—foundational in the Andean horizons since pre-Columbian times—it is crucial to recall that Inkan mining restrained its metallurgic practices to the minimum, centered largely on gold and silver mining to make jewelry or tools rather than weapons. Even the pre-Columbian *Wayra*, which in Quechua means "wind," the Andean small-scale artisanal smelting oven invented for metallurgic activity, was mostly used only sporadically and occasionally. The Inkan *Mit'a* system of corvée labor was taken up and adapted by the Spanish to satisfy their labor demands at Potosí and Huancavelica, where the main silver, gold, and mercury mines were located. Suddenly smelting, using pre-Columbian paleotechnology *Wayra* ovens, expanded during colonial times, intensifying a process of disturbance. Gold (Au) was ransacked early on, but silver (Ag) became dominant for at least three centuries, until its collapse in the international market by the end of the nineteenth century.

The world system next assigned Bolivia the task of focus on newly discovered tin (Sn2) deposits in the 1890s, the only sources found in the Americas. Tin became Bolivia's "wage," its main export, until the bust-and-boom cycle came to an end with the collapse of the market and the International Tin Council in 1986. Andean minerals, it can easily be said, contributed to the making of Europe, the classic example of the "accumulation by dispossession" process that David Harvey (2003) talks about. When the international tin market collapsed due to the emergence of cheaper substitutes (freezing, plastics, aluminum, chemical and nuclear arms, and, nowadays, drones), miners returned en masse to the fields and were said to have been "relocated." By this time, the world's economic appetite was, graphically speaking, addicted to a new subproduct, which originated in the harvesting and processing of coca leaves (*Erythroxylum sp.*) into cocaine, a substance that provoked its own world economic disturbance and toxicity (see also the introduction to this volume).

Surprisingly, that same year, the "relocated" former "industrial" miners energized a process of re-peasantization in Bolivia, challenging the evolutionary view of societal change that Marx had portrayed as "progress," passing from a nomadic to an agriculturalist to an industrial-urban stage (Delgado-P. 2018). Undoing Marx's prediction, the miners returned to the land as peasants, answering to yet another demand in the world system by harvesting coca leaves, the raw material that is processed into a white dust that is eagerly consumed in affluent societies and often serves to add speed to the system. But it was uncontrollable speed in the form of *flash trading* that triggered the market crash of 2008. Some say that the consumption of cocaine was behind it. Charlie Chaplin had already portrayed this issue in his silent film *Modern Times* of 1936, the very first cinematic reference to cocaine consumption and criminalization in popular culture, which was inspired by another crash, the 1929 collapse of Wall Street.[18]

But once again, a pre-Columbian story about coca leaves, called *Mama Qüqa* by Andean Indigenous peasants (dangerously misidentified as a "narcotic" or *aphrodisiacum* by Sigmund Freud in his *Über Coca* [1885]), represents it as being nurturant and healing to Andean Indigenouss but a scourge to those who abuse it. *Qüqa* has been central to Andean spiritual and ritual life since ancient times. The story says that *Qüqa* was an offering of such leaves, a *Q'intu*, made by *Mama Qüqa* herself to a lost herder who was dying of thirst and hunger. No wonder, then, that miners view *Qüqa* as a main offering to *Tío* or *Supay*. The leaves are spread out on a *Tari*, a square cloth representing the four cardinal points, to be shared with the telluric numen as well as chewed by miners themselves to quench their thirst and hunger and, like the lost and hungry herder of the past, receive *K"ama*, reviving energy, from the *Wak'a* shrines. *Qüqa* becomes central to the life-affirming rituals of Andeans and Andean ecology. The physiological explanation is that *Qüqa* chewing adds extra oxygen to the red cells, which Andean people need in larger quantities to avoid pulmonary edema or hypoxia while living at such high altitudes—after all, we are talking about human settlements found between twelve and sixteenth thousand feet above sea level. The highest mining settlement I ever visited, the camps of Mina Alaska, Caracoles and Pakuni, is located at 17,000 feet above sea level. There, tungsten and antimony miners often say that they live inside the *Jach'a Achachila Illimani* Mount.

The coca leaves, as a symbol of green exuberance, warm valleys, and living off the land, are thus transferred to mining environments as part of the "industrial" practices of miners, who often talk about ores as having the qualities of tubers,

Llallawas, Iraqhas, Ch'uqi. They appreciate these ores because of their beautiful shapes and talk about the taste, color, and sweetness of the rocks; some have told me that ores are like the large pool of potato varieties harvested in the Andes. Samples of these *T'urus* or *Tinkas* as the most beautifully capricious *Ispallas* often end up at the local church, where they symbolize seeds (*Muju*) deposited to the feet of Christian saints, in the same way that *T'urus* are offered to *Supay* or *Tío*—to reproduce themselves (like seeds) with their theurgic assistance. As *Qüqa,* the *Ajayu*'s persistence parallels the Indigenous demographic resilience that today has led to the population regaining the same size that Europeans found when their first contact with Andeans occurred five centuries ago. Despite the initial demographic decimation of Indigenous peoples that followed, contact did not kill the *Ajayu*, which, as in the case of *Ispalla, Imillwawa, Phurixa Yuqalwawa, Malki*, survived and spread as rhizomes by way of feeding the shrines of Andean *Ayllus* that are like rhyzomatic plants.

Tukunanpaq: Conclusion—The "Pachacene"

In the 1970s, when James Lovelock and Lynn Margulis (1973, 1) proposed the Gaia hypothesis,[19] they argued that: "the Earth regulates itself, and responds to change, in the same sort of way that a single living organism does." However, the Earth's synergistic self-regulation does not necessarily "benefit" humans; we are just part of it, and not the most important one, although the most damaging. Contextualizing the discussion today, Indigenous peoples have always offered a critique of Western rapacity by acknowledging socionature and proposing the pluriverse for a posthumanistic ethics. As one can see, Andean (or, for that matter, Indigenous) synergistic ritual work faces an uphill battle against the Anthropocene. Their ritual and sustained spiritual activity could also be named the Pachacene, just as some Indigenous peoples in the Maya area propose the Mayacene, both pointing to the renewed cycles of spiritual regeneration. This regeneration implies the restoration of the multiple forms of intricate life: nature, humans, and other-than-humans acknowledging each other as equal, possible "alter-Natives" to the Anthropocene.

As a step in this direction, a small but significant sustainable paleotechnique is being retrieved and restored, one that pays attention to ancient systems for regenerating nature, centered in the *Yana Allpa*. This is a proposal to revive the use of the ancient Indigenous *terra preta* of the Amazon Basin, the main theme of the persistent work of Sach'amama, the project that Frédérique Apffel-Marglin directs with the collaboration of her Quechua counterparts in Peru. Their goal involves restoring this *Yana Allpa* and recreating

> a special type of charcoal in it . . . called *biochar* . . . a porous and, in human terms, it never decomposes . . . . Its porosity and its durability meant that nutrients attach to it forever and are not washed away by the torrential Amazonian rains, thus explaining the truly astounding longevity of the productivity of these soils. (Tindall, Apffel-Marglin, and Shearer 2017, 79)

Old ceramic offerings found in the *Yana Allpa* assure the intricate regenerative process of this embodied *Pacha*.[20] This rediscovered paleotechnique challenges the self-inflicted damage caused by the conscious practice of cosmic destruction that is sponsored by the current notions of progress as ecocide and development as lethal extractivism and burning, which result in entropy, crises, and excess. In fact, plastic, the main oil derivative culprit of the Anthropocene, has infested land and oceans, provoking alarming eutrophication, endangering the very life of water and the creatures in it, since the resistance of plastic lasts virtually forever. Plastic is not biodegradable but disintegrates into microscopic detritus that ends up in ocean ecologies; it has even been found in the stomachs of marine animals that are eventually killed by the hypoxia it provokes.[21] The stratum of plastic refuse in the Andes menaces llamas and alpacas (Geyer 2017), while birds and fish are endangered at Lake Titiqaqa by the same disturbance. The modern teleology behind notions of progress and development along with the risk and excess they entail, has reached a point of no return (Jenkins and Rondón 2015; Delgado-P. 2014, 125–36). Indeed, both alternate contributing in a process of conscious entropy.

Nevertheless, in dispersed nuclei of inhabitants such as the Andes or the Amazon, human, nonhuman, and other-than-human come together, offering respite through the hope that such *Illa* dwelling in *Pacha* and in her *Yana Allpa*, mountains, *Jillaratanaka*, *Machulas*, and *Ajayu* continue to radiate healing sunrays of coexistence and regeneration. Mediated by rituals of cosmic restoration, it is above all this regenerative, cosmogenetic perspective that can be called an emergent anthropology beyond the human.

Acknowledgments

For sharing materials and intellectual exchange, I thank Silvia Rivera Cusicanqui, and Elvira Espejo. I am also grateful to my UMSS colleagues Walter Sánchez, Eduardo Córdova, Fernando Garcés (Museo de Arqueología), and Mónica Navarro, her students at PROIEB (pers. comm., Cochabamba, August 2015) as well as Alvaro and Dacio Ríos Delgado who guided me at Qollqapirwa. For suggestions and feedback on this article, I am indebted to Marcia Stephenson, Norma Klahn, Frédérique Apffel-Marglin, John Brown Childs, Stefano Varese, and graduate student Rolando Vargas (Media and Film Department), all of whom offered insightful dialogue and critique. Silvia Rivera Cusicanqui, Frédérique

Apffel-Marglin, Karen Barad, Donna Haraway, and Anna Tsing have always been inspiring. I remain responsible for the contents.

NOTES

1. Frédérique Apffel-Marglin (2011, 30) offers a seminal contribution to this discussion when she states, "The category of hylozoism gathers a motley group of Renaissance movements, occult philosophies, and peasant practices. What they all have in common is a nondualist view that matter could move itself, that humans were part of nature and part of the cosmos, and that God pervaded the material world as well as the souls of humans." Eduardo Kohn (2013, esp. chap. 2) dwells extensively on this issue.
2. On the pre-Columbian record, the archaeologist John R. Topic (2015) offers a reliable summary of this issue. Despite centuries, several features that Topic mentions can be detected in the contemporary *Wak'a.*
3. Claudette Kemper Columbus (2012) studies the diachronics of this key term of Andean cosmicentricity and hylozoism when she engages the work of Quechua anthropologist and novelist José María Arguedas and his philological analysis of the onomatopoeic Quechua term *Yllu* and *Ylla.* Marcia Stephenson (2010, 29–39) discusses and complements the deep and complex symbolism of *Illa* and *Enqa.*
4. Bruce Mannheim and Guillermo Salas Carreño (2015, 47–72) clarify the linguistic, historical, and ethnographic record on the concept of *Wak'a* and its relationality to the term "entification." See also the work of Ricardo Valderrama and Carmen Escalante (1988).
5. On the issue of "knowledge accumulated for centuries by ethnic communities" and "economies of reciprocity," see Zerda-Sarmiento and Forero-Pineda (2002, 99–114). See also Luis Morató Lara (2005, 193–213), a poet who dedicates a language of telluric healing in his Quechua hymn "*Kallawaya simi mancharisqata jamp'inapaq.*" The German anthropologist Ina Rösing (1995) compiled Aymara healing ceremonials and sacred recitations. See also Apaza (1999, 118–53).
6. The *Amawt'a* Policarpo Flores Apaza, along with Fernando Montes, Elizabeth Andia, and Fernando Huanacuni, offers a detailed dialogic autoethnography on Aymara thought, hylozoism, and perspectivism (Apaza 1999).
7. The term has been written in three ways: *Wak'a*, *Huaca*, or *Guaca.* The first corresponds to the Aymara and Quechua pronunciation, which I use in this chapter. The latter two are Hispanicized versions.
8. Located at 18° 33'S 67° 05'W and south of the city of Oruro, Bolivia, its primary inflow is the Río Desaguadero, approximately three meters deep, with an extension of 1,000 sq. km. Not far from this lake, the Swiss-based multinational Glencore and the Bolivian company Sinchi Wayra operate mines that release heavy silver, zinc, and lead *copagira* (contaminated acidic leach) into the lake. As of May, 2019, Bolivian anthropologists Angela Lara D. and René Orellana M., confirmed the lake's partial resilience due to unprecedented rainfall.
9. I would like to acknowledge dialogues with Javier Sanjinés and Bruce Mannheim on issues pertaining the reontologization of nature (pers. comm., Ann Arbor, November 2014).
10. According to Javier Medina (2001, 138), *Pacha* is a central concept that summarizes Aymara and Quechua archetypical depth and complexity. The word derives from the phoneme *–pa* (which translates as "bi," "two," "duality," "pair") and *-cha* (meaning "strength," "energy in movement").

Recently, the journalist Héctor Tobar (2014) wrote about the thirty-three copper miners who were trapped in the San José mine in the Atacama desert, near Coquimbo, Chile, in 2010. Tobar makes a clear reference to this aspect of nature as alive: "The mine is 'weeping' a lot, the men say to each other. '*La mina está llorando mucho*' .... This thundering wail is not unusual, but its frequency is" (2014, 15).

11. Apffel-Marglin (2011, 191) notes that "*Regeneration* alludes to a non-linear and more cyclical process in which elements circulate, generating, degenerating, and regenerating with the possibility not only of renewal, but also of loss and creation."
12. The terms "healer," "sorcerer," or "shaman" homogenize the multiple curing specialties these men and women of the Andes perform. They are so specialized in the specific rituals they may convoke that, by homogenizing them, their particular forms of knowledge of the sacred are lost or distorted. The researcher Claudia Brosseder identified a list of about forty-four "Andean religious specialists." She affirms that, "The commemoration of the power of *huacas* lost something of their reach to organize a larger community; however they survived intact in small communities and among individuals" (2014, 265). Although her list is not exhaustive, a closer look at each would constitute a formidable study of healing-related bioknowledges. Contemporary ethnography must retrieve and respect such ritual specialties by collecting their proper names or titles. By identifying the lexemic roots of the titles *Yatiri, Ch'amaq"ani, Qulliris, Qapaqchaqëra, Layqas, Paqüs, Jamp'iris,* and *Ch'aqtiris,* their intimate meanings are revealed: for instance, *Yatiri* is associated with the term *Yachaq*, "to know"; *Chamaq"ani* means "to see in darkness"; in *Qapaqchaqëra*, the term *Q'apaq* (to not be confused with *Qapaq*) is associated with the act of burning or producing a purifying, cleansing smoke; *Jamp'i* means "to cure"; and so on.
13. The journalist Héctor Tobar mentions a current example in Chile: "One legend has it that the mountain itself is a woman, and in a sense 'you are violating her every time you step inside her,' which explains why the mountain often tries to kill the men who have carved passageways from her stone body" (2014, 12). Since Chile is a conservative Catholic country, rituals of restoration among miners are now absent, but as we see, hylozoistic elements persist in the minds of Chilean miners.
14. A provocative and complementary reflection on this issue is offered by Javier Sanjinés (2009, 99–153).
15. In some northern Argentine Andean communities, stone heaps or *mojones* similar to *Apacheta*s are considered the place where *diablos* reside. During Carnival time they are awaken and at the end of it, they are "returned" to the *mojones* "to rest" (Costa and Karasik 1996). Carmen Salazar Soler studies this issue in her book (2006), which is a summary of her twenty-year research on Julcani, a silver mine in an area of Peru where lead, gold, tungsten, cadmium, and zinc are also mined.
16. The anthropologist Gerardo Fernández Juárez (1993, 85–115) has written on the symbolism relevant to the *mesa*, the act of feeding the Andean gods and on the meaning it has for the Aymara. The term *mesa* is often mispronounced "misa" by Aymara speakers both entail the holy, *mesa* (an offering on a table) and *misa* (a Christian Mass).
17. The Argentinean ethnographer Rodolfo Kusch (1998, 295–366) noticed this association when he conducted fieldwork in Oruro, located on the Bolivian Altiplano, from 1967 to the 1970s. The root words *Jila* and *Ylla* are also found in the term *Jilaratanaka*, "sacred mountains"; *Jilawarani*, "staff bearer" or "authority"; and *Jilaqallu*, the head representative of a Parciality, a moiety within the Ayllu. See also Ayllu Sartañäni (1992, 198).
18. See the editor's introduction concerning this issue. See also Gold (2018).

19. Lovelock however, rethought the issue and offered to "disenchant" his earlier proposal (see Lovelock [1988] 1995).
20. The notion of biocultural regeneration is extensively discussed in Apffel-Marglin (2011, 167–204) and thoroughly explained in Tindall, Apffel-Marglin, and Shearer (2017, chaps. 2–5).
21. See also Schlossberg (2017): "Scientists estimate that five million to 13 million metric tons of plastic enter the ocean each year . . . New data suggests contamination in rivers and streams, as well as on land, is increasingly common, with most of the pollution in the form of microscopic pieces of synthetic fibers, largely from clothing."

References

Apffel-Marglin, Frédérique. 2011. *Subversive Spiritualities: How Rituals Enact the World.* Oxford: Oxford University Press.

Arnold, Denise Y. 2004. "Midwife Singers: Llama-Human Obstetrics in Some Songs to the Animals by Andean Women." In *Quechua Verbal Artistry: The Inscription of Andean Voices*, edited by Delgado-P. Guillermo and John M. Schechter, 145–79. Bonn, DEU: Bonner Amerikanistische Studien.

Ayllu Sartañäni. 1992. *Pachamamax Tipusiwa (La Pachamama se enoja)*, Qhurqhi, vol. 1. La Paz, BOL: Ediciones Aruwiyiri.

Baer, Hans A., and Merrill Singer. 2014. *The Anthropology of Climate Change: An Integrated Critical Perspective.* London: Routledge.

Barad, Karen. 2011. "Nature's Queer Performativity." *Qui Parle: Critical Humanities and Social Sciences* 19 (2): 121–58. https://doi.org/10.5250/quiparle.19.2.0121.

Barnes, Jessica, Myanna Lahsen, Andrew Mathews, Pamela McElwee, Roderick McIntosh, Frances Moore, Jessica O'Reilly, et al. 2013. "Contributions of Anthropology to the Study of Climate Change." *Nature Climate Change* 3: 541–44. https://doi.org/10.1038/nclimate1775.

Bastien, Joseph. 2012. *People of the Water: Change and Continuity among the Uru-Chipayans of Bolivia.* Salt Lake City: University of Utah Press.

Bovisio, Maria Alba. 2011. "Las huacas andinas: Lo sobrenatural viviente." In *La imagen sagrada y sacralizada*, vol. 1, edited by Peter Krieger, 53–85. Mexico City: UNAM, Instituto de Investigaciones Estéticas.

Bray, Tamara (ed.). 2015. *The Archaeology of Wak'as. Exploration of the Sacred in Pre-Colombian Andes.* Boulder: University Press of Colorado.

Brosseder, Claudia. 2014. *The Power of Huacas: Change and Resistance in the Andean World of Colonial Peru.* Austin: University of Texas Press.

Casey, Nicholas. 2016. "Climate Change Claims a Lake and a Centuries-Old Identity." *New York Times*, July 8, 1, 6–7.

Castro, Victoria, and Carlos Aldunate. 2003. "Sacred Mountains in the Highlands of the South-Central Andes." *Mountain Research and Development* 23 (1): 73–9.

Chaparro, Amanda. 2018. "D'étranges cooperatives de mineurs." *Le Monde diplomatique* 773 (August): 7.

Costa, Mercedes, and Gabriela A. Karasik. 1996. "¿Supay o Diablo? El carnaval en la Quebrada de Humahuaca (Provincia de Jujuy, Argentina)." In *Estudios sobre el sincretismo en América Central y en los Andes*, edited by Bernd Schmelz and N. Ross Crumrine, 275–304. Bonn, DEU: Bonner Amerikanistisches Studien/Holos.

Davidson, Kief, and Richard Ladkani, dirs. 2005. The Devil's Miners. Urban Landscapes Productions. New York: First Run Features. DVD, 82 mins.
Delgado-P., Guillermo. 1996. "K'untur, Jamp'atu y la Boca de Katari." *Lienzo, Revista de la Universidad de Lima* 17: 57–102.
Delgado-P., Guillermo. 2014. "Land, Territory, Entropy." In *Grabbing Back: Essays against the Land Grab*, edited by Alexander R. Ross, 125–36. Oakland, CA: AK Press.
Delgado-P., Guillermo. 2018. "The Radical Anti-imperialist Consciousness of Bolivian Tin Miners in the Early 20th Century." https://www.viewpointmag.com/2018/02/01/radical-anti-imperialist-consciousness-bolivian-tin-miners-early-20th-century/.
Fagan, Brian M. 1995. "Footplows and Raised Fields." In *Time Detectives: How Archaeologists Use Technology to Recapture the Past*, 210–26. New York: Simon & Schuster.
Fernández Juárez, Gerardo. 1993. "*Sullu, mesa* y lógica social aymara." *Revista de Dialectología y Tradiciones Populares* 48 (2): 85–115.
Fernández Juárez, Gerardo. 1995. "Ofrenda, ritual y terapia: Las mesas aymaras." *Revista Española de Antropología Americana* 25: 153–80.
Finnegan, William. 2015. "Tears of the Sun." *New Yorker*, April 20, 55–65.
Flores Apaza, Policarpo, Fernando Montes, Elizabeth Andia, and Fernando Huanacuni. 1999. *El hombre que volvió a nacer: Vida, saberes y reflexiones de un amawt'a de Tiwanaku*. La Paz, BOL: Plural Editores, Universidad de la Cordillera.
Geyer, Roland, Jenna R. Jambeck, and Kara Lavander Law. 2017. "Production, Use, and Fate of All Plastics Ever Made." *Science Advances* (online) 3 (7) (July 19). https://doi.org/10.1126/sciadv.1700782.
Gibson, James William. 2009. *A Reenchanted World: The Quest for a New Kinship with Nature*. New York: Metropolitan Books.
Gold, Jenny. 2018. "Overdosing on the Job." *New York Times*, September 23, B1–B6.
Haraway, Donna J. 2016. *Manifestly Haraway*. Minneapolis: University of Minnesota Press.
Harris, Olivia. 1980. "The Power of Signs: Gender, Culture and the Wild in the Bolivian Andes." In *Nature, Culture and Gender*, edited by Carol MacCormack and Marilyn Strathern, 70–94. Cambridge: Cambridge University Press.
Harvey, David. 2003. *The New Imperialism*. Oxford: Oxford University Press.
Jenkins, Katy, and Glevys Rondón. 2015. "'Eventually the Mine Will Come': Women Anti-mining Activists' Everyday Resilience in Opposing Resource Extraction in the Andes." *Gender and Development* 23 (3): 415–31. https://doi.org/10.1080/13552074.2015.1095560.
Kemper Columbus, Claudette. 2012. "Dos ejemplos del pensamiento andino no-lineal: Los zorros de Arguedas y la *Illa* andina." *Anthropologica* 15: 195–216.
Kohn, Eduardo. 2013. *How Forests Think: Toward an Anthropology Beyond the Human*. Berkeley: University of California Press.
Kusch, Rodolfo. 2003. *Obras Completas*, vol. 4. Rosario, ARG: Editorial Fundación Ross.
Lovelock, James E. (1988) 1995. *The Ages of Gaia: A Biography of Our Living Earth*. New York: W. W. Norton.
Lovelock, James E., and Lynn Margulis. (1973). "Athmospheric Homeostasis by and for the Biosphere. The Gaia Hypothesis." *Tellus* 26: 1–10.
Mannheim, Bruce, and Guillermo Salas Carreño. 2015. "*Wak'a*, Entifications of the Andean Sacred." In *The Archaeology of* Wak'a, edited by Tamara L. Bray, 47–72. Explorations of the Sacred in the Pre-Columbian Andes. Boulder: University of Colorado Press. https://doi.org/105876/9781607323181.COO3.

Medina, Javier. 2001. Suma qamaña: *La comprensión indígena de la vida*. La Paz, BOL: Federación de Asociaciones Municipales de Bolivia, GTZ.
Morató-Lara, and Luis Siwarpuma. 2005. "Quri jarawi, qulqi jarawi: Kallawaya simi mancharispata jamp'inapaq." In *Construcción y poética del imaginario boliviano*, edited by Josefa Salmón, 202–13. La Paz, BOL: Plural Editores.
North, Peter, and Dave Featherstone. 2013. "Localisation as Radical Praxis and the New Politics of Climate Change." In *The Movement of Movements*, part 1: *What Makes Us Move?*, edited by Jai Sen, 3733–854. OpenWord Series on Challenging Empires 4. New Delhi, IND: OpenWord, PM Press.
Orlove, Ben. 2002. *Lines in the Water: Nature and Culture at Lake Titicaca*. Berkeley: University of California Press.
Orlove, Ben. 2016. "Climate Refugees from the Peruvian Andes." *Glacier Hub*, January 28. http://glacierhub.org/2016/01/28/climate-refugees-from-the-peruvian-andes/.
Robins, Nicholas. 2011. *Mercury, Mining and Empire: The Human and Ecological Cost of Colonial Silver Mining in the Andes*. Bloomington: Indiana University Press.
Rösing, Ina. 1995. *Diálogos con divinidades de cerros, rayos, manantiales y lagos: Oraciones blancas kallawayas*. Translated by Rafael Puente and Ginés Pastén. La Paz, BOL: HISBOL.
Rösing, Ina. 2010. *White, Grey, and Black Kallawaya Healing Rituals*. Madrid, ESP: Iberoamericana.
Salazar-Soler, Carmen. 2006. *Supay muqui, dios del Socavón: Vida y mentalidades mineras*. Lima, PER: Fondo Editorial del Congreso del Perú.
Sanjinés C. Javier. 2009. *Rescoldos del pasado: Conflictos culturales en sociedades poscoloniales*. La Paz, BOL: PIEB, Plural Editores.
Santos-Granero, Fernando. 2005. "Writing History into the Landscape: Yanesha Notions of Space and Territorial0069ty." In *The Land Within: Indigenous Territory and the Perception of Environment*, edited by Alexandre Surrallés and Pedro García Hierro, 170–98. Copenhagen, DNK: IWGIA.
Stephenson, Marcia. 2010. "From Marvelous Antidote to the Poison of Idolatry: The Transatlantic Role of Andean Bezoar Stones during the Late Sixteenth and Early Seventeenth Centuries." *Hispanic American Historical Review* 90 (1): 3–39. https://doi.org/10.1215/00182168-2009-089.
Tindall, Robert, Frédérique Apffel-Marglin, and David Shearer. 2017. *Sacred Soil: Biochar and the Regeneration of the Earth*. Berkeley, CA: North Atlantic Books.
Tobar, Héctor. 2014. *Deep Down Dark: The Untold Stories of 33 Men Buried in a Chilean Mine, and the Miracle That Set Them Free*. New York: Picador.
Topic, John R. 2015. "Final Remarks: Catequil as One *Wak'a* among Many." In *The Archaeology of* Wak'a: *Explorations of the Sacred in the Pre-Columbian Andes*, edited by Tamara L. Bray, 369–96. Boulder: University of Colorado Press. https://doi.org/10.5876/97816073231181.CO12.
Tsing, Anna Lowenhaupt. 2015. *The Mushroom at the End of the World: On the Possibility of Life in Capitalist Ruins*. Princeton, NJ: Princeton University Press.
Wachtel, Nathan. 1992. *Dieux et vampires: Retour à Chipaya*. Paris: Éditions du Seuil.
Zavaleta Mercado, René. 1986. *Lo nacional-popular en Bolivia*. Mexico City: Siglo XXI.
Zavaleta Mercado, René. 2018. *Notas para una historia natural de Bolivia*. La Paz, BOL: CIDES AEB, Plural.
Zerda-Sarmiento, Alvaro, and Clemente Forero-Pineda. 2002. "Intellectual Property Rights over Ethnic Communities' Knowledge." *International Social Science Journal* 171: 99–114.

CHAPTER EIGHT

Mapping the Chiasmus

Liberating Patterns in a Planetary Mandala

NEELA BHATTACHARYA SAXENA

> Stare at the canvas as if it is your own blank mind and locate the center. Concentrate and you can even see the grains on the page. Your eyes operate as your will as you find the point of composition. Put a dot there with your pencil. Then measure the distance from the sides, up and down. Depending on the state of your internal environment, the dot will either be left or right leaning, pushed up or drawn down. The canvas is your own body, so draw yourself; discover whether you are comparatively emotional or are stuck in brain chatter. Using your compass, draw a circle from the dot, which is the center of your *anahata chakra*, the heart center. (Satpurananda 2013, pers. comm.)

These were some of the instructions given by Kulavadhuta Satpurananda, my guru, as he taught a class recently in a small art school in Kolkata on the art and mathematics of Tantric mandalas. Satpurananda (also known in the *sangha* community as Baba or Khyapababa, "Crazy Father") demonstrated how to make a Tara mandala through playful and yet serious instructions.

In 2013, when I first attended such a class that, in a strange synchronicity, my guru taught at my home in the United States, I had no clue what was going on. However, almost right away, I started to see the geometric shapes inside my own body. I was perplexed by their vivid luminosity at that time; however, consequent transmutations of my own inner life following Baba's initiation slowly created a profound stillness in me. This is the silence and alertness that the introduction speaks of. Like the shamans of the world, Tantric masters can unlock that receptive quietness and awaken the journey toward cosmic consciousness with ease and grace. I realized that, although I had been reading books on Tantra for decades and had a basic knowledge of yantra*s* and *mandalas*, I knew nothing about their precise methodology or application. Sitting in Baba's class once again, having spent four years of basic practice under his tutelage, this time I began to recognize the real meanings of "as above so below" and the unity of macrocosm and microcosm, a universal staple of esoteric paths.

Humans, as witnesses of the drama of life, have always wondered about their role in this cosmic play. Many scientific, philosophical, psychological, and religious studies have focused on the fundamental patterns that both reveal themselves in nature's structures and appear in the visions of seers and mystics. Some people have delved beyond ordinary observation and into the depths of their own bodies, allowing them to discover the patterns of the planetary mandala and liberate themselves from the relentless turning of the mandala wheel. Using the metaphor of chiasmus from the Greek *X* (chi) and examining the Tantric tradition of India, I will explore some of those engagements to present the World Soul or *Anima Mundi* as the matrix within us.

When we learn to map the chiasmus through embodied meditative practices, we can align ourselves with the planetary mandala, dissolving our own egocentric prison of self that alienates us from life's pulsating grandeur. I am playing with the idea of mapping the chiasmus as structuring the cosmos. For me, the chiasmus, or crossing, metaphorically represents the cosmos where opposites coexist and dance. I am looking at geometrical patterns that mark fundamental dualities as road maps that lead to the understanding of self in relationship to emptiness, or *shunyata*. In the process, from a Buddhist Tantric perspective, the not-self is realized as radical freedom. In this chapter, I explore the liberating patterns in sacred geometry and in Tantric thought that are designed to equate the polarities in our being so that we can come alive to the joy of existence.

Shapes, shapes, and more shapes inhabit our universe, and they inhabit our internal being. A simple observant walk in the woods can reveal nature's graceful dynamic forms. If we see a flower closely, a snowflake through a microscope, or an atomic structure, we can see the Star of David or the central intertwined triangular dyad of a Tantric mandala. Nautilus shells with spiraling patterns fascinate us, making us think about galaxies spiraling in and out of existence. Our eyeball appears as a sphere with a singularity of bottomless depth. A fallen tree in the forest is cut by a woodcutter, and a circle with lines pointing to its center appears that seems to be like the wheel of life, *dharmachakra,* or the spokes of the wheel that Lao-Tzu in *Tao Te Ching* (2009) speaks of. Perhaps a meditative walk around a Chinese sacred mountain made the sage Lao-Tzu recognize the profound power of emptiness inherent in the shapes:

> Thirty spokes meet in the hub. Where the wheel isn't is where it's useful.
>
> Hollowed out, clay makes a pot. Where the pot's not is where it's useful.

Cut doors and windows to make a room. Where the room isn't there's room for you.

So the profit in what is is in the use of what isn't. (Lao-Tzu 2009, 17)

A chiasmus is usually understood as a literary device, a grammatical figure,[1] a sort of "inverted parallelism" used by ancients to communicate complex thoughts. It is a rhetorical device in which the second part of a clause is syntactically inverted to balance the first, creating a figure analogous to the Greek letter *X* or chi, which indicates a crisscross arrangement of words or clauses. The word is derived from the Greek word *khiasmos*, which means "crossing" and is in turn derived from the Greek word *khiazein*, meaning "to mark with an *X*." The related word "chiasma" is a neurological term describing the part of the brain that allows our vision to work in perfect harmony between left and right hemispheres. Thus, the optic chiasma is an intersection or crossing of two tracts, the crucial union of the optic nerves. Perhaps the geometric shapes are a production of the optic chiasma in the brain as a mirror of the two hemispheres of the globe or a cosmic polarity. Finally, Chiasm is the name of a musical group started by Emileigh Rohn, a scientist who finds in her music the creative expression of herself as sound. In this and various other esoteric contexts, a chiasmus operates as a yantra or mandala.

The unitive drive of the human self recognizes the play of dualities, overcomes conflict, and understands the nondual universe that elsewhere I have called a "Gynocentric" matrix (Saxena 2004) or the Mother Principle.[2] Ancient thought contained in the *Rigveda* or other texts with yogic connotations in India is meant to induce harmony and to balance opposites. Similarly, Tantric mandalas and yantra*s*, both Hindu and Buddhist, are meditational tools containing fundamental creative dualities, male and female, marked as triangles pointing upward and downward. These are not agonistic, hierarchical, or dueling dualities; rather, they designate creative play that I have called dancing dualities (Saxena, 2012). When mediated by the third element of *shunyata*, the dualities dissolve.

When looked at from another angle, the figure of chiasmus (*X*) turns into Shiva's *damaru* (drum) or a *vajra*, a symbol of Vajrayana. It signifies an equated pulsation of the two hemispheres of the brain, the Shiva/Buddha and Shakti/Prajna in unison, bringing us to an easeful spontaneity at the point of intersection between time and eternity. Geometric thinking is the figurative and symbolic understanding of the matrix: in Tantric thought, Kali and Shiva (Prajna and Karuna in Buddhist parlance) are marked as two intersecting triangles. Yantras represent the bare lines, while mandalas represent the expanded colorful forms. Symbolically, we map the chiasmus as we create the mandala. This allows us to see the universe in its seemingly oppositional forces within ourselves and to recognize its specular and spectacular mirroring quality.

Why do we all see these patterns? Beyond religions, sects, and ideologies there exists a numinous geometry that comprises descriptive and illustrative mathematics. Curiously, it relates to the way cultures understand the Mother Principle that represents the Void, the womb, the Creatrix. Female deities of the Hindu and Buddhist pantheons, both gentle and fierce ones, play various roles in self-understanding of the collective psyche. While theologians may speak of one God and philosophers speak of monistic substance or Vedantic nonduality, they all seem to get lost in abstractions and thus rarely take the opportunity to be transformed that involves the body. Geometry presents an interesting difference: it is vivid, sensual, and eminently usable.

In Plato, a *chora* is made up of triangles within triangles, very much like the Kali yantra. The mind sees these shapes and makes sense of the world and its apparent chaotic nature. The figure of the chiasmus can also be conceptualized as a crossroad of two triangles meeting at the head. Is it possible that *X* (chi) marks the two roads diverging, one going to the acropolis and the other to the necropolis, to height and depth, to life and death? It is easy to misread the sign as perpetual conflict, but when unsullied by conditioned thought and viewed with clarified vision or cleansed perception, the sign can lead to an understanding of the complementarity of life also represented by the balance of *yin* and *yang*.

There are many instances of Biblical chiastic structures that scholars have discovered and that seem to express both mystical and theological meaning. While that scholarship has defined chiastic structures according to its own theological needs, I am interested in the ancient manifestations of this geometric thinking, especially in the understanding of how patterns are woven into the very structures of the cosmos. Great masters of all cultures have developed elaborate methods and practices to realize that structure within one's own body. This can potentially lead to a balancing of polarities and an equipoise of external and internal realities. I will span the globe by looking briefly at examples of patterns in the West and in the East and then discuss the contemporary scene, where a return of the Mother Principle might show us the way out of our current crisis. The action of this principle is perhaps visually attested to on June 21, the United Nations International Day of Yoga, when millions of people around the world breathe, meditate, and create shapes with their bodies, aligning themselves with the cycle of the solstice.

Patterns West: Sacred Geometry

Since I know very little about Western esoteric traditions, I will examine a few texts that create a fascinating design for me to understand the diverse ways that many ancient paths have worked. I already spoke of the Greek *X* that apparently was a part of the Eleusinian mysteries, but it is extraordinary how ancient esoteric traditions are resurfacing today. Thanks to the virtual space of the internet, we can see how many people in the current world order are playing with these ancient systems. Many of them are dismissed under the blanket term "New Age," and some are perhaps defrauding others and themselves, given the ability of consumer capitalism to commercialize everything.

Yet something seems to be afoot as desperate people afflicted with addictions of every kind search for something that traditional religions no longer provide. One such person is Drunvalo Melchizedek, whose name invokes the mysterious Hebrew master, the spiritual teacher of Abraham. Melchizedek speaks of sacred geometry and merkaba meditation, among other things. Nassim Haramein, an intriguing scientist who has been meditating since he was eleven, has created quite a stir with the film that features him, *The Connected Universe* (Carter 2015). Looking from outside, some people might think that Melchizedek's and Haramein's activities are just as contrived or meaningless as any mathematical or chemical formula seems to noninitiates. I am not a scientist and thus cannot judge their work, but concrete experiences in the inner realm that lead to profound quietude, clarity, and radical peace are all the proof that some of us need.

The artist Jay Hambridge might have popularized the phrase sacred geometry; however, it is "a relatively modern term for the study of archetypal patterns that create everything in the material world" (Gilbert 2006, 15). Robert Lawlor elucidates further:

> Ancient geometry rests on no a priori axioms or assumptions. Unlike Euclidean and the more recent geometries, the starting point of ancient geometric thought is not a network of intellectual definitions or abstractions, but instead a meditation upon a metaphysical Unity, followed by an attempt to symbolize visually and to contemplate the pure, formal order which springs forth from this incomprehensible Oneness. (Lawlor 1982, 16)

Lawlor's book is quite extraordinary; it includes workbooks and elaborate illustrations. I am fascinated by his work and the Pythagorean music of the spheres because I had a very peculiar experience during my early life with visions of Pythagorean triangles; however, not only did I not know anything at all about them, but I was also not oriented toward mathematics. I enjoyed geometry and physics as a child but veered toward literature as my main passion. Later, forced to reveal my personal experiences as I encountered monotheistic prejudices against Indic religions in the United States, I ended up writing about Kali and mother goddesses (Saxena 2004).

Given that I was always fascinated with the sacred feminine and had grown up with Sarasvati, the goddess of all learning, what struck me the most was this declaration in Lawlor's book:

> Geometry as a contemplative practice is personified as an elegant and refined woman, for geometry functions as an intuitive, synthesizing, creative yet exact activity of mind associated with the feminine principle. But when these geometric laws come to be applied in the

> technology of daily life they are represented by the rational, masculine principle: contemplative geometry is transformed into practical geometry. (Lawlor 1982, 7)

He cites a fascinating composite image of even arithmetic personified as a woman, with Pythagoras on her left and Boethius on her right, although he portrays this arithmetic as being of a lower order than geometry. His image is a symbolic representation of the Mother Principle that inhabits both male and female aspects. Incidentally, Tantra is profoundly gynocentric, representing even the ordinary woman as a more complete being than a man. As we know, in the science of the ancient world, philosophy and religion were not compartmentalized. These were undifferentiated activities, perhaps best exemplified by Hypatia, the last integral thinker of the ancient world. It is perchance no accident that these cultures worshipped powerful female divinities.

There was a good reason why esoteric paths including Tantra had elaborate tests for eligibility and were kept out of public knowledge, as powerful universities today ensure tightly organized gatekeeping. We all know how geometry, sacred or not, is used in architecture and many other technological innovations, but its use in psychic revelation is much more complex and could potentially be dangerous. Egyptians, of course, understood it as they created the great pyramids and had the most elaborate alchemical systems of transmutation. Greeks acknowledged their debt to Egyptian geometry. Even today, the geometric solids in Plato's *Timaeus* are re-emerging and countering the overemphasis on Greek rationality, the bugbear of Western enlightenment. Peter Kingsley (1995) has been arguing that Greek thinkers from Parmenides, who was a priest of Apollo, to Empedocles, who understood the elements, were deep meditators and, under the guidance from goddesses like Persephone, they participated in mystical Greek paths of healing and wholeness.

It is perhaps on the heels of patriarchal monotheism[3] and its latest incarnation in modernist reductive scientism that the integral paths were sundered, reaching the point of the severe imbalance that characterizes today's world. Lawlor traces the roots of the practices of geometry, the "measure of the earth," to Egypt and the yearly flooding of the Nile, which

> symbolized to the Egyptian the cyclic return of the primary watery chaos, and when the waters receded the work began of redefining and re-establishing the boundaries. This work was called geometry and was seen as a re-establishment of the principle of order and law on earth. Each year the area measured out would be somewhat different .... This activity of "measuring the earth" became the basis for a science of natural law as it is embodied in the archetypal forms of circle, square and triangle. (Lawlor 1982, 6)

As a cognate to geometry, the Sanskrit word *jyamiti* relates to the earth and, together with the matrix and maya, denotes the limited, circumscribed, and manageable world—although in some more "patriarchal" philosophies, maya is distorted as an illusion. Lawlor's exhaustive and sweeping work moves from Plato to

Zen via Indic mandalas, and thence to Islam, Tao, mystical Christianity as well as modern science, often noticing the integral role of the Mother. He spends considerable amount of space on the problem of "zero" and its various readings and misreadings.

I found it interesting that Lawlor argues that what the Greeks called pure reason is what Indians meant by "heart-mind"; hence, the making of the yantra/mandala is centered in the heart. Lawlor discusses an impressive diversity of subjects, such as numbers, the Fibonacci sequence, the human form in Leonardo's Vitruvian Man, dancing women of Indian temples, and geometry becoming music is extremely impressive. I was not surprised to discover that he became a student of yoga and Sri Aurobindo.

In a recent article, Jerry Gin (2016) asks questions about fundamental patterns, studies the occurrence of number 64 around the world[4] and, following Nassim Haramein's work, speaks about the myriad shapes elaborated in sacred geometry. He writes that

> from the Flower of Life came the figures of the Tree of Life, Seed of Life, Fruit of Life, Vesica Piscis, and Metatron's cube. There are only 5 basic solid shapes in addition to the sphere and torus: cube, tetrahedron, octahedron, dodecahedron, and icosahedron. Everything else is made up of combinations of these 5 basic shapes which are known as Platonic shapes. The shapes all arise from connecting the intersecting points found in Metatron's cube. The patterns formed by the Flower of Life describe how the pattern of life is formed. As an example, after fertilization and cell division, the embryo forms a tetrahedron, then an octahedron, and eventually, the torus pattern emerges. The mystery schools taught this study of sacred geometry. (Gin 2016, 103)

Another fascinating book is *The Myth of Invariance: The Origin of the Gods, Mathematics and Music from the Rig Veda to Plato* by Ernest G. McClain (1984), which focuses on music more than Gin's work does. Following the example of Antonio de Nicolás (1976), McClain, too, bridges the East and the West. He elaborates on de Nicolás and finds musical correspondence in the language of existence and nonexistence, embodied vision, and what he calls the "pre-conceptual science of the Rig Veda." When I was organizing a panel for a 2005 conference on geometric thinking and the chiasmus, I was intrigued to read the following:

> A century ago, Albert von Thimus called attention to many arithmetical and graphical structures in the tonal imagery of the ancient world, particularly to the Greek chi *X*, Plato's symbol for the "World-Soul," the Hindu "Drum of Siva," understandable as the intersection of powers of 2 with powers of 3, or of sequences of octaves and fifths. (McClain 1984, 5)

I marveled how, instinctively and in visions, I had recognized the connections between the Greek *X* and Shiva's *damaru* and was exploring geometric thinking around the world without knowing any of this. McClain recognizes that

> The logic of India is profoundly geometric. Its mandalas and yantras present the observer with static forms which could only be achieved by dynamic processes. Our problem here is to learn to see those forms as Socrates yearned to see his own ideal forms, "in motion." (McClain 1984, 6)

There are many overlapping references in McClain, from the star hexagon to the drum of Shiva, but he goes deeper into music. My own musical training and understanding of primal sound from Shiva's drum make me understand the basic idea, although the mathematical details in his book elude me. He says it was Carl Boyer who "pointed out that Neolithic man (tenth to fourth millennia BCE) 'paved the way for geometry'" and draws attention to pre-Vedic Harappan civilization, when pottery with many geometric patterns abounded (McClain 1984: 54). He also discusses Mary Danielli's work, which argues that the mandala is discovered by each individual deep within and is a wordless symbol that speaks for itself. She also thinks that no one knows how early human beings became conscious of the mandala (Danielli 1974, cited in McClain 1984, 55).

Patterns in Plants Meet Patterns in Mind

Music and rhythm are not too far from our experiments with pharmaceuticals. As a corollary to contemporary youth culture's immersion in psychedelic adventures and musical groups[5] that seem to gain access to those fundamental patterns via dangerous drugs, I would like to focus now on the connection between sacred geometry, expanded consciousness, and entheogens. I have wondered about the prohibition against eating mushrooms in India and the loss of the soma plant that is now being rediscovered in Central Asia.[6] Some people have conjectured that perhaps misuse of the plants led to such prohibition. McClain discusses the use of soma in the Rig Vedic rituals and establishes a profound connection with music (1984, 49), which plays an important role in *samveda*. However, what interests me the most is William Blake's insight that "If the doors of perception were cleansed, everything would appear to man as it is, infinite" (cited in Smith 2000, 74).

Since in the Tantric tradition, cleansing our perception and deconditioning our minds play a vital role, for which sacred plants are used provisionally with great care and respect, I will now explore in some detail Huston Smith's engagement with entheogens, which led him to make the startling assertion that "Plato's basic outlook derived from his Eleusinian initiation," (Smith 2000, 113). Smith speaks eloquently of his experiments with psychedelic substances that he encountered in his quest "to know the final nature of things: reality's deepest structure and what follows from that structure for maximizing the human potential" (2000, 3). He struggles with nomenclature, since these substances have attached themselves to a complicated history. The etymology of the term "entheogen" has been traced to

"God-containing," but Smith thinks "God-enabling" is more accurate (2000, xvii). Although he wants to avoid the word "psychedelic," its meaning as "mind-manifesting" is the most appropriate for my purposes, as it avoids the even more controversial term "God." This eminent scholar's profound experiences speak to my sense that the sacred plants reveal the hidden layers of human consciousness where the plant world meets the human. Perhaps a mirroring happens, and the two patterns meet, merging human consciousness with that of plants.

Smith writes about his awakened encounter with five bands of consciousness:

> It was as if the layers of the mind, most of whose contents our conscious mind screens out to smelt the remainder down into a single band we can cope with, were now revealed in their completeness—spread out as if by spectroscope into about five distinguishable layers. (Smith 2000: 10)

Smith uses terms like "empirical metaphysics" and "psychological prism" as he contemplates "Plotinus's emanation theory, and its more detailed Vedantic counterpart." However, as he experienced them, they were no longer just conceptual theories: he "was *seeing* them" (2000, 11). He marvels, "How could these layers upon layers, these worlds within worlds, these paradoxes in which I could be both myself *and* my world and an episode could be both momentary *and* eternal—how could such things be put into words?" (2000, 11). He clarifies hastily, lest we assume otherwise, that this encounter was not a pleasurable one but, rather, a terrifying experience of what Otto Rank calls the *mysterium tremendum*. Having gone through such terror myself in my twenties without any substance use whatsoever,[7] I can readily understand his dread. Smith gives some description of shapes and colors that undulated like "writhing serpents," yet he concludes the chapter by saying "but my head was clear" (2000, 12–13).

Heeding Alan Watts's dictum "when you get the message, hang up the phone," Smith has stayed away from these entheogens with a healthy respect and fear (2000, 130–131). My guru warns of potential drug addiction as well. Smith's vast experiences with diverse religions have given him deeper insights, such as Zen, which, like many others, "recognizes that unless the experience is joined by discipline, it will come to naught" (2000, 31). While yoga has become the focus of a massive industry in the Western world, Smith reminds us of the indispensable prerequisites of *yamas* and *niyamas*, which are the preliminary rules of discipline and purification (2000, 42).

Smith also shows how misperceptions of these experiences have historically led to "sexual irresponsibility, anarchy, and lethargy" (2000, 39) as well as the self-glorifications that Carl Jung describes as inflation.[8] However, given the epidemic of drug addiction and profiteering that have desacralized these plants, it is perhaps time to re-evaluate not only their medicinal value but also their consciousness-expanding abilities under strict guidance from enlightened teachers.[9]

Many people are recognizing the healing power inherent in the nonhuman world and remember the ancient wisdom of sacred geometry that

> also finds its echoes in the deep cellular structures of human beings, animals, plants and minerals, especially gemstones; colored gem shapes in their crystalline matrixes represent healing waves of light that correspond to the 5 Elements and assist in the ritual of healing. (Wakefield 2009, 27)

With regard to Carl Jung, it is perhaps a sign of the times that, in 2009, after extensive editing by Sonu Shamdasani, Jung's descendants agreed to allow the publication of *Liber novus*, or *The Red Book*, which had a far-reaching impact on psychology and psychotherapy. Jung, who was in many ways a modern reincarnation of Gnostic Christianity's ancient adepts, understood the relationship between alchemy and psychotherapy. His archetypes and use of mandalas are quite well known, but until the publication of *The Red Book*, most people were not aware of his actual experiments and boundary-breaking experiences. Jung spoke poignantly about his unconscious flooding him with the numinous and how he came to the recognition of the meaning of mandalas:

> Only now did I gradually come to what the mandala really is: "Formation, transformation, the eternal mind's eternal recreation." And that is the self, the wholeness of the personality, which, when everything is well, is harmonious, but which can bear no self-deception. My mandala images were cryptograms on the state of my self, which were delivered to me each day. (Jung 2009, 43–44)

While the book is replete with beautiful figures, drawings, images, and calligraphy, I wish to draw attention to the image of a Celtic cross inside a flaming circle (2009, 125). Like the Greek *X* (*chi*), this cross represents the balance of polarities and has been forgotten or repressed by patriarchal Christianity. It is also similar to the Buddhist Tantric *ativajra*, in which two *vajra*s crisscross each other, denoting balance in the brain. Other parallels can be found in the Kabbalistic tree (discussed in Saxena 2016) and in works about the geometry of mystical Judaism (such as Gershom Scholem 1991).

Post-Jungian James Hillman also speaks of reanimating and "en-souling" the world in *The Soul's Code*, in which he develops what he playfully calls his "psycho-daimonics," an approach he uses to return to the ancient understanding of the soul (Hillman 1996, 144). His engagement with polytheistic consciousness to escape tyrannical monotheism and monocentric consciousness brings him to Greece. His return to the ancients represents an effort to create a more serious engagement with polytheistic imagination that takes us to India and its Tantric traditions, where ancient paths and lineages remained seamlessly alive.

PATTERNS EAST: TANTRIC MANDALAS

The word Tantra invokes all kinds of emotions and reactions in people who have only a vague notion of it. For a long time in India, the word has been associated with forms of occult and black magic; in the United States, it gained a certain notoriety in relation to its sexual rituals. Both reputations contain partial but misguided truths; however, the real science of Tantra can be quite mind-boggling, but there should be nothing mystifying about it. "Mysticism" is a term that has been abused by so-called scientific rationalists, but the eminent scholar Frits Staal has insightfully noted that it may be more "scientific" than faith-based religions: "Among the great religions of mankind, Christianity stresses faith and irrationalism to an extent that the others (including Islam) never even considered" (Staal 1975, 23). He argues that the charge of irrational mysticism is mistakenly hurled at Eastern religions because

> in many forms of mysticism, uniqueness is rejected in favor of repeatability, commonality, and generality, just as the personality of God is often rejected in favor of an all pervading impersonal divine absolute . . . . Mystics tend more to generally to reject the irrational need for faith and dogmatism. (Staal 1975, 30)

Tantric tradition includes texts, methodologies, as well as precise elucidations regarding what extends or actualizes theoretical knowledge. Many scholars have been attempting to break its secret code through intellectual means, while others have been exploring its remarkable history and liberating potential as part of their genuine search for truth. A few things are certain about it: it is profoundly centered on the Mother Principle; there is no access to its real mystery without a proper teacher; and its highest goal goes beyond the Vedantic realization of Brahman, or expanded consciousness, dissolving into the very womb of existence.

In the following section, I will give a brief description of the science and art of this knowledge system, which, when practiced under a guru, can radically change our brain chemistry, profoundly reorienting our lives. I will focus primarily on the structure of Tantra, for which my main source of knowledge is the work of my guru, Kulavadhuta Satpurananda, although I will also cite a few other texts.

Satpurananda is a direct lineage holder of Indian Vajrayana which is slightly different from its better-known Tibetan version.[10] He is one of the contemporary Indian Mahasiddhas, known in the Western world as the "crazy masters" of Tibetan Buddhism. A descendant of renowned Bengali Tantric Atisha Dipankar/Srigyan from the eleventh century, Baba belongs to Tarakula or Tara family. He was initiated into Tantric yogic paths at the age of five and, by age twelve, he went through the final initiation of the Aghora Way. For the next eighteen years, he practiced and attained the perfection of wisdom, and at the age of thirty-two, he was given the highest title of Kulavadhuta, one who has gone beyond and returned

(*Tathagata*) to the world. The word *kula* means family and is derived from the word *ku* (womb) and *laya* (dissolution); *kulachar* means practices of the "Buddha" family that includes other *agama* groups such as Shaktas, Shaivas, Jainas, Vaishnavas, Baul, and even Sufis.

Satpurananda also received the final empowerment of the way of *sahaja*, or "spontaneity." He is a painter, sculptor, poet, Baul singer, meticulous cook, designer, iconologist, Sufi master, and Gnostic chalice holder. He does not belong to any religion and has not established any specific school. He has been recognized as a Dzogchen-pa by the great Tibetan master Chatral Sangye Dorje Rinpoche. His robed disciples follow the yogic path known as *Ngakpa* in Tibetan. Ecstatic and yet profoundly cosuffering with humanity, he radiates profound *maitri*, or "deep friendliness." Affectionately addressed as "Baba," he teaches those who seek him without any barrier of caste, creed, gender or religion.

According to Baba, Tantra is the cream churned out of the ancient knowledge system known as *agamas*. There are three main strands of Tantrism in India that were given to three different regions based on their geographical significance and the specific karmic makeup of the people who lived in each. The first and the root source is *Tarakula*, the Himalayan or northern strand, a form of Buddhist Tantra whose origins go back to antiquity. Second is *Kalikula*, which encompasses the middle of the country. The third is *Srikula*, which was given to the south. Tantra is highly individualized, and, like Indic religions, there is no one size that fits all.

I learned a great deal from Baba's teachings in which he explicated Buddhist Tantra and its female deities. In one of his lectures, Baba said that the Buddha discovered the structure of the mind and dissolved it. With Nagarjuna and Mahayana, we see further elucidation and techniques, but it was the second Buddha, Padmasambhava or Swayambhunatha (also called Guru Rinpoche), who developed the science to the fullest and spread it in many lands. *Vajrayanatantr*a recognizes suffering as the cause of all enlightenment leading to *nirvana*, that is, the realization of the Void. The Void is the primordial nature of both being and becoming. Void-ness is not empty or vacant; it is the middle point between the polarity of all positives and negatives. It is the neutrality of existence not disturbed by any factors involving cause and effect. It is the understanding of Being-but-Nothing. This is the core of Buddhist philosophy of truth, the practice of which is Tantra.

Baba says that the Void-ness or "Suchness" is the spontaneity of Being-but-Nothing happening, which is personified as Tara the Saviouress. She is the balance of the paradox of the dependable origination, *nirvana* and *samsara*, the parallel truths between which the perfection of wisdom, Prajnaparamita, stands as the saviouress. The gradual evolution of Buddhist philosophy matures as adepts discover cosmic reality. It corresponds to realization practiced through body-mind-speech by enlightened beings who call for the elemental realities personified as five

Taras.[11] In India, the deities, *murtis*, or "icons," are actually mandalas dressed up and beautified in anthropomorphic or zoomorphic forms.

Let us now return to my guru's class and learn a bit more about mandalas. Baba began the class by explaining why such knowledge is beneficial for lay practitioners as well as useful for artists, since Indian art is fundamentally Tantric in its conception and execution. He then began to play musical notes using his own breath and a curious flute made of bone. We heard all seven notes of Indian classical music. He described them as *nada*, meaning "sound gives birth to forms." The sound also helped us relax and come into the present moment, leaving behind brain chatter, anxiety, and other disturbances of the mind.

Baba meticulously taught us about the theory of mandala making and then held a workshop in which we created the Tara mandala using profound meditation techniques. Staring at the empty white sheet and finding the dot in the middle where the center of the mandala dwells made the theory come alive. It is crucial to experience this profound dissolution into *shunyata* of all our perceptions and the very self that perceives the world in order to understand the Tantric science that slowly developed in the Indic landscape. The expanded consciousness that is the knowledge of *Brahman*, or the expanding universe known to the Vedic rishis, may witness the vastness of the lighted world, but it is the experience of dissolution in the womb of the dark Mother that is the hallmark of all the diverse Tantric paths in India.

The term Tantra was not used until the first Buddhist text *Guhyasamaj* mentioned it. *Agamas* preceded the tantra*s*, which are a refinement of *agamas* as a science; what is known as Kashmir Shaivism is the Kaula order of Tantra. The word Tantra also brings the complex term *kundalini* to mind. As Mookerjee and Khanna (1977, 22) put it, "Kundalini's rising, in the language of modern science, means the activation of the vast dormant areas of the brain." People of all times and places have experienced some aspect of this unsettling experience. It is a kind of sudden immersion of the soul into the "World Soul," as Paulo Coelho depicts it in his novel *The Alchemist* (1998).[12] All kinds of intense emotions or peak experiences can induce this experience, and sometimes it happens quite spontaneously.

The development of the science of this experience, which is activated by a guru through *shaktipat*, has a traceable history. Like an expert physician, a guru decides when a student is ready for this medicine of life-changing experience. To induce it, the guru creates an environment involving intense meditation, along with the help of tools like mandalas and yantra*s*. This experience is dangerous enough to require the student to stay with the guru, who administers precise herbal remedies and prescribes a strict yogic discipline of diet and exercises. These practices are not taken lightly, since any wrong application risks leading to madness or brain disorders. On the other hand, when the Great Mother, in the form of the guru, provides guidance for the student, the result is no less than an awakening into Buddhahood.

The word yantra has come to mean "machine," but etymologically it means that which saves one through the power of three *gunas*: the qualities of *satva* ("clarity"), *rajas* ("energy"), and *tamas* ("inertia"). These *gunas* constitute the world and condition us toward behavior patterns (*sanskaras*, or "karma imprints") that create repeated grooves in the psyche, leading to suffering.

In the Indic scene, yantra*s* are extremely common, often being used for material gain and unconsciously leading to unsavory consequences. In some cases, unscrupulous people have used them to wield power in India for unethical ends, since the exact science and ethical use of this radical knowledge system has been suppressed for many reasons. A mandala is the more expansive use of yantra*s*, since it represents the very structure of creation. Some people may be familiar with the ways that the Dalai Lama ritually makes and unmakes the Kalchakra mandala. Apparently, our bodies have a blueprint of the entire evolutionary journey, so following this journey up to the advanced point of evolution represented in the mandala can radically change us. It reminds me of the theory of ontogeny and phylogeny, which holds that the fetus replicates the entire evolutionary process during the gestation period.

Madhu Khanna presents a wide variety of images of yantra*s* in her work *Yantra: The Tantric Symbol of Cosmic Unity* (2003). She carefully portrays an array of fierce deities,[13] from the simplest to the most esoteric. Here, she invokes the familiar metaphor of the spider web:

> In the *Brhadaranyaka Upanishad* (2, 1, 19) there is the metaphor of a spider sitting at the center of its web, issuing and reabsorbing its threads in concentric circles, all held at one point . . . . The spider's threads symmetrically expand into a visible circumference, and though there are divergent lines in between and varying distances to be spanned they can all be traced back to the central point of the web. (Khanna 2003: 9)

This movement from the *bindu*, or "central point," to the periphery and back again is the most elemental structure of the path. Indeed, it brings us back to Baba's class.

Let us now enter the circle drawn by some of Baba's students, which represents the sky element with rainbow colors. This lighted, manifested cosmos stands upon the unmanifested *shunya* or *Mahavyom* ("great sky"). Inside this circle is drawn the square that represents the earth element. It has four doors, which are the four goals or *purusharthas*: *dharma* in the south, the discipline that holds us; *kama* in the east, that which makes us desirous of discovery; *artha* in the west, that which denotes significance of being; and *moksha* in the north, that which cleanses our habitual patterns. Our spinal column is the axis mundi, the vertical line running through the middle of the circle. Baba also explained how every drawing involves creating eight spokes of the wheel, which denote the eightfold path of the Buddha from the "right view" (*samyak drishti*) to *Samadhi* (*samyak sambodhi*).

As the students slowly enter the interior of their beings, the making of the mandala helps them encounter the water element within another circle, which introduces them to their *prana* or "nervous systems." The water element is colored red. The eightfold lotus represents the *prakriti*, "one's nature"; its outer petals denote the practitioner's external nature while the inner ones represent the internal. These are drawn differently for each kind of mandala. The students then enter the intertwining equilateral triangles, the star hexagon that sits in the middle, which is colored white. This is the fire element or mind. Triangles are the most important Tantric symbols, representing the threefold activity—of *icchha*, or "will"; *jnana*, or "gnosis"; and *kriya*, or "activity"—of the feminine principle of the triangle that points downward.

The triangle pointing upward denotes *purusha*, or "realizer," the masculine principle, with the three sides representing the *shunya* (voidness), *dvaita* (duality) and *advaita* (nonduality). Other meanings include body, speech, and mind; the three *gunas* of creation, sustenance, and annihilation; and a Buddha's *dharmakaya*, *sambhogkaya*, and *nirmankaya*. The circle inside the triangle represents the air element, shown in blue/black, and the consciousness of self, which vanishes into the dot (*bindu* or *shunyatma*) where we began. Those who enter these depths recognize that they are witnessing their consciousness. Going beyond this involves entering into contact with the principle of nirvana. This is the most rudimentary understanding of what it entails; more complex yantras and mandalas, such as the most famous Sriyantra ones, involve equally complicated paths that cannot be elucidated in this short paper.

I have been practicing *spanda vipassana* meditation taught by Baba for several years. It has calmed my interior and given fascinating insights into myself. This meditation includes the journey into the mandala of our body, as described above, in a curious way. It seems to me that with every breath drawn from the heart center of our body, the mandala expands into the lighted cosmos and returns to the point of singularity in the middle. This is the very pulsation and rhythm of our life and death. Perhaps the collapsing of the wave of supreme consciousness creates each being, which eventually returns to the wave of existence we call death. Beginning the journey from yellow earth (Yellow Tara) and returning to green earth (Green Tara), when *samsara* and *nirvana* are equated, denotes the path of a Buddha.

According to some texts, certain kinds of Tantric practices known as *Chinachar* came from China; however, practices of *Mahachinachar* were taught by Padmasambhava in Tibet which was known as Mahachina. Although this story is beyond the pale of this paper, I want to draw attention to the famous *yin/yang* symbol as I conclude this discussion of patterns from the East. This Tai Chi symbol, a kind of mandala, perhaps most visually represents the dynamic aspect of dualities moving in and out of each other. They are often imagined as two black-and-white fish. Balancing and harmonizing these principles is the bedrock of ancient Daoist practices in China, which both enriched and learned from Buddhism's journey there through many centuries of exchange. The Zen and Chan arts of calligraphy and gardening are also meditative practices and form parts of our global engagement with various ways of balancing our beings.

Conclusion

Now is the perfect moment to revive *Anima Mundi*, which was destroyed under a vast array of linear and supersessionist mythologies and theologies. The world's spiritual paths speak of the fragmentation of the primal unity in the very process of its manifestation, but the "murder of *Anima Mundi*," as Apffel-Marglin notes in the introduction of this volume, is of a different order altogether, since it culminated in nothing less than outright war against life itself. The original totality of the world was sundered to the point of atomization, leading to the nuclear age and its devastating consequences for the world's diverse cultures. Yet ironically, in the very attempt to grasp the fundamental unit of the God particle of creation, scientists discovered a sort of Buddhist emptiness at the heart of what they had previously considered to be solid and immutable "objective" matter.

Nassim Haramein draws on both his Bedouin and European heritage when he offers reflections on atoms. Since we are made of atoms, we are mostly space, yet we do not speak about this, focusing instead on the atom as an infinitesimally small part of existence, a perspective that leads to pathologies of all kinds. Haramein's work is quite startling, and some of his claims are controversial and problematic to apply. He clearly shows how physicists, by focusing only on expansion of the universe (*prakasha*), have missed the opposite movement of contraction (*vimarsha*). When Haramein draws polarized equilateral triangles and creates the star hexagon, he introduces the missing feminine element, thereby redressing the imbalance. His work is vehemently opposed by "rationalists," but Baba says Haramein is using mathematics to show us what Tantras teach, which Baba has experientially recognized through his *sadhana*. However, inner truths are not self-evident in the external world; in addition, whether this knowledge will be used for enlightenment or domination is yet to be seen.

In this respect, I want to draw attention to the experiences of a Nobel-Prize–winning scientist, Wolfgang Pauli, who went through analysis with Carl Jung. Physicist F. David Peat writes about Pauli's dream, which was "a dream of 'the most sublime harmony,' his 'vision of the world clock'" (Peat 2014, 47). It is striking to see the accompanying image (2014, 46), which looks like a mandala. Peat writes:

> Jung identified the point of rotation of the discs with the mystical *speculum*, for both partake of the movements and stand outside it. The two discs belong to the two universes of

> the conscious and the unconscious, which intersect in this speculum. The whole figure with its elaborate movements is therefore a mandala of the Self, which is at the one time the center and the periphery of the world clock. (Peat 2014, 47)

It is not an accident that all such claims, including Tantric inner "technology," are dismissed by reductionist "rationalists." In a remarkable book presenting a sweeping study of the Western world and modernity, physician Iain McGilchrist (2009) examines the role that our emphasis on using the left hemisphere of our brains has played in creating the imbalance of the current world: "There has been a relentless growth of self-consciousness, leading to increasing difficulty in cooperation," which has led to an "increasingly mechanistic, fragmented, decontextualized world, marked by unwarranted optimism mixed with paranoia and a feeling of emptiness," which is a product of "unopposed action of a dysfunctional left hemisphere" (McGilchrist 2009, 6). However, he also says that "the kind of attention we pay actually alters the world; we are literally partners in creation," and he believes "it to be profoundly true that the inner structure of our intellect reflects the structure of the universe" (2009, 5, 460). Ancient mind–body sciences had come to the same conclusions. Since we can change the world through our attention and behavior, collective action worldwide can have salutary effect. We need to turn our focus away from that particle of self-consciousness and explore the waves that occur within our own spaciousness.

Reticence about one's inner visions can be both individually and collectively fatal. The question becomes, how do we walk the fine line between reductive rationality and consumerist science, on the one hand, and blind belief in religious ideologies, which are mostly patriarchal, on the other? These have not only pitted humans against other humans, but in the ultimate grandstanding of bloated egotism, they have pitted humans against the living cosmos itself, ironically producing more alienation than all pharmaceuticals put together can handle.

Yet there is hope. It seems to me that we may be resurrecting *Anima Mundi* and reentering the womb of the Mother to regenerate the earth. Curiously, this is happening outside traditional religions; postmaterialist science has been uncovering the quixotic nature of reality, which cannot be pigeonholed under the label of "rationalism" or any ordinary theoretical knowledge. James Hillman (1996, 6) proclaims that "we have been victims of theories," and Baba Satpurananda believes that the root word of "theory" is *theos*, or "god," which is a mere concept devoid of experience. The difficult problem of consciousness is now being widely discussed, and many scientists themselves are revealing their true beliefs about subjective experiences.

I have found that ordinary people, along with many contemporary thinkers, artists, writers, and scientists, seem to be mapping the chiasmus in their own way to find liberating patterns. People in every possible field are engaged in reawakening the World Soul and the Mother Principle within themselves. We are journeying into the circle and discovering the *Anima Mundi* in our own bodies as we discard the deeply entrenched dualistic metaphysics, philosophies, and theologies. In the process, we find the axis mundi in our own spine, which has been represented in innumerable images, from Jacob's ladder to Mount Meru. Techniques are being discovered by masters to ascend and descend this tree of life, whether it is portrayed as the Kabbalistic tree, the Bodhi tree, or a banyan with roots branching skyward.

The ancient staff called a caduceus, using serpent energy as a guide, is bringing back the regenerating form of knowledge. Everything indicates that contemplative and integral geometry under the sign of the Mother is returning. Like Einstein, who "geometrized" time and space, we are now recognizing that we have fluid structures and are fractal geometries ourselves, having infinite potential and possibilities. Numerous books are sounding the call for us to immerse ourselves in the World Soul, including F. David Peat's work (2014), mentioned earlier, and David Fideler's *Restoring the Soul of the World* (2014). A journal edited by Fideler called *Alexandria* (Fideler 1991–2000), dealing with cosmology, philosophy, myth, and culture, may inspire modern Hypatias as we re-evaluate our relationship with all ancient cultures and learn to respect them for their liberating patterns.

Indic paths had somehow managed to resist at least partially the relentless march of modernity and extreme self-consciousness. Polycentric Indic *dharma* scenes are fundamentally nontheistic since they mostly see provisional creator "agents" who dissolve with the dissolution of each cycle of time. These "deities" also need to be dissolved in meditation. They are created by humans and given shape, energy, and power through the power of our beings. I would like to close by invoking Kulavadhuta Satprananda's Tantric *vigyan*, a newly emerging postclassical science, and propose that we decouple religion from structures of knowledge and awaken our depth intelligence through practices that are ubiquitous in "mystical paths" and sacred geometries in the world.

NOTES

1. The Oxford English Dictionary (OED) defines a chiasmus as "a grammatical figure by which the order of words in one of two parallel clauses is inverted in the other."
2. See my blog *Stand Under the Mother Principle*, http://neelabsaxena.blogspot.com/.
3. For more on this subject, see the chapter on matricide in Saxena (2016).
4. He mentions the Kamasutra but is perhaps not aware of 64 yoginis and Tantric temples in India (Gin 2016). Satpurananda (pers. comm.) explains that the number 64 refers to parts of the brain.
5. Musical groups like *Contortionist, Tesseract, Uneven Structure,* and *Animals as Leaders* use fascinating geometric symbols and patterns in their videos and logos.
6. See Jeremy Jeffs's video series (2007) on India.

7. See Saxena (2004), where I tried to describe my experience of being devoured by Kali, whom I called "pregnant-nothingness." Decades later, my guru Satpurananda explained it as spontaneous *samadhi*. Kali devours time, which is the experience of *samadhi* and an ego death. This initial experience, perhaps a consequence of *shaktipat* by my family's guru Shiddha Ma, slowly led through many difficult years to my current practices, which I learned from Satpurananda.
8. I recall my guru's comment about why ganja as a ritual entheogen is prohibited to be used in the Southern Tantric tradition of India while it is appropriate in the North due to the climate and makeup of the people of these regions.
9. See Frédérique Apfell-Marglin and Stefano Varese's references in the Introduction to this volume.
10. See Saxena (2016) for a discussion of the complex history of Vajrayana Buddhism, which has a different trajectory than Vedic Brahminism in India.
11. To understand the deities of Vajrayana, see Bhattacharya (1958).
12. In the novel, the character named Santiago speaks of this sudden immersion into the World Soul when he finally becomes the alchemist transmuting himself. When we look at Paulo Coelho's life and his psychiatric episodes, it is possible to surmise that he had such an experience, which may have been suppressed by powerful medicines until he escaped and made his way to become a renowned writer.
13. Before I met Baba, I had written about my encounter with Chinnamasta, a fierce Mahavidya and a form of Kali (Saxena 2011). Although I wrote it as though the figure spoke to me (an experiential approach to writing that Baba appreciated), I knew nothing about the precise meaning of the figure, who is also a mandala.

References

Bhattacharyya, Benoytosh. 1958. *Indian Buddhist Iconography*. Calcutta: Firma, KL Mukhopadhyaya.

Carter, Malcolm, dir. 2015. *The Connected Universe*. Film (98 mins.). Distribber. http://getconnected.resonance.is/. (accessed on April 15, 2017)

Coelho, Paulo. 1998. *The Alchemist*. San Francisco: Harper Collins.

Danielli, Mary. 1974. *The Anthropology of the Mandala*. Amherst, NY: Center for Theoretical Biology.

de Nicolás, Antonio. 1976. *Meditation through the Rig Veda: Four-Dimensional Man*. York Beach, ME: Nicolas Hayes.

Fideler, David, ed. 1991–2000. *Alexandria: The Journal of Western Cosmological Traditions*, vols. 1–5. Grand Rapids, MI: Phanes Press.

Fideler, David, ed. 2014. *Restoring the Soul of the World: Our Living Bond with Nature's Intelligence*. Rochester, VT: Inner Traditions.

Gilbert, Robert J. 2006. *Shift: At the Frontiers of Consciousness* 11: 15–19.

Gin, Jerry. 2016. "Fundamental Pattern and Consciousness." *Cosmos and History: The Journal of Natural and Social Philosophy* 12 (2): 99–113.

Hillman, James. 1996. *The Soul's Code: In Search of Character and Calling*. New York: Warner.

Jeffs, Jeremy, dir. 2007. *The Story of India*. Six-Part Series. BBC Home Video.

Jung, Carl. 2009. *The Red Book: Liber Novus*. Edited with an Introduction by Sonushamdasani. New York: Norton.

Khanna, Madhu. 2003. *Yantra: The Tantric Symbol of Cosmic Unity*. Rochester, VT: Inner Traditions.

Kingsley, Peter. 1995. *Ancient Philosophy, Mystery and Magic: Empedocles and Pythagorean Tradition.* Oxford: Oxford University Press.
Lao Tzu. 2009. *Tao Te Ching.* 2009. Translated by Ursula K Le Guin. Boston: Shambhala.
Lawlor, Robert. 1982. *Sacred Geometry: Philosophy and Practice.* London: Thames and Hudson.
McClain, Ernest G. 1984. *The Myth of Invariance: The Origin of the Gods, Mathematics and Music from the Rig Veda to Plato.* York Beach, ME: Nicolas-Hayes.
McGilchrist, Iain. 2009. *The Master and His Emissary: The Divided Brain and the Making of the Western World.* New Haven, CT: Yale University Press.
Mookerjee, Ajit, and Madhu Khanna. 1977. *The Tantric Way: Art, Science, Ritual.* London: Thames and Hudson.
Peat, F. David. 2014. *Synchronicity: The Marriage of Matter and Psyche.* Pari, Italy: Pari Publishing.
Saxena, Neela Bhattacharya. 2004. *In the Beginning IS Desire: Tracing Kali's Footprints in Indian Literature.* Delhi, IND: Indialog.
Saxena, Neela Bhattacharya. 2011. "Mystery, Wonder and Knowledge in the Triadic Figure of Mahavidya Chinnamasta: A Shakta Woman's Reading." In *Woman and Goddess in Hinduism: Reclamations and Re-envisionings*, edited by Rita Dasgupta Sherma and Tracy Pintchman, 61–75. New York: Palgrave Macmillan.
Saxena, Neela Bhattacharya. 2012. "Gynocentric Theology of Tantric Hinduism: A Meditation Upon the Devi." In *The Oxford Handbook of Feminist Theology*, edited by Mary McClintock and Sheila Briggs, 131–56. Oxford: Oxford University Press.
Saxena, Neela Bhattacharya. 2016. *Absent Mother God of the West: A Kali Lover's Journey into Judaism and Christianity.* Lanham, MD: Lexington Books.
Scholem, Gershom. 1991. *On the Mystical Shape of the Godhead: Basic Concepts in the Kabbalah.* New York: Schocken Books.
Smith, Huston. 2000. *Cleansing the Doors of Perception: The Religious Significance of Entheogenic Plants and Chemicals.* Boulder, CO: Sentient Publications.
Staal, Frits. 1975. *Exploring Mysticism: A Methodological Essay.* Berkeley: University of California Press.
Wakefield, Mary Elizabeth, and Michel Angelo. 2009. "Acupuncture and Gem Essences: An Ancient Healing Ritual Using the Five Elements and Sacred Geometry." *Oriental Medicine Journal* 17 (1): n.p.

CHAPTER NINE

Shamanic Archaeology at Chavín de Huántar

ROBERT TINDALL

ONE

My first visit to Takiwasi, the center for the treatment of addiction that utilizes the methods of Amazonian shamanism along with Western psychotherapy, and its host town, Tarapoto, Peru, was many years ago, in a quieter age.

My partner, a Chilean therapist, had already developed a strong affinity with Takiwasi's surprisingly effective approach to treating addicts and the unique character of its founders, the doctors Jacques Mabit and Rosa Giove. When I joined Susana there, she was doing psychotherapy with the addicts in treatment in the ample, tree-shaded grounds of the center, conducting her dissertation research, and soaking up the accumulated knowledge about traditional plant medicines and shamanic techniques utilized at Takiwasi to heal—especially the psychodynamic effects of plants such as chiric sanango, azucena, rosa sisa, tobacco, tamamuri, and came renaco used in *dietas* and *purgas* there.[1]

At that time, we rented a rustic but very *cargado* (i.e., spirit-filled) house, around the corner from Takiwasi, for a hundred bucks a month. We slept on borrowed mattresses, cooked on a borrowed stovetop, and invested in a few pots and spoons. We were on pilgrimage at that time. When we left the center, we simply put all our accumulated possessions in the back of a pickup truck and drove to the entrance at Takiwasi, where we gave them away to the staff.

Yet the hundreds of details which we thought we knew about the *mestizo* ("mixed heritage," i.e., of European/Indigenous descent) culture of Tarapoto were sadly out of date upon our return to Takiwasi in 2012. An economic boom had altered the landscape, so prices were far higher. And by arriving with our three-year-old daughter Maitreya with the intention of staying—involving schooling, decent housing, local community and friends, language issues, reckoning with local diseases such as parasites, and so on—we were no longer pilgrims, skimming over the landscape. We had become immigrants, putting down roots.

The cultural honeymoon in Tarapoto was short-lived. Indeed, our arrival could not have been timed better for triggering the extremes of culture shock.

Even under normal conditions, Peruvians in frontier towns live by the motto, "I make noise, therefore I am." Tarapoto could be loud. Very loud—with both the chainsaw grind of the constant *motocarros* (three-wheeled rickshaws drawn by motorcycles) and the blasting of the rhythms of Peruvian dance music late into the night.

Yet now Tarapoto was celebrating its anniversary. Like an earth-shaking bombardment, a week of beer-sodden, music-blasting parties began, lasting into the early hours of the morning. Susana had been called away by a family emergency, so Maitreya and I, living in a tiny bungalow on the edge of the town, passed evening after evening surrounded by the surreal cacophony, like a war against the air. Moreover, my daughter had been mauled by insects during our time in the rainforest, so she tossed and turned scratching all night as I applied lotions and tried to catch some sleep. An epidemic of dengue was also a threat. One day we watched a parade of schoolchildren following a slow ambulance. It was scary. We were vulnerable.

Finally, a knocking on my door came one evening just as I was putting Maitreya to bed with her bottle. "They are here to spray against the spread of dengue," the worried owner of the hostel told me. Men with backpacks full of chemical pesticides were roaming the streets and entering homes with blowers, filling the night with toxins against the mosquito population. There had been no advance notice, no opportunity to plan to escape from the fumes. Without even a chance to dress my daughter, I rushed like a refugee into the crowded streets with her in my arms.

In the mornings, after watching Maitreya bravely walk in her Catholic school uniform into her kindergarten class, dogged in her determination to learn Spanish, I would run errands through the sordid and ugly streets in the center of town. I kept repeating, like a mantra, W. B. Yeats's lines from one of his final poems:

> Those masterful images because complete,
> Grew in pure mind, but out of what began?
> A mound of refuse or the sweepings of the streets,
> Old kettles, old bottles, and a broken can
> Old iron, old bones, old rags,
> that raving slut who keeps the till. Now that my ladder's gone,
> I must lie down where all the ladders start.
> In the foul rag and bone shop of the heart. (Yeats 1996, 489)

It was my way of keeping heart.

In the midst of a total recalibration of my senses to the mestizo culture of Tarapoto, I hit bottom. Choking on pollution and noise and plastic and scattered beer-bottle caps littering the streets, I knew where I was on the anthropologist's map: I was moving through the phase of acculturation where the honeymoon of fresh, new impressions ends, and the trapdoor of reality opens beneath your feet.

With the return of Susana, the hardest part of what I called the "Tarapoto Mestizo Blues" began to pass. We found a house on the outskirts of town (what the motor-car drivers call *la jungla*) and moved into its cavernous spaciousness. There we began luxuriating in the sound of authentic quiet—the chirping of birds in trees. Monkeys even clambered through the open grounds.

We bought a refrigerator. A double-sized mattress. A set of dishes, pots and pans. We began decorating the house and giving it a spirit of home. We had some furniture made of bamboo. Around the corner, we found a hotel with a pool that, by a stroke of luck, we were given free access to use. My daughter began waking up speaking in Spanish or firing off complex grammatical constructions. She even began correcting my pronunciation: "It's 'Perú,' Dad, not 'Péru.'"

While Susana engaged in individual therapy with the patients and supervised the therapeutic team, I began leading meditation in the exquisite Zen Buddhist dojo that Takiwasi, in a gesture of true ecumenicism, had built some time ago. The patients took to the practice with gusto and began requesting more focused instruction. I had also begun doing another kind of work with the patients—ceremonies with ayahuasca—and that is where this story really begins.

I had begun toying with the idea of a work of fantasy, an alternative history in which the first European immigrants to encounter the Amerindians were not Spanish conquistadors but instead Celts fleeing the Anglo-Saxon invasions of the British Isles. Feeling that the pre-Christian Celts and the pre-Colombian peoples of the Amazon rainforest and the Andes mountains shared a certain kind of worldview, I reckoned a Celtic invasion would not have led to genocide but, rather, to an accommodation and eventual creative fusion of the two cultures. I wanted to explore what that synthesis would have looked like a generation after the landing of the Celts coming from the West, using an old tale about the archetypal figures of a king, a servant, and a white snake to do so.

Yet I needed a place to work. I contacted a friend of mine, Andres, a member of the cultural aristocracy of Lima, who thought about it and said, "You know, Robert, I think Chavín de Huántar might be a good place for you to go." Curious, I went online and found the sparse images and encyclopedic entries about the site. A ceremonial complex at 10,425 feet above sea level, located at the confluence of the Mosna and Wacheksa rivers, at the geological and cultural crossroads between the mountains, jungle, and sea, Chavín had flourished as a sacred site for five hundred years (800–300 BCE). Its light had been eclipsed long before the rise of the Incan empire. Its art, carved in stone, was shocking, outlandish, as if from an alien civilization. I had no idea what to make of it.

Except for a couple of visits to Cuzco to walk its colonial streets and visit the shattered remnants of the Incan empire, I had not had much contact with the traditions of the Andes. My knowledge of the lore and practices of rainforest shamanism was respectable, but my understanding of the time depths upon which that world floated like flotsam and jetsam was scant. Scattered, persecuted, competitive, tending toward witchcraft and charlatanism, *vegetalista* shamans were survivors, inheritors of rich seams of cultural traditions transmitted during the brazen thuggery, enslavement, murder, and dislocation of the previous century's rubber boom. As far as I knew, there was no Mecca, no Jerusalem, no Eleusis to which pilgrimage had once been made, nor could now be made.

That sacred locale, the *axis mundi* of pre-Colombian peoples, had once existed, however, and my friend had pointed the way to me.

A couple of nights later, the path opened up during an ayahuasca ceremony at Takiwasi. Well launched into the visionary sea of the vine of spirits, I was called forward by the Colombian *taita* who was coleading the work that evening. Sitting before him, I felt his song pour over me like pelting rain and then was surprised when the *taita* spread his arms like a hawk, and, gathering in the wind, blew something into me in a single burst. At that very instant, I saw the *huaca*, the serpent-festooned, great-eyed figure with a crocodile's grin from the heart of the labyrinth of Chavín, and I knew I was called upon to make my pilgrimage there.

Two

The town of Chavín, with its colonial square, church, and market where the Indigenous tongue of Quechua mixes with Spanish, lies adjacent to the archeological site of Chavín de Huántar, like a house cat beside a sleeping jaguar. I settled in at one of the little hotels popular with archaeologists during their work season. Unaccustomed as I was to the Andes, the uncanny resemblance of the people, the rarified air and light, the valley walls and snow-capped mountains, the stench of urine and stale beer, and other less tangible resemblances to the Himalayan region where I once traveled kept playing tricks on my memory. Stepping out onto the sidewalk in the morning, I kept finding myself in Nepal.

At first sight, the temple was something of an anticlimax. Accustomed to the evocative romance of archeological sites like Machu Picchu in the Sacred Valley, the temple had the air of an ancient warehouse stripped of its goods. Wandering into its square plaza, still perfectly framed by its low walls and megalithic stones of black granite, I beheld the lineaments of the temple, partly collapsed, as an evocative rubble quarry. Puzzled, solitary, and alone, I crossed the expanse, climbed the stairs out of the sunken court, and ascended to the cold stone. At the center of the massive structure stood a great portal, framed by two columns carved with anthropomorphized, staff-bearing birds of prey. Like an Escher painting, their geometrical shapes struck my untutored eye as both wildly psychedelic and profoundly sophisticated. The temple loomed above me like an alien obelisk, even in the warm sunshine that bathed the verdant green fields of the site. Yet there was no ominous, contained threat in it; it was simply a puzzle. Ascending further, I passed the roped-off circular plaza and, climbing the fragments of a flight of stairs, arrived at the entrance to the *huaca*, the "deity" of the ceremonial complex, a fifteen-foot-tall granite carving of a figure surpassing description.

I evoke this literary trope because, truly, all attempts to map this image, to depict its topography with any degree of objectivity, would miss the point. I now believe the *huaca* can only be understood through intimate communion, an intense, engulfing subjectivity that removes, even if only temporarily, the boundary line between subject and object. Most scientifically trained researchers are not comfortable with the old ways of *seeing*, fearing what poet Gary Snyder described as the "passage into that myth-time world," and "the unsettling vision of a natural self" (Snyder 2004, 15). What dreams may come when they release their hold on Cartesian dualism, they fear, will derail their capacity to interpret the meaning of their "artifacts."

Yet not all researchers shun the Indigenous worldview. Anthropologist Hillary Webb, after some time apprenticing with the Native healers Amado and Juan Luis,[2] approached the *huaca* and gives this account of her unsettling vision:

> For me, it was like seeing something out of a nightmare. Logically, I couldn't say why it had such a profound effect. Certainly, the Lanzon's face with its fangs and strange hybrid expression was eerie, but my reaction was totally unexpected and completely nonrational. The only way I can describe what it was to sit there in front of it is that looking at it was like trying to stare directly into the sun. Kneeling there, I had the sense that if I were to spend too much time with it, something inside me might be obliterated, collapse from the power of it. An eclipse of the soul might occur that would be too much for my psyche to handle. (Webb 2012, 101)

I had no such epiphany on my first visit. Making my way down the well-worn stone steps into the Old Temple was like descending into an ancient barrow—a musty inertness, the fug of exhaling earth, muteness of chambers. Gazing down one passageway after another, I caught sight of an illuminated statue at the end of one, and, picking my way the length of the narrow, low tunnel, I arrived before the *huaca*, nicknamed the Lanzón for its resemblance to a spearhead.[3]

I gazed upon it in profound incomprehension: a perfect "duh." I saw neither menace nor benediction. All I saw was an inexplicable megalithic figure, as alien and remote as the surface of the moon. Eventually I got up and stumbled out.

Wandering the streets of the town of Chavín that evening, I glanced into a family tienda from which light spilled out onto the dark cobblestone road and saw a table covered with local artisanry—fine miniature reproductions of the temple art carved in stone. Stepping forward to purchase a stylized jaguar head, it occurred to me to inquire for a guide that my friend from Lima had recommended, an elusive figure named Esteban. They smiled, said, "Of course," and dialed their phone. Within a minute, the diminutive, muscular figure of Esteban stood framed in the doorway.

Of African and Indigenous descent, Esteban was a talented artist and a raconteur. The following day, he took me on a rambling tour, narrating the history of the temple. A meeting ground for the spiritual traditions of the Amazon rainforest and the Andes mountains, Chavín's iconography is dominated by four figures: the jaguar, the harpy eagle, the crocodile, and the anaconda—four powerful animals whose provenance lay far away in the rainforest to the East, yet whose visionary presence migrated to the heart of the temple. As Esteban guided me through the underground labyrinth, grunting and chuckling like a jaguar in the darker, roped-off areas, a sense of topography began to take shape for me. When Esteban introduced me to the *wachuma* shaman, a human figure with a serpent headdress and distinct jaguar fangs protruding from his mouth and clawed hands and feet, I was hooked. I had been under the lingering misapprehension that shamanism, such as I had seen it practiced by *vegetalista* shamans, was an exclusively seminomadic, hunter-gatherer tradition that faded into obscurity once the hierarchies of priesthoods, governance, agriculture, and the like emerged with the coming of civilization. Yet here was something remarkable: a high civilization, geographically far-reaching, enduring for centuries, which, according to archeological evidence, eschewed warfare, blood sacrifices, and political coercion. Indeed, if the archaeologists are correct, within the region of the "Chavín-complex" peace reigned for hundreds of years. Chavín had an enviably high-functioning cosmology.

At the heart of this realm was a vigorous practice of shamanism. By the time Esteban walked me over to a massive standing grove of *wachuma* cactus and introduced me to one of the visionary plant medicines that informed the spirituality of the temple, I knew I would need to commune with it to understand something of the temple's inner life.

To ingest a "hallucinogen" is admittedly a grave undertaking. Understood by Native peoples throughout the world as divine gifts, entheogenic plants "reveal the divine within," opening up a blissful and fearful prospect. Entering our frail human constitution with the vigor of a wild serpent, they can define who we are in ways beyond our ken. For Westerners, the experience of intimacy created by these plants, the enchanting, all-subsuming "Secondary World" (as Tolkien [2008] describes encounters with the realm of Faerie) within them can also be psychologically derailing. Of such potency and beauty are the visions, it is tempting to subscribe a status of "Primary World" to them and start seeing the ordinary world as a shadow of the spiritual realm. There is also the ever-present danger of psychological damage and abuse—as is now commonly reported within the world of drive-by (and occasional hit-and-run) ayahuasca shamanism. Participants in ceremonies can find themselves trapped in very serious breakdowns once the "shaman" has disappeared over the horizon and can no longer offer any assistance—if he or she ever could.

In short, a ceremony utilizing a sacred plant is, like heart surgery, dangerous. And like heart surgery, such interventions into our deep selves require great skill on the part of the leader. During my years of residence in the Amazon, I had become rather discriminating toward the shamans under whose guidance I took traditional medicines, inquiring about their lineages, their years of experience, their reputation among the Native communities, the purity of their brews, and so on.

As is usual for guides, Esteban knew a local *wachumero* who, he said, prepared an excellent brew and could lead us in a ceremony in the temple. My guide spoke with the oratorical flourishes and smiling assurances that have accompanied traveling medicine shows for centuries, yet despite his jaguar chuckling, I liked Esteban and agreed to meet the shaman, called "Jack," the next day.

Instead of the local Andean shaman with a wide-brimmed hat and sandals, Jack turned out to be a gringo wearing jeans, hiking boots, and a baseball cap. As he shook my hand and introduced his Peruvian partner Yngrid, he spoke with a Midwestern twang and seemed a character more likely to be met at a truck stop than at an ancient ceremonial site in the Andes. After a brief chat, we clambered into his four-wheel drive pickup, crossed the bridge under which the Mosna River travels through the heart of modern Chavín's market, and drove a short distance back into the valley. There we descended to some land Jack had acquired directly opposite the temple compound, separated only by the river, which runs down, vigorous and cold, from the melting snowcaps above.

It was a good place to do a ceremony. Rounded by blossoming San Pedro cacti (as *wachuma* is called in Spanish), an altar stone and a fire pit were set in the middle, with a circular maze of stone leading to it clockwise. Little *chozas*, or huts for sleeping, were set back up the slope. There was a covered eating area, and as we descended the meticulously laid stone steps the local workers, whom Jack introduced by name, walked up and shook my hand in welcome. Jack pointed out some truly megalithic carved stones set higher up the hillside. When we climbed to these stones, we saw they were clearly part of the original temple construction, carved and placed in perfectly alignment with the geometry of the compound across the river. I realized that beneath our feet was likely a lattice of temple construction about which nothing was known, with only these two surviving stones being the jagged remnants that protruded above the earth. This was especially likely, since Andean peoples, rather than dismantle their sacred places, would bury them as offerings to Pachamama.

"Would you believe the farmer that I bought this land from pulled down some of the stones that were part of this alignment?" asked Jack. "He said it got in the way of his grazing sheep." Seeing the righteous indignation in his eyes, I realized that Jack, far more than the archaeologists who have been steadily draining Chavín of its riches for over a century, was deeply committed to the preservation and restoration of the temple's original integrity. This was good. I chose to join him and the small group of other "pilgrims" from afar in the ceremony the following day.

Three

At midmorning, a small group of us gathered and opened the ceremony with conches sounding to the four directions. We gathered around the altar and drank the *wachuma* medicine. Jack's preparation was as thick as a milkshake but not as bitter as other Native medicines I have drunk. We then loaded up and drove back across the river to the archeological site, where we purchased our tickets from the guards and entered.

Finding our way down to the opposite bank of the river from where we had begun, we spread out and sat either quietly chatting or alone among the trees. Unlike Esteban, who affected the mannerisms of a jaguar shaman, Jack spoke intently about the spiritual dimensions of the temple, the portals to be encountered within it, and its cosmic function in the seen and unseen worlds. I listened to his metaphysics, only half-attending, as the world began to take on another hue and my breath began transpiring with the trees and river. I was also nauseous and had to lie on my back staring into the deepening blue sky and hoping the dizziness would pass.

What followed involved such an intensification of consciousness that ordinary narrative ceases around the moment I got up and began to walk the path to the great sunken square plaza lying before the New Temple. As I crested the hill and descended the precisely cut, massive granite stairs now warm in the sunshine, my awareness began to both telescope and microscope. Time sped on, lurched drunkenly, or hung serenely suspended. At the same time, my habitual waking orientation, all that wearisome analysis, measurement, and surveillance which so clogs up awareness, receded, as if that little figment, that insignificant knot of neural activity that sustains the illusion of an external, objective world "out there," had crashed. My habit of self, in short, went off-line.

Who was I, then? That was the fear and the adventure I now faced, gazing in awe upon this terrible place, Chavín de Huántar. Where was I? Whose dream was I?

This is fantastic! I thought, gleefully rubbing my hands together. Or at least I would have if I could have thought at all. Instead, I sat upon the ancient stones in a deep muse. I knew something momentous was at hand, that I had come to Chavín for whatever this birth was I was about to undergo. Other participants in the ceremony moved around me. I chatted with them with the surface of my mind. I performed feats of ratiocination. Meanwhile, the temple pounded upon my deeper awareness even more powerfully than the Andean sun above upon my wisp of a body.

There is, believe it or not, a handy little map that can be consulted for the stages of the journey that I was undergoing and still had to undergo on that day. It's helpful to locate yourself on a map, because, of course, it's encouraging to know you aren't crazy and that, in fact, you are following in the footsteps of your ancestors. Furthermore, this map validates a certain social function for shamanic states—but more on that later.

This map occurs in *The Mind in the Cave*, where cognitive archaeologist David Lewis-Williams (2002) offers a neurophysiological model for the process of entering that "mental vortex that leads to the experiences and hallucinations of deep trance," which, for Upper Paleolithic people, he claims, was psychologically indistinguishable from their actual entrance into "the subterranean passages and chambers ... the 'entrails' of the nether world" of the prehistoric painted caves (Lewis-Williams 2002, 209).

What, then, is this hardwiring we all share? According to Lewis-Williams, the ordinary spectrum of consciousness of *Homo sapiens* ranges from a waking orientation, to dreaming, and beyond the edge of the psychic world, unconsciousness, as expressed graphically in Figure 9.1.

As we can see from Lewis-Williams's schema, our ordinary experience of consciousness ranges from wakeful problem-solving (most likely the state you are in right now as you read this) to daydreaming, dream imagery, the dreams of sleep, and then deep unconsciousness.

Imagine, however, what it would be like to dream while being fully awake.

To achieve that end, intensified consciousness takes a different fork in the road before entering the realm of hypnagogia. Rather than seeing dancing sugarplums, we may encounter "entoptic phenomena," that is, geometrical will-o'-the-wisps that begin to dance before our eyes and "flicker, scintillate, expand, contract, and combine" to draw us on—just as the "embellishing images" of the prehistoric caves once "blazed a path into the unknown" (2002, 209).

In the second stage of this intensified trajectory, a struggle to focus occurs, and "the brain attempts to decode these forms as it does impressions supplied by the nervous system in an alert, outwardly directed state" (2002, 128). This stage of construal is normally rather brief, a final preparatory moment before the leap into what Lewis-Williams, with his Cartesian worldview, characterizes as

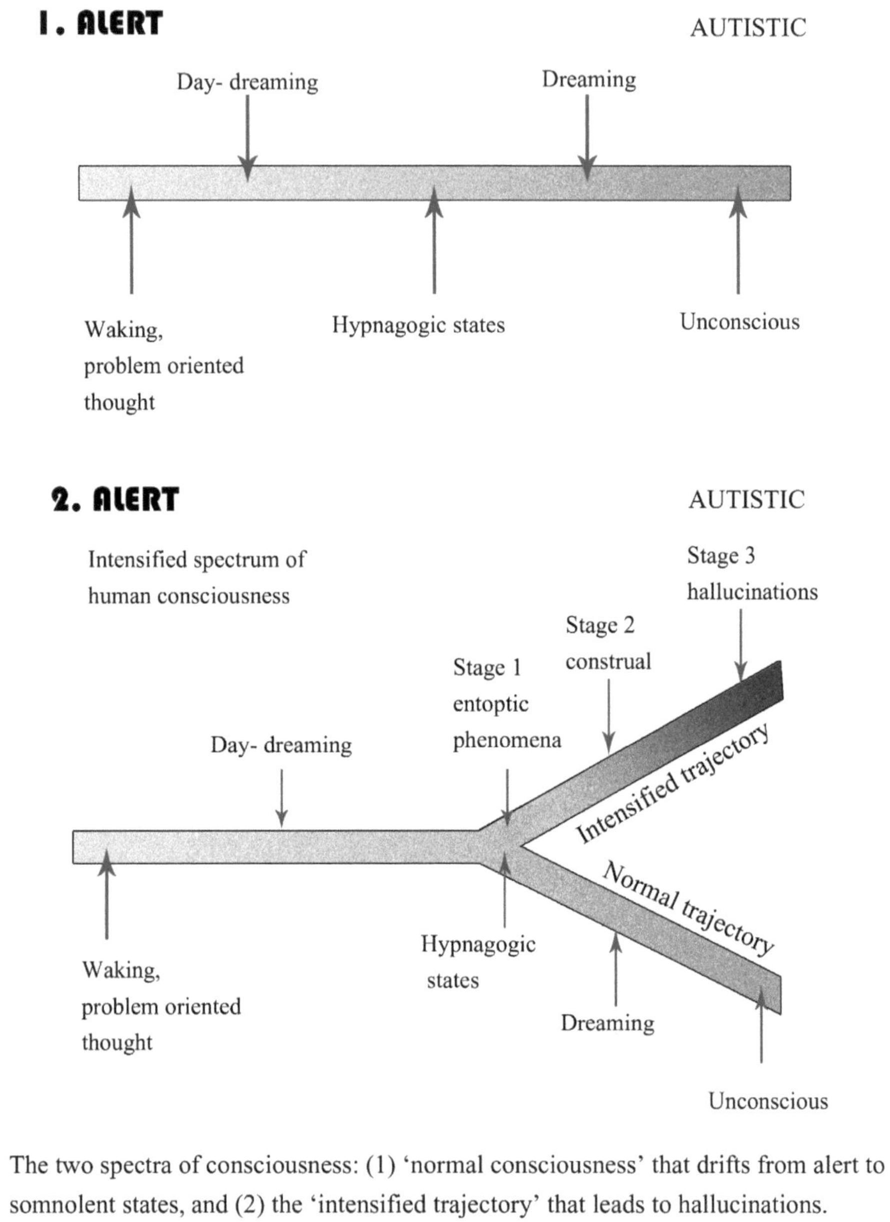

Figure 9.1. The two spectra of consciousness, according to Lewis-Williams (2002) (Reproduced from *The Shamanic Odyssey* by Robert Tindall with Susana Bustos, Ph.D., published by Inner Traditions International and Bear & Company, © 2012. All rights reserved. http://www.Innertraditions.com. Reprinted with permission of publisher).

"hallucinations," but which the rest of humanity understands as visionary and as having the greatest worth.

According to Lewis-Williams, in this third stage of the unfolding visionary state, "Many people experience a whirling vortex or rotating tunnel that seems to surround them and draw them into its depths" (2002, 128–29), and they are drawn into full intra-action with the worlds that lie beyond the range of ordinary consciousness.

The virtue of Lewis-Williams's model is its elegant simplicity. Yet he may be making too much of neurological hardwiring, which offers an objective-sounding mechanistic model for transformations of consciousness. The Native voice of a Huichol shaman gives a better idea of this leap into the visionary stage:

> There is a doorway within our minds that usually remains hidden and secret until the time of death. The Huichol word for it is *nierika*. *Nierika* is a cosmic portal or interface between so-called ordinary and non-ordinary realities. It is a passageway and at the same time a barrier between worlds. (Halifax 1979, 1)

Eventually, my mind began to move out of construal and I began to *see*. This *seeing* was with an inner eye, the one I suppose the Huichol shaman was describing as the doorway that opens at the time of death. It wasn't that the features of the landscape around me changed in their form. The llamas grazing on the hillside were still llamas, the valley walls rising high above us were still perfectly valley walls, and the rest of the buzzing world continued with its colorful, precise multiplicity. No trouble there. What changed was the depth: the frontier between life and death was down, so when I walked, I seemed to penetrate realms. I knew where I was now, and it was time for me to go and discover what it was I already knew.

I approached Jack, who, in his sunglasses and black Gore-Tex jacket, had the aspect of an astronaut, and told him I would like to depart from the group and move up higher into the temple. He studied me and then sent me onwards.

It was fortunate for me that no busloads of tourists had yet been disgorged at the site that day. I had the way up to the Portal of the Falcons to myself. As I slowly ascended the curving path toward the looming walls of cut granite above, I was aware of the millennia-old stream of pilgrims flowing around and through me. By the time I placed my foot upon the final step up to the New Temple (carved, in the playful spirit of medieval cathedrals, with two serpents), I felt overwhelmed by a pounding wave of energy, titanic and sacred, coursing through the pillar and lintel construction of the portal. I walked into it as into a mighty river and fell onto my knees before the carvings of this ancient world and wept.

I had come home. I knew this place in the deepest recesses of my soul. After—could it be centuries of transmigration?—of barely, of barely surviving this lifetime's experience as a street kid and prisoner of the urban United States, I had returned. An embrace engulfed me, a gratitude welled up in me, and I knelt, clutching my prayer beads, holding on as the dawning revelation crashed over me: I was known and loved in this place. The temple, I realized, was no archeological ruin: it was a sacred abode. When I finally stood up, it was with the posture of a beloved son.

Of course, even in the midst of the visionary experience, I was aware of how remarkable it all was. My work with medicines such as ayahuasca and peyote in traditional contexts down through the years had somewhat tempered my naïveté and bright faith—as well as developed my ability to discern within the intensified trajectory of consciousness. Not only had I known in myself and others profound healing of both psychological trauma and physical disease utilizing shamanic trance, but I had also acquired a navigator's sense of how to move through the deeper, perilous waters of the psyche and distinguish reality from figment and even downright mischievous trickery—or worse.

This was no trickery, no light show. In fact, strictly speaking, there had been no visual "hallucinations" at all but, rather, what prehistorian Jean Clottes, in Werner Herzog's (2010) film *Cave of Forgotten Dreams*, characterizes as "permeable consciousness." Nor was it unconditional love or a narcissistic regression to a uterine existence. The time depth, the sense of awakening memory, the power of the catharsis, all strongly suggested that my vision was an ancestral, Indigenous one.[4] In short, it was a very clean burn—and it began to open my eyes to the inner workings of the temple.

After making full prostrations in the verdant green grass before the portal, I shouldered my bag and prepared to move higher up into further unexplored regions. I glanced below. The pilgrims were just beginning to gather to leave the plaza. Turning away, I clambered up like a mountain goat, eager to continue my solo communion with the temple.

Rounding the sharp edge of the New Temple, I climbed past the roped-off circular plaza where, in the oldest spiritual strata of the site, before the creation of the New Temple with its baroque visionary art, pilgrims and suppliants had gathered and danced before ascending to the *huaca* or had received benediction from temple priests who emerged from the Old Temple.[5] My feet then carried me to the entrance to the labyrinth, the many-chambered, three-thousand-year-old Old Temple, wherein resided the *huaca* that had so unsettled anthropologist Hillary Webb.

I greeted the guards, who lounged about the entrance, and requested permission to enter. With the natural kindness of Andean people, they welcomed me and gestured to the threshold. With a tingling sensation, I carefully descended the stairs, placing my hands upon the stone protrusions worn smooth by the passage of generations, and set foot upon a landscape as strange and silent as the lunar surface.

In my absorption in the moment, I had forgotten about the *huaca*, so it wasn't with my thinking self but my heightened senses that I bumped into it. Catching sight of something unaccountable at the end of a passageway, I crouched like a cat, all my senses on alert. I began to stalk, every nerve aquiver, toward this inexplicable being grinning at me out of the darkness, my eyes, I reckon, having gone predatory—thought being rejected as too slow and clumsy for this contingency. Yet reason did eventually dawn on me. I was gazing upon the *huaca*, I realized, lit as if by a light spilling in from an opening in the temple ceiling above. As I approached nearer, my wariness lessened and sheer wonder came over me. I was gazing into a face of extraordinary, benevolent love. My legs gave way before it, and I sat gazing into the carven image's oceanic expression—a crocodile grin, fangs, serpents, jaguar heads—the whole cacophony of beings intertwined within the carving—and saw unworldly compassion and wisdom embodied there, the workings of the mind of the heavens itself.

Eventually, I began to form conclusions. I was not gazing upon a sculpture, a representation of something in the modern way of conceiving of representational art. This *was* the being itself, not a carbon copy. The veils had been torn from my eyes before the Portal of the Falcons, and I was beholding the deity of Chavín that was said to have come from the stars and whose presence had brought hundreds of years of peace upon the land.

It takes a sense of humor and bemused acceptance for a simultaneous plurality of utterly contradictory worldviews to tolerate this sort of stuff, an experience that contradicted every particle of received wisdom in my own culture and, indeed, one that is violently condemned by my own religious background. Fortunately, my gratitude and wonder, my sense of engulfing love far surpassed the consternation that might have arisen. I sat bathing in that opening into timeless apprehension that I was given that day.

When I finally emerged from the labyrinth, the group was gathered in the foyer, courteously awaiting my emergence before entering themselves. Esteban warbled upon a bone flute. Seeing a conch sitting at his side, I asked if I might sound it. He handed it to me, smiling. Walking to the edge of the stone platform, gazing out into the valley, I leaned out and blew the conch. The valley walls formed a perfect amphitheater, and the sound echoed back, resonating through all the land, welling up as rich and intimate as joy itself.

That afternoon, I came to know Chavín informed by the consciousness of *wachuma*, growing accustomed to wearing *wachuma*-tinted glasses, as it were. I became able to discern more and more specificity of detail within the temple, and in one wonder-struck moment, I stood before a section of wall in the New Temple, my hand upon the stones, utterly convinced that I had participated in laying that wall in a bygone age: I knew the texture of those rocks with the same intimacy as I knew the fretwork upon my guitar.

Four

So began my adventures at Chavín. Not heroic adventures but, rather, pilgrimages in the old sense as described by Ian Baker:

> Tibetan tradition speaks of *kha sher lamkher*, "Whatever arises, carry it to the path"—a Buddhist injunction to abandon preferences and integrate all experience beyond accepting and rejecting. Without that dynamic openness to adventure (from the Latin *ad venio*, "whatever comes"), Tibetans say, pilgrimage devolves into ordinary travel and the hidden lands—both physical and metaphysical—will never open. (Baker 2006, 187)

This practice of "shamanic archaeology," as I've coined it, obviously puts in abeyance the Cartesian dualism underlying ordinary archeological investigation and embraces instead a "dynamic openness." Yet the mode of perception that splits the world into subject and object is always available, contrary to what many scientists fear. Rather than derailing the finely honed, discriminating mind that scientific investigation relies upon, visionary experience commonly makes the mind more flexible, more insightful, and more adaptable. It simply requires the intense subjective participation of the investigator in the process of inquiry to the point where the investigator *becomes* the object of investigation rather than standing outside and opposed to it. Subject and object lose their defining stances, and the investigator's outlook shifts from anthropocentric to *cosmocentric*. We *are* a community of beings.

In short, the fundamental tenet of humanism, that *Homo sapiens* is the measure—or even capable of taking the measure—of all things, ends up looking absurdly myopic.

Although we can argue that the dreams that may come during this intensified trajectory are of great spiritual and psychological worth, that they are "ontologically rooting" in the words of philosopher Simon May,[6] the question arises: Does this visionary work disclose or transform realities normally unavailable to ordinary consciousness? Is it, in some sense, measurable?

It turns out that it is indeed, in two significant ways: through the healing of diagnosed physical diseases and through archeological discovery.

Marion Jackson arrived in Chavín de Huántar on a pilgrimage of her own, in her case seeking healing for stage-four colon cancer. This cancer had metastasized to her liver through which, after surgery and an intense round of chemotherapy, tumors were still laced. Before departing New York, her doctors had told her she had only months to live and no hope for any successful intervention by Western medicine. They expected her to return from her travels to die, nothing more.

I spoke with Marion over the phone, catching her in a break between her yoga teaching in New York and a trip to teach at a monastery in Nepal. She is still young, in her late thirties, with a fruity laugh that is contagious. She described to me how she had been convinced her healing required her to engage in an inner work as well as an outer treatment, so, while undergoing treatment from Western medicine, she had followed her intuition and embarked upon the path that brought her to Chavín. "All of my friends were saying, 'You're not going to Machu Picchu?' Although I can't explain why, and no one I knew had ever been there, I felt strongly drawn to the temple. I simply knew I had to go to Chavín."

Marion told me that upon arriving, she had "a feeling of deep remembering, as if I had been in the temple many times before. I felt so at home, so safe and cradled in its embrace." During her first day wandering the temple grounds, she encountered Jack, who could only speak briefly with her. The next day, they bumped into one another again near his land across the river. After Jack showed her his *wachuma* garden, she described her condition to him. Although Jack was about to leave town, he gave her a bottle of *wachuma* medicine with a single instruction: "Drink this and do not leave the temple complex while the medicine is strong in you."

Marion was cautious and drank only half the bottle the following day when she entered the temple, but it was sufficient. She has had many subsequent ceremonies in the temple and her more powerful visionary experiences were to follow. But in that first ceremony, as the medicine was strongest, she entered the Old Temple labyrinths and found her way to a subterranean room, likely once designated for the priests or priestesses of the temple. There, in the "womb of the Earth," it was if she had arrived at the axis mundi of the world. Light radiated in from one of the mysterious narrow openings that lead out of the temple, and gazing down its length, she knew that she was seeing the successful healing of her condition. "I was filled with that light and I saw that I was going to live. I sobbed in gratitude and humility. As I lay my head on the cold ancient stones of the labyrinth walls, I gave thanks to the powerful temple, to the ancestors of Chavín who felt so close to me, and to the intelligent plant spirit that was working to cleanse and heal me."

Marion left the Old Temple as if reborn, with a deep knowing that she was going to survive the cancer. Returning to the square plaza, she danced in the wildflowers with sheer joy.

When she returned to the New York hospital, the doctors were surprised to see her looking so strong. "What happened to you down there?" they asked. When PET and CT scans were run, it was found that, while she wasn't entirely free of cancer, the tumors in her liver had not grown at all in her months in South America. In fact, the ones which had been deemed inoperable were now calcified or calcifying. Surgical intervention was possible.

Marion chose to undergo surgery and additional chemotherapy, yet she now wonders if the treatments were necessary. As she continued doing ceremonial work with *wachuma*, the "plant spirit" appeared to be playing the strongest role in her physical, emotional, and psychic healing—although she didn't dare share that information with her doctors. One day, she disclosed her plant medicine use to her doctor and received a brusque ultimatum in response: "I forbid you to ingest that while you are my patient! If you want to ruin your liver when you are through with treatment, that is your choice."

"I learned to stay quiet after that!" said Marion, and then laughed as she described how Western medicine unwittingly validated the traditional healing approach she was taking. As part of her treatment, she would regularly report to the hospital to have comprehensive blood work and tests of her liver enzymes. "Whenever I would arrive just after a *wachuma* ceremony," she said, "they would run the tests and say, 'Wow, your blood work and enzymes are perfect today!'" She continued:

> My doctors were always very impressed with my physical and emotional strength. I owe a big part of that strength to what the plant was teaching me and to how it was working to create balance within me. I felt that I was carrying the medicine and the energy of the temple within me. My spirit would return to the temple, to those labyrinths, even while I was in the hospital in New York. It was as if I were connected by my umbilical cord to that place. In conjunction with the medicine, this connection kept me strong, resilient, and optimistic through what was potentially a very difficult time. (Jackson, pers. comm.)

Marion is now entirely recuperated from her "terminal" cancer. She credits her healing to the sacred realm, the spiritual topography, of Chavín de Huántar. "I feel as if I was reborn in the womb of Chavín. I am eternally thankful and connected to this place and will continue to return to the temple for many pilgrimages."

FIVE

> Our revels now are ended. These our actors,
>
> as I foretold you, are melted into air, into thin air.
>
> (Shakespeare, *The Tempest*, Act 4, Scene 1)

The practice of archaeology, as in all other sciences, is as much about *seeing* as is the practice of shamanism.

Europeans knew about the Paleolithic temple caves for centuries before recognizing their actual time depth and inestimable cultural worth. However, saddled with a worldview that disallowed the existence of human culture before 4004 BCE (that is, the date of Bishop Ussher's calculation for the creation of the world), visitors to the caves simply could not perceive the vast antiquity of the iconography from 25,000 years before. The graffiti they scrawled upon the walls testifies to this. It was not until European science had tentatively begun to establish the hoary time depths of the geological processes of the earth that the possibility of evolution, fossil records, and human culture of anything other than superficial depth could be perceived.

As I gradually accustomed myself to *seeing* through the worldview of the pre-Colombian high cultures of the Andes, I began to perceive features in archeological sites that I previously had not been able to *see* at all. At Ollantaytambo, for example, after ascending to the Sun Temple, I found that the features of the severely defaced jaguars carved upon the Wall of the Six Monoliths were far clearer, lying just beneath the scars left by Spanish vandals.

Yet the revelation of the deep past is a cat-and-mouse game, our unseen inheritance a plaything in the hands of the industrial forces unleashed upon Peru. This fact was driven home to me some time after my return from Chavín when, in the darkness of the wee hours of the morning, I walked out of Takiwasi after an ayahuasca ceremony. The jungle roads were utterly silent, the stars gleamed down, and I needed to get back to my home a couple of miles away across the Shilcayo River in the section of town called La Banda. The road passed through Tarapoto's version of a red-light district (comprised of a couple of wooden shacks by the bridge), where I had been advised not to attempt the passage at night. I therefore called a friend of mine who drove a *motocarro* and requested he come to pick me up. Restless, I had set off down the road beneath the palm trees in the direction I knew he would be approaching from, when suddenly a voice said to me, "Slow down and pay attention!"

Arrested in my movement, I halted and looked. Across the road was a boulder, one that I and the other workers at Takiwasi passed by every day and which had never excited comment, yet this evening it seemed to fairly shout out at me. As I walked toward it across the road, I was electrified to see the faintest yet perfectly discernable lineaments of a jaguar head. I stood before it musing when my local mestizo friend pulled up, a Tarapotino who had never drunk ayahuasca in his life. I asked him, "Do you see a jaguar head in this stone?" and pointed to the rock. "Yup," he said, nodding his head, conviction written all over his features.

I lay awake until early morning, turning over and over the implications of my "discovery"—should it end up being something other than an insubstantial pageant faded by the morning. Archeological remains are few in the San Martin region of Peru, and there is no evidence for or memory of megalithic cultures that carved in stone in that region of the high Amazon. Was it reliably, physically real with a historical "aura," or had I seen a spirit, a protector of the grounds of Takiwasi?

The next day, I avoided the stone. I wanted to digest the experience and was, moreover, afraid that when I approached it again, my vision might turn out to have been a "hallucination." After all, a couple of days wouldn't matter if the stone had sat there for centuries.

I was wrong. A couple of days later, I could not find the stone in its usual setting. My pulse quickening, I cast about and spotted it turned upside down in a ditch, all vestiges of the jaguar head vanished from sight. A vehicle, no doubt a huge truck, had dislodged it from its position.

Whatever glimpse I may have been given of an unknown pre-Colombian people had dissolved, leaving not a trace behind.

Notes

1. Among Westerners caught up in the current *ayahuasca*-tourism boom, it is little recognized that the "vine of the spirits," as it is translated from Quechua, is only one medicinal plant among many in the *vegetalista* tradition.
2. Speaking of Chavín, Webb's informant Juan Luis stated,

   In all of the Americas, this is probably the most ancient culture, the Chavín people. All our great-grandchildren will always know that that was the center, that that was the place where these lineages could meet again. [Chavín] is the meeting point. Our planet was basically created with the wisdom and the life that was brought from the cosmos in [Chavín]. That's the center. It is the strongest place of *yanantin-masintin*, for it is capable of creating new humanity if we are talking about humanity, or new life, or new levels of life in the planet, in the cosmos. (Webb 2012, 91)

   Webb's other informant, Amado, added this:

   Chavín is the place where a cosmic seed was planted—where the Medicine was brought from the cosmos to this earth. You see, we have legends that say that the Medicine was brought from the stars . . . the original center of spiritual centers was Chavín. This is where the Medicine began. All the work with ayahuasca, huachuma, vilca—all these sacred medicines that allow you to enter into higher-level dimensions of light began here. (Webb 2012, 92)

3. I was later informed by a *curandera* who grew up in the region that whenever her family encountered a stone shaped like a *lanzón*, they would thrust it as a blessing into the fields. "The *huaca* is from the stars," she told me. "It fertilizes the earth."
4. One criteria for distinguishing authentic visionary material is that, in its wake, one is left with a deep sense of cleansing, gratitude, and communion. Such experience does not give answers so much as free up the possibilities of life, which I felt in plenitude upon arising.
5. All the practices at the site over the centuries are largely speculative, but it's certain that highly developed rituals were practiced there, based upon the archeological evidence, including the discovery of a sizable stash of conches in a buried storeroom, all inscribed with the motifs of the temple.
6. Here is how May explains "ontological rootedness":

   My suggestion is that we will love only those (very rare) people or things or ideas or disciplines or landscapes that can inspire in us ontological rootedness. If they can, we will love them regardless of their other qualities: regardless of how beautiful or good they are; of how (in the case of people we love) generous or altruistic or compassionate; of how interested in our life and projects. And regardless, even, of whether they value us. For love's overriding concern is to find a home for our life and being.

Love, I will argue, is the rapture we feel for people and things that inspire in us the hope of an indestructible grounding for our life. It is a rapture that sets us off on—and sustains—the long search for a secure relationship between our being and theirs.

If we all need to love, it is because we all need to feel at home in the world: to root our life in the here and now; to give our existence solidity and validity; to deepen the sensation of being; to enable us to experience the reality of our life as indestructible (even if we also accept that our life is temporary and will end in death). (May 2013, 5)

References

Baker, Ian. 2006. *The Heart of the World: A Journey to Tibet's Lost Paradise*. London: Penguin Books.

Halifax, J. 1979. *Shamanistic Voices: A Survey of Visionary Narratives*. London: Penguin.

Herzog, Werner, dir. 2010. *Cave of Forgotten Dreams*. New York: IFC Films. http://www.ifcfilms.com/films/cave-of-forgotten-dreams.

Lewis-Williams, David. 2002. *The Mind in the Cave*. London: Thames and Hudson.

May, Simon. 2013. *Love: A History*. Princeton: Yale University Press.

Snyder, Gary. 2004. *The Practice of the Wild*. Washington, D.C.: Shoemaker and Hoard.

Tindall, Robert, with Susan Bustos, Ph.D. 2012. *The Shamanic Odyssey*. Rochester, NY: Inner Traditions International and Bear & Company.

Tolkien, J. R. R. 2008. "On Fairy-Stories." In *Tolkien on Fairy-Stories*, expanded ed., edited by Verlyn Flieger and Douglas A. Anderson. New York: HarperCollins.

Webb, Hillary S. 2012. *Yanantin and Masintin in the Andean World: Complementary Dualism in Modern Peru*. Albuquerque: University of New Mexico Press.

Yeats, W. B. 1996. *The Collected Poems of W. B. Yeats*. New York: Scribner.

Contributors

D. Ahmed is Chairperson and Senior Research Fellow, Center for the Study of Gender and Culture, Lahore, Pakistan. From 1976 to 2009, she taught at the National College of Arts in Lahore, where she was Professor of Psychology and Communications and Director, Graduate Program in Cultural Studies. She is also a practicing psychotherapist. She is the author of *Masculinity, Rationality and Religion: A Feminist Perspective*, and editor and contributing author of *Gendering the Spirit: Women, Religion and Postcolonial Response.*

Frédérique Apffel-Marglin is Professor Emerita, Department of Anthropology at Smith College. She founded Sachamama Center for Biocultural Regeneration (SCBR) in the Peruvian Upper Amazon in 2009. She has published fourteen books and circa seventy articles. Her most recent book, written with Robert Tindall and David Shearer, is *Sacred Soil: Biochar and the Regeneration of the Earth* (North Atlantic Books, 2017). The Spanish translation was published in Lima, Peru, with the title of *Yana Allpa* (Apus Graph, 2019).

Guillermo Delgado-P. is a Bolivian anthropologist and Quechua linguist. He obtained his doctoral degree at the University of Texas at Austin. He teaches at the University of California-Santa Cruz in the Departments of Anthropology and Latin American Studies. He is coeditor of *Indigeneity: Collected Essays* (Santa Cruz, CA, 2012), *Identidad, ciudadanía y participación popular* (Plural Editores, 2003), and *Quechua Verbal Artistry* (Bonn, 2004).

Fern Feldman is a rabbi, nurse practitioner, and hypnotherapist living in Seattle, Washington. Her academic background includes a degree in French literature. Her chapter "Gender and Indeterminacy in Jewish Mystical Imagery" was published in *Unsettling Science and Religion: Contributions and Questions from Queer Studies* (Lexington Books, 2018). Her interests include exploring the sacred dark in Jewish tradition, gender queer imagery in the Kabbalah, and working in the field of harm reduction.

Inés Hernández-Ávila is Professor of Native American Studies at University of California-Davis and a scholar, poet, and visual artist. She is Nimipu (Nez Perce), enrolled on the Colville Reservation in Washington, and Tejana. She is one of founders of the Native American and Indigenous Studies Association (NAISA). Her most recent publication, coedited with Norma E. Cantú, is *Entre Guadalupe y Malinche: Tejanas in Literature and Art* (University of Texas Press, 2016).

Jacques Mabit, a French medical doctor, first came to Peru with the organization Doctors Without Borders in 1980. From 1986 onward, he began participative and experimental research on traditional Amazonian medicine in the Peruvian Upper Amazon, which led him to the founding of the Takiwasi Center in 1992. The Center offers an original therapeutic protocol for the treatment of drug addiction and mental health, for which it has gained the attention of both the public and the scientific community.

Neela Bhattacharya Saxena is Professor of English and Women's Studies at Nassau Community College, New York. She has published two books, *Absent Mother God of the West: A Kali Lover's Journey into Christianity and Judaism*, and *In the Beginning IS Desire: Tracing Kali's Footprints in Indian Literature*, as well as several articles. She writes a blog called "Stand Under the Mother Principle." More information can be accessed at her website, neelabhattacharyasaxena.com.

Robert Tindall is a writer whose work explores the crossing of frontiers into other cultures, time depths, and states of consciousness. He is the author of three books on shamanism and indigenous lifeways: *The Jaguar that Roams the Mind*, *The Shamanic Odyssey: Homer, Tolkien, and the Visionary Experience*, and *Sacred Soil: Biochar and the Regeneration of the Earth* (with Frédérique Apffel-Marglin and David Shearer). He works as a professor of literature and divides his time between the San Francisco Bay Area and Peru.

Stefano Varese, is an Italian/Peruvian anthropologist, Professor Emeritus of Native American Studies at University of California-Davis, and former director/founder of the Indigenous Research Center of the Americas (IRCA). He received his PhD from the Pontifical Catholic University in Lima, Peru in 1967. His research began in the Peruvian Amazon in the mid-1960s with the publication of his book *Salt of the Mountain: Ashaninka Resistance and Utopia in Amazonia*, (with five Spanish editions and translations into English and French). He has published seven more books and over one hundred articles.